The Theory of Coherent Radiation by Intense Electron Beams

Particle Acceleration and Detection

springer.com

The series *Particle Acceleration and Detection* is devoted to monograph texts dealing with all aspects of particle acceleration and detection research and advanced teaching. The scope also includes topics such as beam physics and instrumentation as well as applications. Presentations should strongly emphasise the underlying physical and engineering sciences. Of particular interest are

- contributions which relate fundamental research to new applications beyond the immediate realm of the original field of research
- contributions which connect fundamental research in the aforementioned fields to fundamental research in related physical or engineering sciences
- concise accounts of newly emerging important topics that are embedded in a broader framework in order to provide quick but readable access of very new material to a larger audience

The books forming this collection will be of importance for graduate students and active researchers alike.

Series Editors:

Professor Alexander Chao
SLAC
2575 Sand Hill Road
Menlo Park, CA 94025
USA

Professor Christian W. Fabjan
CERN
PPE Division
1211 Genève 23
Switzerland

Professor Rolf-Dieter Heuer
DESY
Gebäude 1d/25
22603 Hamburg
Germany

Professor Takahiko Kondo
KEK
Building No. 3, Room 319
1-1 Oho, 1-2 1-2 Tsukuba
1-3 1-3 Ibaraki 305
Japan

Professor Franceso Ruggiero
CERN
SL Division
1211 Genève 23
Switzerland

Vyacheslav A. Buts   Andrey N. Lebedev
V.I. Kurilko

The Theory of Coherent Radiation by Intense Electron Beams

With 42 Figures

Professor Vyacheslav A. Buts
32 Ubrorevicha Str.
61336, Kharkov
Ukraine
E-mail: vbuts@kipt.kharkov.ua

Professor Andrey N. Lebedev
Department of Theoretical Physics
PN Lebedev Physical Institute
Leninsky Prospect 53
119991, Moskva
Russia
E-mail: lebedev@sci.lebedev.ru

V.I. Kurilko
(1932–2001)

Library of Congress Control Number: 2005939175

ISSN 1611-1052
ISBN-10 3-540-30689-7 Springer Berlin Heidelberg New York
ISBN-13 978-3-540-30689-4 Springer Berlin Heidelberg New York

This work is subject to copyright. All rights are reserved, whether the whole or part of the material is concerned, specifically the rights of translation, reprinting, reuse of illustrations, recitation, broadcasting, reproduction on microfilm or in any other way, and storage in data banks. Duplication of this publication or parts thereof is permitted only under the provisions of the German Copyright Law of September 9, 1965, in its current version, and permission for use must always be obtained from Springer. Violations are liable for prosecution under the German Copyright Law.

Springer is a part of Springer Science+Business Media
springer.com
© Springer-Verlag Berlin Heidelberg 2006
Printed in The Netherlands

The use of general descriptive names, registered names, trademarks, etc. in this publication does not imply, even in the absence of a specific statement, that such names are exempt from the relevant protective laws and regulations and therefore free for general use.

Typesetting: by the authors and TechBooks using a Springer LATEX macro package
Cover design: *design & production* GmbH, Heidelberg

Printed on acid-free paper   SPIN: 11017370   54/TechBooks   5 4 3 2 1 0

Preface

This book is intended neither as a manual for electrodynamics nor as a monograph dedicated to specific problems of vacuum electronics. It would be naive for the authors to attempt that after numerous brilliant courses of studies, already classical, had been published and after incredibly large number of works, dedicated to research and development of microwave devices, their operation, optimization, etc., had appeared in literature.

One can hardly add anything to the classical theory of the radiation emission by a point charged particle. Almost all the possible configurations of external fields with various boundary conditions for the microwave radiation field had already been investigated. The fundamental effects – Vavilov–Cherenkov radiation, transition radiation in a system with inhomogeneous parameters, and Doppler effect in the case of a relativistic particle moving with acceleration – have been investigated in detail. One can easily find the description of these problems in a large number of specialized monographs and reviews.

At the same time, it is necessary to take a very important logical step to apply the theory to microwaves generation and amplification. The point is that the spectral density of the radiation emitted by a single particle is very low. Multiplication of this density value even by a huge number of individual emitters yields the result of practical interest only in the case of very short waves (e.g., synchrotron radiation) because of the absence of alternative methods. In this situation, coherence of emitters plays the decisive role. It leads to a sharp increase in the spectral–angular brightness of radiation (to be more precise, the field mode composition is implied). Respectively, the efficiency also increases. Surely, coherence, imposed by any of initial conditions, finally has to vanish – at least because of an increase in entropy during the irreversible process of radiation emission. However, in systems, essentially disequilibrium from the viewpoint of thermodynamics, a stage of self-organization can precede self-destruction of the coherence. At this preliminary stage, the coherence is being maintained – or even heightened – because of the radiation reaction (i.e., due to the radiation field backward influence on the motion of particles). As regards low-frequency systems, smaller than the wavelength, the problem of the coherence maintenance does not arise. However, this problem becomes urgent if one deals with the TWT-type microwave systems with distributed wave–particle interaction.

As a rule, classical electrodynamics of the point particle may be described according to one of the two patterns: it is either "a charged particle in the given field" or "the radiation emission by the charged particle when its motion is prescribed." One can get the total physical picture by combining these schemes, which is more or less artificial. Applied to a single charged particle, this approach is justified because of the weakness of the radiation reaction. It causes only inessential changes in the particle motion parameters during a time interval in question. However, this method does not fit strong coherent proper fields. There were attempts to elaborate a "self-consistent" electrodynamics of point charged particles. Unfortunately, even the most promising one remained unfinished (P. Dirac, 1940s). In particular, there does not still exist any Lagrangian description of mutual influence of several relativistic particles radiation field taken into account. Therefore, we believe that it would be worthwhile, even if qualitatively, to extend the notions concerning the radiation reaction on a single charge to the case of an ensemble of interacting particles.

Similar difficulties in many particles theories (e.g., the plasma theory) are successfully overcome in the self-consistent field approximation, when the totality of particles is regarded as a charged medium (either hydrodynamic or kinetic continuum). Appearance of the microwave field (i.e., the radiation emission) is regarded then as collective instability of internal degrees of freedom or self-excitation of negative energy proper waves. However, the drawback is that within this approach the spontaneous radiation emission is not taken into account. Besides, in the plasma theory, little attention is paid to relativistic effects and, generally speaking, to the problem of generating microwave fields with prescribed characteristics.

As regards investigations in vacuum electronics, they are aimed, by definition, to optimization of a particular device. Naturally, one is principally interested then in relative advantages of the given construction, while the general physical picture is not being discussed in detail. Surely, there exist many excellent monographs in this field, where physics of the process is profoundly discussed. Notwithstanding this fact, these works, on our opinion, still use a rather specific theory for various devices. We really believe that a common approach, for example, to investigations of Vavilov–Cerenkov radiation, principles of operation of TWT, and Landau damping in collisionless plasmas is not just an attempt to find effective physical parallels but can also be of scientific value. Probably, the most convincing evidence of that is the up-to-date concept of stimulated radiation emission. It combines not only quantum theory of the black body equilibrium and quantum lasers but also purely classical devices of vacuum electronics with distributed interaction.

In more practical sense, this book was stimulated by the quick development of high-current relativistic electronics. By itself, this field is a natural continuation of traditional vacuum electronics – it just so happened that demand for higher powers stimulated the use of higher currents and higher energies of electron beams. Besides, the advance to shorter wave ranges has conditioned giving traditional slow-wave structures up and using Doppler relativistic effects. All these factors have caused changes in many concepts. For instance, the phenomena, previously treated as unpleasant space charge effects, at present make sometimes the basis of the device operation; the substitution of strictly specified beam quality accelerators for traditional electron guns has cardinally changed the device geometry. In addition, there has arisen a necessity of using open optical cavities or leaving the radiation free at all, etc. Beside, principally new results have been achieved – i.e., the development of the high-current beam technology has enabled advancing into the gigawatt power range, while elaboration of devices working on the basis of Doppler deep transformation (the so-called free electron lasers or FELs) has provided the possibility of stimulating the monochromatic tunable radiation emission even in the soft x-ray band. Because of all of these factors, a large number of specialists in various branches gathered together within this field. All of them had specific concepts, their own experimental and theoretical approaches, different terminology and even their own prejudices. It was our impression that the first discussions somewhat reminded the construction of the Tower of Babel. Surely, later on a mutual understanding was somehow achieved but it is still to be formulated. Periodic literature is of a little help in this aspect.

These factors have determined both the book's composition and the selection of material. In its essence, the book is divided into three parts. The first one is dedicated to the radiation emission by a single relativistic particle. When dealing with the problem of the controllable generation of narrow-band high-power microwaves, the authors have not considered the effects such as wide-band bremsstrahlung which is typical, for instance, of x-ray tubes, while focusing their attention on the prolonged interaction of relativistic particles with a copropagating wave. Based on simple and clear reasoning, this approach enables getting an important piece of information about the field spectral–angular distribution in free space and about the mode composition in an electrodynamic structure. In particular, avoiding Maxwell equations, one can trace the common nature of Cherenkov radiation emitted in media and in slow-wave electrodynamic structures – such as periodic waveguides or diffraction lattices. Similar prolonged interaction might be achieved when a particle is moving along a helix in longitudinal magnetic field or passing through the undulator – a system where the transverse magnetic field alternates in space. In these cases, Doppler normal and anomalous effect plays an essential role because it determines the beam optical activity in the short-wave range, even if the particle is passing through macroscopic structures.

By the way, the synchrotron radiation emission fits the same scheme due to the deep Doppler effect – notwithstanding the fact that for the cyclic motion the wave accompanies the particle only within a short section of the curvilinear trajectory. This fact is rather important because it reveals the common character of the synchrotron and undulator radiation emission, used in FELs.

Presentation of the short-wave undulator radiation emission as a result of the wave scattering by a moving charged particle and prospects of the coherent backward scattering by an intense beam have required the preliminary dwelling on the theory of scattering by a charged particle in the magnetic field – all the more so this problem can be analytically solved under rather loose conditions.

In concluding this section, we present a problem, classical in electrodynamics of the point charged particle – the radiation reaction in relativistic and nonrelativistic cases, also dwelling on the corresponding well-known paradoxes. Paying some attention to the radiation reaction influence on the prolonged particle dynamics, we bear in mind mainly a sequent application of this concept to the case of coherent radiation reaction in a many-particles system.

In fact, similar reasons have dictated our selection of all material for this part. For instance, here the reader can find the total field expansion in potential and solenoidal modes of an arbitrary structure. At the same time, we have left aside the traditional expansion in multipoles because, from the viewpoint of physics, it hardly has any meaning in distributed microwave systems.

The second part – the radiation emission by an ensemble of charged particles – could be regarded as the keystone one. At its beginning, we have presented certain general notions concerning partial coherence of the radiation emitted or scattered in regular structures of various dimensions. Furthermore, we have attempted to describe the stimulated radiation emission as a process of the emitting system self-organization. At the dawn of quantum mechanics development, the "stimulated radiation emission" had been defined as the process reverse to radiation absorption. Later on, the notion of stimulated emission, applied to classical systems with linear spontaneous spectrum, has been regarded as autophasing of individual emitters under the influence of their proper radiation field. It is worth mentioning that both approaches yield completely identical correlation between the spontaneous radiation spectrum and wave amplification under conditions of inverted population. However, the classical approach, which implies mutual autophasing of the particles, is much more illustrative and corresponds better to physics of the process.

Besides, in the second part we have also traced the correlation between discrete and continuous models of the beam. The latter permits applying such an effective tool as the hydrodynamic and kinetic self-consistent equations. It also justifies the use of the concept of negative-energy proper waves and their interaction with electromagnetic waves of the "cold" system. These aspects have been minutely described by an example of the typical problem of an electron beam propagating along a waveguide in a longitudinal magnetic field.

Finally, the third part deals with applying the general ideas to specific schemes. We have presented there beam–plasma systems, gyrotron, and FEL (in spite of our desire, the FEL chapter turned out to be rather bulky, which is conditioned by the novelty and unusual nature of the device). In accordance with the reasons given above, we tried to avoid coming into details and specificities of the devices schemes. An exception has been made only in cases of absolute necessity – e.g., when we had to explain briefly the principle of operation of the open cavity because of dwelling on the diffraction effects.

The book is written by physicists, for physicists, and about physics. To understand the mathematics involved, one has to handle Fourier and Laplace transforms. The general theory of functions of complex variables is also necessary (within the framework of the university course). The authors tried to describe the models, which can be described analytically, as strictly as it was possible. We do believe that even a limited analytical model is more illustrative than just the results of numerical simulations. Of course, the appropriateness of the model choice is another thing. For those who are ready to take the calculations for granted, "hand-waving" arguments could be sufficient. We tried to use them as often as possible – even taking the risk to sound simplistic.

As regards the references, we can just give our apologies. It is evident that neither the authors nor the reader can physically make acquaintance – even cursory – with all works on the subject. There were even poorer chances to arrange the list of references according to priority – if the latter can be established at all. Therefore, the authors have referred only to the most known manuals and reviews (within the limits of the possible) available both in English and in Russian. We hope that the reader will find useful information in this literature. Original papers from journals are mentioned only in cases of absolute necessity, without giving any priority to them. An excuse, somewhat poor, is that no exception has been made for the authors' own works.

The book deals with the problems that were being discussed by the authors with many of their colleagues during decades. Thus, it is only fair to consider that these people have also contributed to the concept presented. We are sincerely grateful to all of them, but, unfortunately, it is almost impossible to mention all the names here. Besides, it would be tactless to make the people who have helped us responsible – if even partially – for the authors' possible omissions or errors. Instead, we would like to pay our greatest respect to those whom we consider our teachers: V. L. Ginzburg, Ya. B. Fainberg, A. V. Gaponov-Grekhov, A. A. Kolomensky, A. I. Akhiezer.... We also must mention names of our colleagues: B. Bolotovskiy, Ph. Sprangle, A. Rukhadze, A. Sessler, M. Petelin, J. Nation, N. Ginzbburg, V. Bratman, A. Agafonov, and many others. We owe a great deal to them for the scientific exchange and their friendship. The original idea of this book belonged to our late friend V. I. Kurilko, and we dedicate it to his memory. E. Bulyak and I. Bogatyreva have rendered an invaluable contribution to the technical work with the text.

The typescript has been prepared with the support of the scientific fund STCU (grant #855).

Kharkov, Moscow *Vyacheslav Buts*
February 2006 *Andrey Lebedev*

Contents

Part I Radiation by Single Particles

1 Synchronous Wave–Particle Interaction 3
  1.1 Basic Definitions and Terminology 3
  1.2 Conditions of Radiation Emission 5
    1.2.1 Classical Approach 5
    1.2.2 Semiquantum Approach 8

2 Radiation Emitted by Particle Moving Uniformly 11
  2.1 Vavilov–Cherenkov Radiation 12
    2.1.1 Radiation in Uniform Dispersive Medium 12
    2.1.2 Cherenkov Radiation in Magnetized Plasma
          Waveguide 19
  2.2 Transition Radiation 22
    2.2.1 Medium Step–Like Inhomogeneity 22
    2.2.2 Smooth Inhomogeneity of Medium 25
  2.3 Cherenkov Radiation in Periodic Structures 27
    2.3.1 Proper Waves in Periodic Structures 28
    2.3.2 Excitation by Moving Charge 33

3 Microwave Bremsstrahlung 37
  3.1 Radiation Field 37
    3.1.1 Proper Waves of Free Space 38
    3.1.2 Proper Waves of Waveguide Systems 41
  3.2 Dipole Radiation 47
  3.3 Undulator Radiation 49
  3.4 Cyclotron Radiation 52
    3.4.1 Cyclotron Radiation Emitted in Waveguide 53
    3.4.2 Synchrotron Radiation 55

3.5	Scattering by Free Charged Particle	59
3.6	Scattering and Absorption by Bound Particle	65

4 Radiation Reaction ... 71
- 4.1 Conservation Laws ... 72
- 4.2 Radiation Reaction and Emitter Field ... 76
- 4.3 Radiation Friction and Charged Particle Dynamics Radiation Cooling ... 79

Part II Radiation by Particles Ensembles

5 Coherence of Individual Emitters ... 89
- 5.1 Spatial Coherence ... 89
- 5.2 Interference in Regular Lattices ... 92
  - 5.2.1 One-Dimensional Distribution ... 93
  - 5.2.2 Multidimensional Lattices ... 93

6 Spontaneous and Stimulated Emission ... 97
- 6.1 Semiquantum Interpretation ... 98
- 6.2 Classical Limit ... 102
- 6.3 Stimulated Emission and Beam Phasing ... 105
  - 6.3.1 Phase Dynamics in Quasi-Synchronous Wave ... 105
  - 6.3.2 Phase Bunching by External Wave (Low-Gain Regime) ... 107
  - 6.3.3 Spatial Amplification in Particles Flow (High-Gain Regime) ... 110
- 6.4 Dynamic Chaos ... 114
  - 6.4.1 Resonant Perturbation Theory ... 115
  - 6.4.2 Randomization of Motion ... 118
  - 6.4.3 Criteria of Dynamic Chaos ... 119

7 Proper Waves in Flows of Charged Particles ... 123
- 7.1 Proper Waves in Beams of Interacting Particles ... 123
  - 7.1.1 Dispersion Relations ... 124
  - 7.1.2 Partial Beam Waves ... 130
  - 7.1.3 Proper Waves in Flow ... 132
  - 7.1.4 Proper Waves in Transversely Limited Beam ... 134
- 7.2 Negative Energy Waves ... 139
- 7.3 Kinetic Effects ... 144
  - 7.3.1 Kinetic Equation ... 145
  - 7.3.2 Dispersion Relations ... 147
  - 7.3.3 Kinetic Effects and Landau Damping ... 148

Part III Certain Modern Applications

8 Cherenkov Radiation in Beam–Plasma Systems 159
   8.1   Dispersion Equation 160
   8.2   Cold Beam Instability...................................162
         8.2.1   Absolute Instability 162
         8.2.2   Convective Type Instability 167
   8.3   Warm Beam Instability 169

9 Cyclotron Resonance Masers (CRM) 173
   9.1   General Principles 173
   9.2   CRM in Small–Signal Approximation 176
         9.2.1   Dispersion Equation for Ribbon Beams 176
         9.2.2   Bunching of Particles in CRM 183
   9.3   Particle Interaction with Large Amplitude Wave...........185
         9.3.1   Averaged Equations of Motion 185
         9.3.2   Qualitative Analysis 188
         9.3.3   Stochastic Regime 196
   9.4   Nonlinear Regime of Operation...........................201

10 Free Electron Lasers (FELs) 207
   10.1 FEL–oscillators. Low-Gain Regime 209
         10.1.1 Optical Cavity.................................210
         10.1.2 Gain 213
         10.1.3 The Self–Excitation Threshold 215
         10.1.4 Steady–State Oscillations and Output Power 217
         10.1.5 Beam Quality...................................220
   10.2 FEL–Amplifier: High-Gain Regime 221
         10.2.1 Dispersion Relation 222
         10.2.2 Two-Dimensional Effects 231
   10.3 SASE Mode of Operation..................................238
         10.3.1 Amplification of Spontaneous Radiation
                 in Uniform Flow 239
         10.3.2 Amplification and SASE Mode in Short Bunch....245

11 Blowup Effect in Linear Accelerators 251

References..257

Index..261

Part I

Radiation by Single Particles

1
Synchronous Wave–Particle Interaction

Electromagnetic radiation emitted by charged particles is an indispensable part of majority of courses on general electrodynamics. There are special monographs available, dedicated to some of the typical cases – such as synchrotron, transition, and Vavilov–Cherenkov radiation. As a rule, it is usually emphasized that the radiation field is the self-field of a charged particle in the far-field zone, where it decreases in inverse proportion to a distance from the emitter so that the total energy flow through a closed surface remains constant while the surface expands to infinity.

However, this statement is not always correct (e.g., when longitudinal oscillations are excited in a dielectric dispersion medium or if the emitter motion trajectory is infinite). In this chapter, we will describe certain general conditions for the radiation emission by charged particles, moving uniformly and rectilinearly in spatially homogeneous (or periodic) electrodynamic systems in this direction.

1.1 Basic Definitions and Terminology

First, let us define the physical notions that are to be used not only in this chapter but all through the monograph.

The term "elementary emitter" refers to a point (zero–dimensional) charged particle, characterized by finite magnitudes of its charge (q), rest frame mass (m), and the total energy (the relativistic factor γ).

The term "coherent" when used to describe the radiation emitted by an ensemble of identical individual charged particles, refers to situations where the radiation characteristics change qualitatively with enlarging of number of these emitters.

Finally, the term emission of radiation is used to describe the processes where either the microwave field "is moving away" from an elementary emitter that excites it or the emitter "deserts" the region of electromagnetic oscillations excited by itself. The latter conditions occur in cases of exciting either longitudinal (polarization) waves in a medium or of proper oscillations in resonant elements of the system. Both of the processes are characterized by the emitter energy losses due to the excitation of the fields and the emitter deceleration by these fields.

It should be noted that, as a rule, we do not specially emphasize the difference between the Coulomb and radiation components of the microwave field excited by the individual charged particle. Strictly speaking, it is possible to distinguish between these components through dependencies of their amplitudes on a distance from the source only in the far–field zone of the source [1], where the strong inequality

$$2\pi R \gg \lambda$$

is satisfied (here R is the distance from the emitter to the point of observation; λ is the characteristic length of the wave). However, for taking into account interaction of many radiating particles, the near–zone field can be of the same importance.

The self-field structure of a point charged particle, moving with some acceleration in vacuum, presents the simplest illustration of this statement. The electric field component, expressed via the retarded Lienard–Wiechert potentials, may be presented as a sum of two addenda [1]: decreasing in different manner in the region where $2\pi R \gg \lambda$. However, we will have to consider the source near–field zone as well, where the inequality does not hold and both fields are indistinguishable. In particular, in the simplest case of the dipole microwave bremsstrahlung, contributions of both terms to the deceleration force differ only in a numerical coefficient of the order of unity [1].

As regards the term "microwave bremsstrahlung" this term implies the quasi–monochromatic microwave radiation emitted because of acceleration of electrons, whose velocity varies periodically (in magnitude and/or in direction) in external fields.[1] Here and below the term "microwave bremsstrahlung" is to be used only in this sense.

It should be emphasized that we do not apply the widely used label spontaneous to the process of radiation emission by an individual charged particle. The terminology spontaneous transition, introduced by A. Einstein in the examination of a quantum two–level emitter, was defined as "radiation that takes place without any external influence on the emitter" [2]. Literally (i.e., in the sense of "not caused by anything outside"), this definition contradicts the causality principle: the zero matrix element, describing the emitter interaction with external forces, yields a zero probability of transition (and corresponding emitter has an infinite lifetime in the excited state). At the same time, it is known in classical electrodynamics that an oscillator with a finite amplitude does always emit radiation – even in the absence of external microwave fields [1]. This paradox has been explained in the course of further

---

[1] In nuclear physics it usually defines the broadband hard x-ray radiation under conditions of electrons scattering at nuclei.

development of quantum theory of radiation [3]. In particular, the notion of zero electromagnetic oscillations has permitted demonstrating that these are the very oscillations that stimulate "spontaneous" transitions [4].

Hence, because of these reasons, it would be pointless to use the notion of spontaneous emission of radiation in microwave electronics with respect to the very process of emitting radiation by an individual charged particle.[2] At the same time, it should not be forgotten that, as a rule, the terminology historically refers to the processes of radiation emission by ensembles of charged particles. However, here it is used in the meaning opposite to "stimulated radiation." In fact, one implies that the total radiation is emitted by systems of uncorrelated individual oscillators, and their electromagnetic field phases are random. A more detailed examination of the subject will be given in Part II.

A reader will find that we use throughout the book electric and magnetic fields rather than traditional electromagnetic potentials. The latter would not simplify wave equations in our models, mainly one–dimensional, while the field description makes particles interaction mechanisms more transparent. A necessary exclusion is made only for discussing general properties of vector eigenfunctions in periodic electromagnetic systems.

1.2 Conditions of Radiation Emission

Naturally, the possibilities of essentials altering the particle energy, due to its interaction with the radiation field, are of principal interest for applications. The process of interaction involves two subsystems. One of these is represented by free electromagnetic waves propagating in a "cold system" (i.e., a system that does not contain emitting particles). The second subsystem includes particles moving along fixed trajectories in the absence of waves. To save their individual physical identity, the subsystems will be regarded as weakly interacting. It is not necessary, of course, to regard the wave as a plane one – generally speaking, its phase velocity and fields configuration prescribed by the system properties can be arbitrary. Below, we will consider the conditions for an effective energy exchange between a charged particle and the microwave field.

1.2.1 Classical Approach

If the energy of the wave quantum ($\hbar\omega$) is considerably smaller than that of the particle (γmc^2) and the effectiveness is supposed to be high enough a number of the quanta emitted during the interaction time must be large as well. Consequently, such conditions must be provided, under which interaction lasts for a large number of the wave periods. During this time, the average

---

[2] Modern quantum mechanics texts refrain from applying this notion in the theory of radiation.

field work on the particle remains nonzero. It is easy to notice, that the same considerations may also refer to the radiation field emitted by an oscillator. To make the long story short, if the radiation field has to perform some negative work on the particle, the field temporal change must be synchronous with the particle motion.

Let us consider a harmonic wave with the electric field vector

$$\mathbf{E} = \mathbf{E}_0 \exp[i(kz - \omega t)]$$

and a moving oscillator with the velocity

$$\mathbf{v} = \mathbf{e}\beta c + \tilde{\mathbf{v}},$$

where $\mathbf{e}$ is a unit vector in z–direction. For the varying component of the velocity $\tilde{\mathbf{v}}$, with the frequency Ω in the laboratory frame of reference, the synchronism (i.e., the field phase perpetuation at a point of the particle location) takes place under the conditions

$$\omega - k\beta c - s\Omega \approx 0 \,; \qquad s = 0; \pm 1; \pm 2; \ldots \,. \tag{1.1}$$

This condition can be presented as

$$\beta_{\text{ph}} = \beta + s\frac{\Omega}{kc}\,. \tag{1.2}$$

Here $\beta_{\text{ph}} c \equiv \omega/k$ is the wave phase velocity in z–direction.

The relations (1.1) determine the frequency spectra of waves that essentially interact with an oscillator:

$$\omega_s = \frac{s\Omega}{1 - \beta/\beta_{\text{ph}}(\omega_s)}\,. \tag{1.3}$$

For example, a moving relativistic oscillator effectively exchanges energy with a copropagating electromagnetic vacuum wave if the wave frequency is

$$\omega_s = \frac{s\Omega}{1 - \beta} \approx 2s\Omega\gamma^2, \qquad \beta \to 1\,. \tag{1.4}$$

The approximate equality, valid for the ultra-relativistic case ($\gamma \gg 1$), indicates that the wave frequency ω_s can essentially exceed the oscillator frequency Ω.

The possible cases of the synchronism, dependent on a value of the integer s, may be conceptually classified as follows.

Vavilov–Cherenkov Effect

If $s = 0$ for $\omega \neq 0$, the condition of synchronism (1.1) is identical to the condition of the Vavilov–Cherenkov radiation, characterized by the wave-phase velocity in a medium equal to the velocity of the particle:

1.2 Conditions of Radiation Emission

$$\beta = \beta_{\text{ph}} = 1/n\cos\vartheta \ .$$

Here n is the medium refractivity, and the angle ϑ indicates the wave propagation direction (with respect to the particle velocity vector).

However, the point to be made is that this case is not the most typical and, in addition, a strongly degenerated one if n is independent of ω. As a rule, an electron beam is moving in a vacuum channel surrounded by metallic and/or dielectric walls, which provide a certain dispersion of the system proper waves (i.e., there exists a specific dependence $\omega(k)$). If some of the proper waves are slow (i.e., their phase velocities in the direction of the particle motion coincide with the particle velocity during a sufficiently long period of time), exactly these waves will be excited. In this case, the relation (1.1) should be regarded as an equation determining frequency spectra of the emitted waves. The field configurations and amplitudes of the waves depend on a type of the system.

Sometimes, as in the case of linear accelerators, it is useful to consider the influence of an external accelerating field on a particle from the same viewpoint (long–term wave–particle interactions). A linear accelerator is designed so that its electrodynamic structure (e.g., an iris–loaded waveguide, a chain of cavities, etc.) enables the accelerating wave propagation with a phase velocity equal to that of particles. Thus, the process of the long–term acceleration becomes inverse to the Cherenkov radiation emission and should be accompanied by the latter. The radiation field, coherently detracted from the external field, decreases its amplitude. This corresponds to a load provided by the current of the accelerated particles.

Doppler Effect (Normal and Anomalous).

If $s \neq 0$, the synchronism condition naturally yields the Doppler formula, determining the frequency of the wave emitted (or absorbed) by an oscillator with the frequency Ω in the laboratory reference frame (Ω is not equal to the oscillator proper frequency in the rest frame!). In an isotropic nondispersive medium, this expression gives the following dependence of the emitted frequency ω_s versus the emission angle ϑ and the medium refractivity n:

$$\omega_s = \frac{s\Omega}{(1 - n\beta\cos\vartheta)} \ . \tag{1.5}$$

If $s > 0$ (normal Doppler effect), the particle interacts with a fast wave ($\beta < 1/n$) that outruns the oscillator by s wavelengths during one period of the particle oscillation. If $s < 0$ (anomalous Doppler effect), the particle interacts with a slow wave lagging behind the emitter ($\beta > 1/n$). It should be noted that if the emitter oscillations are transverse (i.e., directed along the wave electric field line), this straight–moving oscillator can also interact with free vacuum plane waves as a result of the normal Doppler effect. In particular, at small emission angles, a relativistic oscillator emits very hard radiation at the frequency

$$\omega_{+1} \approx \frac{2\Omega}{\vartheta^2 + \gamma^{-2}} \gg \Omega; \qquad \vartheta^2 \approx \frac{1}{\gamma^2} \ll 1 \,. \tag{1.6}$$

(or the oscillator can be accelerated by a microwave copropagating wave, characterized by the same frequency).

Certainly, if the motion occurs in a waveguide system, it is more appropriate to consider (radiation emitted by the particle in the form of cylindrical waves or other proper waves of the system. Note that a moving oscillator can effectively interact with both the fast and slow waves (even with the backward ones).

The possibilities listed above are illustrated in Fig. 1.1 where intersections of the cold system dispersion curve $\omega(k)$ and straight lines $\omega = k\beta c \pm \Omega$ determine the radiation (absorbtion) spectrum.

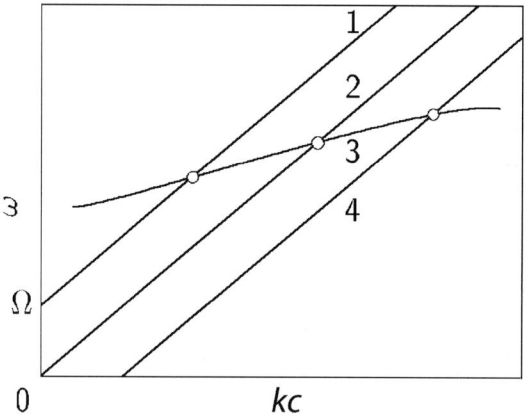

Fig. 1.1. Synchronous wave-particle interaction diagram. Curve 1: $\omega = kv + \Omega$. Curve 2: $\omega = kv$. Curve 3: $\omega(k)$ (electromagnetic wave). Curve 4: $\omega = kv - \Omega$. Circles indicate normal Doppler effect, Cherenkov radiation, and anomalous Doppler effect (from left to right)

1.2.2 Semiquantum Approach

It may seem as if the conditions for Doppler–shifted radiation derived above work only in the case of the emitter nonzero oscillation amplitude. Actually, the latter statement is not always true. To make sure of this, it is enough to consider a charged particle moving uniformly and rectilinearly strictly along a line of force of an external homogeneous magnetic field. Under these conditions, each of the fixed values of the emitter energy $\mathcal{E} \equiv \gamma mc^2$ is split up into two sublevels corresponding to particle spin of different orientations. Transitions between these levels determine the frequency Ω in the numerator of the

1.2 Conditions of Radiation Emission

expression on the right of (1.5). In the given case, this frequency is proportional to the external magnetic field strength and the effect is purely quantum one.

In general, any transition between emitter internal energy levels satisfies the necessary condition for the radiation emission or absorbtion corresponding to normal or anomalous Doppler effect. Following [5], we present below general conditions for the radiation emission by such emitter.

Let us designate the total and internal energy of the emitter in the initial and final states by $\mathcal{E}$ and W with subscripts i and f respectively. We start from the laws of conservation of total energy and total momentum in the "emitter + photon" system:

$$\mathcal{E}_i - \mathcal{E}_f = \hbar\omega > 0; \qquad \mathbf{P}_i - \mathbf{P}_f = \hbar\mathbf{k} . \qquad (1.7)$$

where $\mathbf{k}$ is a wave vector of the emitted photon (we consider a plane wave emission). Using the kinematic relativistic relations,

$$\mathcal{E}^2 = W^2 + \mathbf{P}\cdot\mathbf{P}c^2; \qquad \mathbf{P} = \mathcal{E}\mathbf{v};$$
$$\mathbf{v}\cdot\mathbf{k} = \beta ck\cos\vartheta; \qquad W = \mathcal{E}\sqrt{1-\beta^2}$$

and expanding over powers of $\hbar$ we get in the first approximation:

$$W_i - W_f = -\hbar\omega\sqrt{1-\beta^2}\left(1 - \frac{\beta_0}{\beta}\cos\vartheta\right) . \qquad (1.8)$$

This yields for the emitted photon frequency:

$$\omega = \frac{W_i - W_f}{\hbar}\sqrt{1-\beta_0^2}\left(1 - \frac{\beta_0}{\beta}\cos\vartheta\right)^{-1} . \qquad (1.9)$$

For an equidistant system of internal levels

$$W_i - W_f = \hbar\Omega's ,$$

where $\Omega' = \Omega\sqrt{1-\beta^2}$ is the oscillator frequency in the rest frame. Then (1.9) coincides with (1.3). Moreover, it manifests unequivocally the essential difference in change of levels population in the cases of emitting a normal quantum or an anomalous one. Really, as it follows from (1.9), under the conditions of normal Doppler effect ($\beta\cos\vartheta < \beta_{\text{ph}}$), while emitting a quantum, the emitter transits from an upper energy level to a lower one ($s = +1$). On the contrary, in the case of anomalous Doppler effect ($\beta\cos\vartheta > \beta_{\text{ph}}$), the quantum emission is accompanied by the emitter transition from a lower energy level to an upper one ($s = -1$). As it follows from (1.9), the emitter longitudinal motion is a source of an additional internal energy, required for such transition.

In the case of emitting a Cherenkov quantum ($s = 0$), when the emitter internal energy remains constant ($W_f - W_i = 0$), the emitter deceleration provides the only source of the photon energy.

Thus, the semiquantum approach presented above yields the emission conditions identical with those based on the concept of synchronism.

Of course, the concept of the wave–particle synchronous motion as a necessary condition of the radiation emission is of a limited value. For example, it obviously cannot be applied to the case of bremsstrahlung when the particle velocity sharply changes. The arguments about the "long enough" interaction need a quantitative proof. Nevertheless, the concept is simple and can be easily generalized to the case of interacting particles losing, in a sense, their individuality. We delay these problems for Part II and consider now the main emission mechanisms for an individual particle still basing on the considerations above.

2

Radiation Emitted by Particle Moving Uniformly

Even the qualitative considerations given above do indicate that the radiation emission by a charged particle moving uniformly in vacuum is impossible because the conditions of synchronism cannot be satisfied for any of free space proper waves. As is known, these waves propagate with the velocity of light, which always exceeds the velocity of a charged particle. Moreover, the radiation emission is a phenomenon invariant with respect to the choice of a reference frame. Therefore, impossibility of its realization in vacuum becomes obvious when one transfers to the rest frame of the charged particle, where the particle proper field in vacuum is of a purely electrostatic nature. The situation changes radically in the presence of a medium or another environment. This situation is not invariant under transition to new coordinates. In these systems, among their proper waves, there could be at least one wave synchronous with the particle over a long period of time and possessing the electric field component parallel to the particle velocity. The very possibility of the particle energy transfer to this wave—together with the fact that the latter can freely leave the radiation source—guarantees realization of the effect of radiation emission.

These conditions can be satisfied in the classical case of Cherenkov radiation. Transverse proper waves of a medium with a dielectric constant ε have the phase velocity $c/\sqrt{\varepsilon}$, which may be less than the charged particle velocity v. Besides, a real dielectric is characterized by a certain dispersion (i.e., the dependence $\varepsilon(\omega)$) as well as by the existence of its proper resonant frequencies, where $\varepsilon(\omega) = 0$. So, in the total system of the medium proper waves, there also exist longitudinal polarization waves with such frequencies. Although these waves have no magnetic field components, they can propagate freely and absorb the charged particle energy. In its physical nature, the effect of exciting these waves by a particle is very close to the emission of Cherenkov radiation. In the case of a dielectric medium, these types of radiation should be studied together.

Besides, any inhomogeneity of an electrodynamic environment along a rectilinear trajectory of charged particle also causes the emission of radiation of a specific type, defined as "transition radiation." The reference to the proper waves synchronous with the particle is less obvious here. However, formally, it is quite justified—even in the case of a single inhomogeneity, when it would be more appropriate to characterize the radiation by the total energy emitted rather than by the power losses. At the same time, the transition radiation emitted because of periodic inhomogeneities is very similar to the Cherenkov radiation. The common nature of both types of radiation is conditioned by the phenomenon of the synchronous wave–particle interaction.

Below in this chapter, we will examine Vavilov–Cherenkov radiation emission, excitation of polarization oscillations, and the transition radiation conditioned by the presence of isolated and periodic inhomogeneities of the system.

2.1 Vavilov–Cherenkov Radiation

The Vavilov–Cherenkov radiation emission by a charged particle moving uniformly is a complicated many parametric physical process even in the simplest isotropic spatially homogeneous medium. It is always accompanied by excitation of the medium proper longitudinal oscillations (waves). To describe these processes in the direct analytical form, below we will use the simplest model of a dielectric medium proposed in [6].

2.1.1 Radiation in Uniform Dispersive Medium

We now consider a charge q moving uniformly and rectilinearly with the velocity v in a dielectric medium. Following [6], we suppose that the medium consists of an almost transparent homogeneous gas of isotropic oscillators with the proper frequency Ω. The dispersion, corresponding to this medium model, is characterized by the dielectric constant

$$\varepsilon(\omega) = 1 + \frac{\omega_{\mathrm{p}}^2}{\Omega^2 - \omega^2 - 2\mathrm{i}\omega\nu} \ . \tag{2.1}$$

Here ω_{p} is the effective "plasma" frequency (it is proportional to the square root of the oscillators volume density), and ν is a small positive damping constant. As it will be demonstrated below, the latter is required for a correct restoration of the temporal dependence of the radiation field according to the causality principle.

As we take into account the frequency dependence of the dielectric constant, Maxwell equations must be rewritten for monochromatic fields proportional to $\exp(-\mathrm{i}\omega t)$, i.e., for the complex amplitudes of the real fields

$$\mathbf{A}(\mathbf{r},t) = \int_{-\infty}^{+\infty} \mathbf{A}(\mathbf{r},\omega) \exp(-\mathrm{i}\omega t) \, \mathrm{d}\omega$$

(2.2a)

$$\mathbf{A}(\mathbf{r},\omega) = \frac{1}{2\pi} \int_{-\infty}^{+\infty} \mathbf{A}(\mathbf{r},t) \exp(+\mathrm{i}\omega t) \, \mathrm{d}\omega \ .$$

(Where necessary, further we distinguish between the physical quantities and their Fourier transforms by labeling their arguments.)

So, Maxwell equations take the form

$$\operatorname{rot} \mathbf{B}(\mathbf{r},\omega) = \frac{4\pi}{c}\mathbf{j}(\mathbf{r},\omega) - ik_0\varepsilon(\omega)\mathbf{E}(\mathbf{r},\omega) \tag{2.3}$$

$$\operatorname{rot} \mathbf{E}(\mathbf{r},\omega) = ik_0 \mathbf{B}(\mathbf{r},\omega) \qquad k_0 \equiv \frac{\omega}{c}.$$

We now submit the field excited by the particle in the form of an axially–symmetric TM–wave; only this wave possessing a nonzero longitudinal component of the electric field in the particle trajectory can perform a work on the particle. For a point charged particle moving uniformly along the z–axis, the longitudinal current density is

$$j(\mathbf{r},t) = \frac{qv}{2\pi r}\delta(r)\delta(z-vt), \tag{2.4}$$

where $v = \beta c$ is the velocity of the emitter.

Under these conditions, the function $j(\mathbf{r},\omega)$ has the form

$$j(\mathbf{r},\omega) = \frac{q}{4\pi^2 r}\delta(r)\exp\left(i\frac{k_0}{\beta}z\right). \tag{2.5}$$

Respectively, all the components of the solution driven by this current have to depend on z as $\exp(ik_0 z/\beta)$. Hence, the set of Maxwell equations for the complex amplitudes can be reduced to two relations expressing the transverse fields via the longitudinal one

$$B_\varphi = \beta\varepsilon(\omega)E_r = \frac{ik_0}{\kappa^2}\frac{dE_z}{dr}; \qquad \kappa^2 = k_0^2\left(\varepsilon - \beta^{-2}\right), \tag{2.6}$$

and the inhomogeneous Bessel equation for E_z:

$$\frac{1}{r}\frac{d}{dr}r\frac{dE_z}{dr} + \kappa^2 E_z = -\frac{iq\kappa^2}{\pi c r k_0\varepsilon(\omega)}\delta(r)\exp(ik_0 z/\beta). \tag{2.7}$$

For $r > 0$, where the current density (2.5) is zero, the equation (2.7) is homogeneous and its solutions are cylindrical functions of the argument κr. Physically, the radiation field is defined as a wave moving away from the charged particle. Under the real and positive values of κ, the solution satisfying the condition for radiation emission at infinity should be chosen between the two possible particular solutions of this equation:

$$E_z(r,z,\omega) = C\, H_0^{(1)}(\kappa r)\exp\left(i\frac{k_0}{\beta}z\right). \tag{2.8}$$

Here $H_0^{(1)}(u)$ is a Hankel function of the first kind. This function is defined in the complex plane of the argument everywhere except the cut along the negative real semiaxis. It behaves asymptotically as an outgoing cylindrical wave when the argument u goes to infinity [7, 8]:

$$H_0^{(1)}(u) \approx \sqrt{\frac{2}{\pi u}} \exp(iu) ; \qquad u \to \infty . \tag{2.9}$$

If $u \to 0$, the Hankel function has the asymptotic

$$H_0^{(1)}(u) \approx (2i/\pi) \ln u .$$

Hence, (2.8) gives

$$\frac{dE_z}{dr} \to \frac{2i}{\pi r} C ; \qquad r \to 0 . \tag{2.10}$$

At the same time, for small values of r, the direct integration of (2.7) yields:

$$\frac{dE_z}{dr} \approx -\frac{iq\kappa^2}{\pi r \omega \varepsilon(\omega)} \exp(ik_0 z/\beta) . \tag{2.11}$$

Comparing (2.10) and (2.11), one determines the wave amplitude:

$$C = -\frac{q\kappa^2}{2\omega\varepsilon(\omega)} . \tag{2.12}$$

After the inverse Fourier transform with respect to time, we finally get the charged particle total field in the explicit form:

$$E_z(r,z,t) \tag{2.13}$$
$$= -\frac{q}{2c^2} \int_{-\infty}^{+\infty} \left[1 - \frac{1}{\beta^2 \varepsilon(\omega)}\right] H_0^{(1)}(\kappa r) \exp\left[i\omega\left(\frac{z}{v} - t\right)\right] \omega \, d\omega .$$

The physical meaning of this result is clear. Indeed, the Fourier transform with respect to time t depicts a current density harmonic modulation in the form of a traveling wave with the frequency ω and the wave number ω/v. Naturally, the fields excited by these current density waves in the dielectric ought to have the same frequencies and the wave number along z. Their propagation direction is characterized by the angle with respect to the charged particle trajectory

$$\vartheta \equiv \mathrm{arctg}\,(\kappa v/\omega) = \mathrm{arctg}\,\sqrt{\beta^2 \varepsilon(\omega) - 1} ; \qquad \beta^2 \varepsilon \geq 1 . \tag{2.14}$$

These angles actually exist only for those frequencies that satisfy the condition for the Cherenkov radiation. At sufficiently large distances from the particle trajectory (where the strong inequality $\kappa^2 r^2 \gg 1$ takes place), the field has the form of a superposition of cylindrical waves moving away from the emitter trajectory at the angles ϑ.

As it follows from (2.14), for the accepted dispersion model the Cherenkov radiation is emitted outside a conical surface with the apex of cone $\vartheta(\omega_{\min})$ where the frequency $\omega_{\min}$ is given by

$$\omega_{\min} = \begin{cases} [\Omega^2 - \beta^2\gamma^2\omega_{\rm p}^2]^{1/2} & \text{if} \quad \Omega^2 > \omega_{\rm p}^2\beta^2\gamma^2; \\ 0 & \text{if} \quad \Omega^2 < \omega_{\rm p}^2\beta^2\gamma^2; \end{cases} \qquad (2.15)$$

where $\gamma = (1-\beta^2)^{-1/2}$. The angle ϑ monotonously increases with the radiation frequency ω within the range $\omega_{\min} \leq \omega \leq \Omega$. Only in the limiting case of a nondispersive medium, the total radiation is emitted exactly at $\vartheta_{\min}$.

Coming back to (2.14), one can see that the angles ϑ are complex when $\text{Re}\left[\beta^2\varepsilon(\omega)-1\right] < 0$. Physically, it means that at these frequencies the medium provides the total internal reflection of the microwave field as it occurs in optics at the angles smaller than the Brewster angle [9, 10]. At large distances from the particle trajectory, the field amplitudes of these frequency components decrease exponentially with r:

$$H_0^{(1)}({\rm i}|\kappa|r) = \frac{2}{{\rm i}\pi}K_0(|\kappa|r) \approx -{\rm i}\sqrt{\frac{2}{\pi|\kappa|r}}\exp(-|\kappa|r) \ . \qquad (2.16)$$

Here $K_0(u)$ is a MacDonald function [7, 8].

In the particular case of vacuum ($\omega_{\rm p}=0$), the radiation field vanishes, and the integral (2.13) gives the conventional expression for the Coulomb field of a charged particle moving in a free space. In the general case the right–hand side of (2.13) may be somewhat simplified by singling out the emitter Coulomb field deformed by the medium, the field of the Cherenkov radiation, and the field of the excited polarization oscillations. The integration contour in (2.13) should be closed in the upper (if $z > vt$) or lower (if $z < vt$) half planes of the complex variable ω with only integrand singularities making nonzero contributions to the integral. They include the function $\varepsilon(\omega)$ zeros at the points

$$\omega_1^{\pm} = \pm\sqrt{\omega_{\rm p}^2 + \Omega^2} - {\rm i}\nu$$

and the cuts connecting the branch points of the Hankel function, which are $\kappa = 0$ and $\kappa \to -\infty$ (note that the point $\omega = 0$ is regular). The transformation of the cut into the ω plane for the cases of $\Omega^2 > \omega_{\rm p}^2\beta^2\gamma^2$ and $\Omega^2 < \omega_{\rm p}^2\beta^2\gamma^2$ is shown in Fig. 2.1. In the first case, the cuts go from $\omega = 0$ along the imaginary axis to $\pm\infty$ and also join $\omega = \pm\omega_{\min} - {\rm i}\nu$ with $\omega = \pm\Omega - {\rm i}\nu$ just below the real axis overlapping the region of Cherenkov radiation with $\varepsilon\beta^2 > 1$. In the second case, the cut joins $\omega = 0$ with $\omega = \pm\sqrt{\omega_{\rm p}^2\beta^2\gamma^2 - \Omega^2} - {\rm i}\nu$ again below the real axis. The cut along the imaginary axis go then to $\pm\infty$ starting at $\omega = \pm{\rm i}\sqrt{\omega_{\rm p}^2\beta^2\gamma^2 - \Omega^2}$. So far as the vertical cuts in both cases go along the imaginary axis, they correspond to harmonics exponentially decreasing with distance $|z-vt|$ from the charge and, thus, to nonpropagating Coulomb fields. By the way, they are the only fields ahead of the particle (for $z > vt$ when the integration contour is closed in the upper half-plane).

Examining the field behind the charged particle ($z - vt < 0$), one must close the integration contour in the lower half–plane. The residues at the zeros

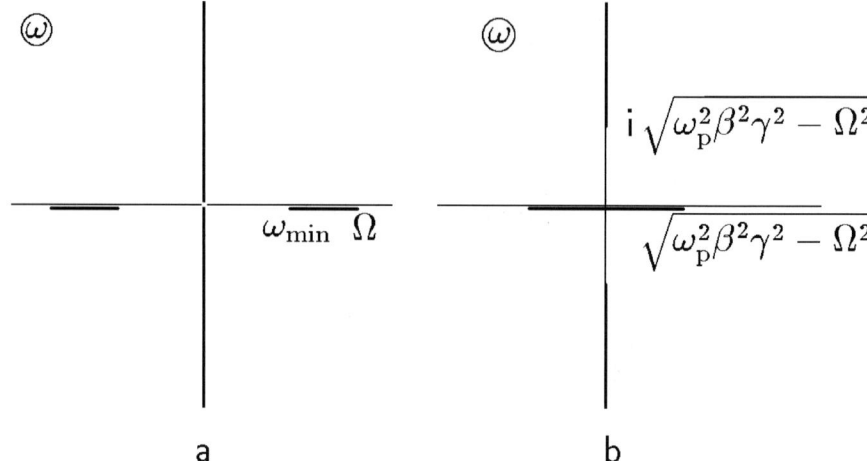

Fig. 2.1. Cuts in the plane of the complex variable ω. (a) $\Omega^2 > \omega_p^2 \beta^2 \gamma^2$, (b) $\Omega^2 > \omega_p^2 \beta^2 \gamma^2$

of $\varepsilon(\omega)$ yield the field of medium longitudinal polarization oscillations excited by the charged particle at the frequency $\omega_1^+ = \sqrt{\omega_p^2 + \Omega^2}$:

$$E_z^{(p)}(r, z < vt) = -\frac{2q\omega_1^2}{v^2} K_0\left(\frac{\omega_1^+ r}{v}\right) \cos\left[\omega_1^+ (t - z/v)\right]. \quad (2.17)$$

One may prove that the polarization of these oscillations is purely longitudinal in the following way. Let us Fourier–transform the first equation in the system (2.3) with respect to the coordinate **r**. Scalar multiplication by the vector **k** of both the parts of the linear inhomogeneous equation, obtained for $\mathbf{E}(\mathbf{k}, \omega)$ yields:

$$\mathbf{E}_l \equiv \frac{(\mathbf{k}, \mathbf{E}(\mathbf{k}, \omega))}{k^2}\mathbf{k} = -\frac{4\pi i(\mathbf{j}(\mathbf{k}, \omega), \mathbf{k})}{\omega \varepsilon(\omega) k^2}\mathbf{k}. \quad (2.18)$$

As the expression indicates, zeros of the function $\varepsilon(\omega)$ really do correspond to the medium free longitudinal oscillations.

In the case examined the excited microwave field has the form of the wave (2.17), running behind the charged particle with the phase velocity v as it should be from the viewpoint of the synchronism condition (see Fig. 2.2). Its amplitude exponentially decays in transverse direction.

As has been mentioned above, the integrals along the cuts located in vicinity of the real axis yield the Cherenkov radiation field:

$$E_z^{(C)}(r, z < vt) \quad (2.19)$$
$$= -\frac{2q}{c^2} \int_{\omega_{\min}}^{\Omega} \omega\, d\omega \left[1 - \frac{1}{\beta^2 \varepsilon(\omega)}\right] J_0(\kappa(\omega)r) \cos\left[\omega(t - z/v)\right].$$

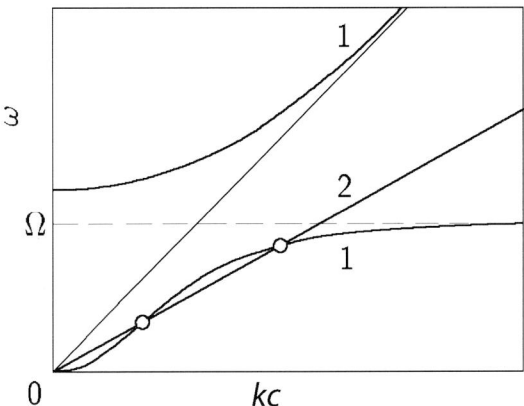

Fig. 2.2. Schematic dispersion for the Fermi model. Curve 1: electromagnetic waves. Curve 2: $\omega = kv$. The circles correspond to Cherenkov radiation and to polarization losses

Here $\omega_{\min}$ is determined by (2.15).

Finally, the integral along the cut in the imaginary half–axis gives the Coulomb field in the back half–space ($z < vt$), which is precisely antisymmetric with respect to the emitter location.

The seemingly formal operations reveal the deep physical meaning of the medium damping constant ν introduced above (finally it may be equated to zero). The infinitesimal small damping creates asymmetry between the past and the future by shifting pole locations into the lower half-plane of ω. From the viewpoint of physics, this asymmetry implies retaining only retarding potentials and excluding those advanced.[1]

The Cherenkov transverse radiation field does not exist in front of the particle because it cannot be in time for getting into the area ahead of the Cherenkov cone: the field propagates along z–axis synchronously with the particle, while along the radius r its phase velocity is

$$v_{\rm ph} \equiv \frac{\omega}{k_\perp(\omega)} = \frac{v}{\sqrt{\beta^2 \varepsilon(\omega) - 1}} \; . \qquad (2.20)$$

Calculating the total longitudinal field at the point of the emitter location, we can find a decelerating force acting on the particle and, thus, determine the radiation energy losses. In the case examined this force is prescribed by half–sums of the total fields in front and behind the particle:

$$F_z^{(1)} = \frac{q^2 \omega_{\rm p}^2}{v^2} K_0\left(\omega_1^+ r_{\min}/v\right) \; ; \qquad (2.21)$$

---

[1] Because of this reason, below we suppose that $\nu = 0$ and integrals over ω in the inverse Fourier transformation are taken along the contour that passes above singularities on the real axis.

$$F_z^{(C)} = \frac{q^2 \omega_p^2}{v^2} \left[\ln \gamma^2 - \beta^2\right] \quad \text{if} \quad \Omega^2 > \omega_p^2 \beta^2 \gamma^2 \tag{2.22}$$

$$F_z^{(C)} = \frac{q^2 \omega_p^2}{v^2} \left[\ln \left(1 + \frac{\Omega^2}{\omega_p^2}\right) - \frac{\Omega^2}{\gamma^2 \omega_p^2}\right] \quad \text{if} \quad \Omega^2 < \omega_p^2 \beta^2 \gamma^2 \;.$$

As it follows from these formulae, the polarization losses increase approximately in proportion to the medium density. If $\gamma \gg 1$, they weakly depend on the emitter energy. As regards the Cherenkov losses, they increase linearly with the medium density only in the range of its relatively small values, where the inequality $\Omega^2 > \omega_p^2 \beta^2 \gamma^2$ holds. In the region of $\Omega^2 \ll \omega_p^2$, they practically do not depend on the medium density, being determined only by a minimum wavelength of radiation emitted in the Cherenkov spectrum

$$\min \lambda_C = \frac{2\pi v}{\Omega} \;. \tag{2.23}$$

For this wavelength the radiation field energy spectral density is maximal.

As regards the parameter $r_{\min}$, formally introduced in the r.h.s. of (2.7), it is determined by the maximal momentum $p_{\max}$ transferred from the moving charged particle to an atom of the medium:

$$r_{\min} = \hbar / p_{\max} \;.$$

If the corresponding energy value exceeds the ionization potential of the atom I_i, the formula (2.7) describes not only the particle energy losses due to exciting the medium longitudinal (polarization) oscillations, but also the losses for the medium ionization. A minute analysis of this problem can be found in [10].

It should also be noted that in the nondispersive dielectric model with $\varepsilon = \text{const}$ the Coulomb fields of the charged particle behind and in front of the latter vanish completely (as well as the field of longitudinal oscillations). In this case the limits of integration in the expression for the Cherenkov radiation field (2.19) extend all over the real half–axis $\text{Re}\,\omega > 0$. The corresponding decelerating force acting on the particle increases indefinitely as a result of the overestimated contribution of the high frequencies ($\omega \gg \Omega$).

As a rule, in real dielectrics a number of the proper frequencies Ω are large, and they are located in frequency bands from microwaves and up to the optical range. If the medium transparency is high enough, the conditions for Cherenkov radiation are satisfied in vicinity of these resonance frequencies. So, the radiation spectrum can take the form of a set of bands, including the overlapping ones. In condensed dielectrics where the strong inequality $\omega_p^2 \gg \Omega^2$ takes place, the polarization losses considerably exceed the Cherenkov radiation losses. Under these conditions, to single out the latter in the pure form, a vacuum channel in a dielectric [11] can be used. Surely, if the channel radius a is small compared with the optical radiation characteristic wavelength

($a \ll v/\Omega$), its presence practically would not exert any influence on the Cherenkov radiation parameters. Besides, if the radius a is sufficiently larger than the characteristic skin depth where the polarization losses occur ($a \gg v/\omega_p$), these losses would become exponentially small. Thus, the Cherenkov radiation in a condensed dielectric is observable in its pure form in a channel of radius

$$v/\omega_p \ll a \ll v/\Omega \,.$$

2.1.2 Cherenkov Radiation in Magnetized Plasma Waveguide

An idealized nature of the medium model used above is rather obvious. Really, the medium was supposed to be boundless in the transverse direction and the specific dispersion law was prescribed. However, if radiation is emitted by an electron in a perfectly conductive circular waveguide filled with a homogeneous plasma and placed into uniform longitudinal magnetic field, these conditions would be somewhat closer to reality, at least within the frames of plasma electronics (this branch of microwave electronics is under intensive development nowadays [12, 13]). In the approximation of a strongly magnetized plasma[2] this system, being anisotropic, may be described by a diagonal tensor of dielectric permeability [14, 15]:

$$\hat{\varepsilon}(\omega) = \begin{pmatrix} 1 & 0 & 0 \\ 0 & 1 & 0 \\ 0 & 0 & \varepsilon(\omega) \end{pmatrix}; \quad \varepsilon(\omega) = 1 - \frac{\omega_p^2}{\omega(\omega + 2i\nu)} \,. \tag{2.24}$$

The qualitative distinction from the previous model is conditioned by the tensor nature of the dielectric constant and by the presence of walls (they exclude formation of the Cherenkov cone, which is a characteristic of a boundless medium).

In the case considered here, the structure of Maxwell equations is the same as for an isotropic medium (see (2.3)). However, the relationships (2.6) should be changed for

$$B_\varphi = \beta E_r = -\frac{i\gamma^2 \beta^2}{k_0} \frac{dE_z}{dr}; \quad \kappa^2 \equiv -\frac{k_0^2 \varepsilon(\omega)}{\gamma^2 \beta^2} \,. \tag{2.25}$$

The corresponding inhomogeneous Bessel equation for E_z takes the form (compare with (2.7)):

$$\frac{1}{r}\frac{d}{dr} r \frac{dE_z}{dr} + \kappa^2 E_z = \frac{iqk_0}{\pi c r \gamma^2 \beta^2} \delta(r) \exp(ik_0 z/\beta) \,. \tag{2.26}$$

Note that for the positive $\varepsilon(\omega)$ (imaginary values of κ) the dependence of the field amplitude E_z on r must be exponential rather than oscillatory as in the previous case (see (2.8)).

---

[2] This approximation is valid when the thermal energy of plasma electrons is smaller than the magnetic field energy.

When $\varepsilon(\omega) < 0$, the solution of (2.26) is a cylindrical function of a real argument

$$Z_0(\kappa r) = J_0(\kappa r) N_0(\kappa b) - N_0(\kappa r) J_0(\kappa b) \ .$$

Here J_0 and N_0 are Bessel and Neumann functions, respectively.

The function $Z_0(\kappa r)$ is chosen to satisfy the zero boundary condition on the waveguide wall ($r = b$). Taking into account that for small u

$$N_0(u) \to \frac{2}{\pi} \ln u; \qquad J_0(u) \to 1; \qquad N_1(u) \to -\frac{2}{\pi u}$$

we restore the field in its explicit form by the same technique as above:

$$E_z = -\frac{iq}{2c^2\gamma^2\beta^2} \int_{-\infty}^{+\infty} \frac{Z_0(\kappa r)}{J_0(\kappa b)} \exp\left[i\frac{\omega}{v}(z - vt)\right] \omega \, d\omega \ . \tag{2.27}$$

As in the previous case, different singularities of the integrand determine the fields that have distinct physical meanings. The cuts on the real axis of the complex variable ω yield the Coulomb field component. The latter, being antisymmetric with respect to the charged particle location decreases on both sides. The radiation itself, that is, the wave field is due to zeros of $J_0(\kappa b)$. Being located on the real frequency axis, these zeros correspond to real values of κ. The fields, determined by the corresponding residues, do not exist in front of the charged particle. Behind the emitter they can be presented as the sum

$$E_z^{(\text{rad})} = -\frac{2q}{b^2} \sum_{n < n_{\max}} \frac{J_0(\lambda_n r/b)}{J_1^2(\lambda_n)} \cos\left[\frac{\omega_n}{v}(z - vt)\right] \tag{2.28}$$

$$\omega_n = \sqrt{\omega_p^2 - \lambda_n^2 \gamma^2 \beta^2 c^2 / b^2} > 0 \ ; \quad z < vt \ .$$

Here λ_n are the roots of the Bessel function J_0 and the summation is extended over those numbers n, for which the values of ω_n are real. In particular, if $r = 0$ and $z \to vt$, the formula (2.26) gives the average decelerating field that performs negative work on the charged particle, that is, determines the energy losses due to the radiation emission.

The physical meaning of the results obtained and their interpretation from the viewpoint of the concept of the wave–particle synchronous motion are illustrated in Fig. 2.3. In particular, one can see the point of intersection of the line $\omega = kv$ with the dispersion curve of a slow TM–wave in a "cold" magnetized plasma waveguide.

For a fixed value of n, the dispersion curves belong to the regions $0 < \omega < \omega_p$ and $\omega > \sqrt{\omega_p^2 + \lambda_n^2 c^2/b^2}$. The upper one relates to a fast TM–wave with a phase velocity that always exceeds the velocity of light:

$$\omega_+^2(k) = \frac{1}{2}\left(\omega_p^2 + \lambda_n^2 c^2/b^2 + k^2 c^2\right) \tag{2.29}$$

$$+ \sqrt{(\omega_p^2 + \lambda_n^2 c^2/b^2 + k^2 c^2)^2 - 4k^2 c^2 \omega_p^2} \ .$$

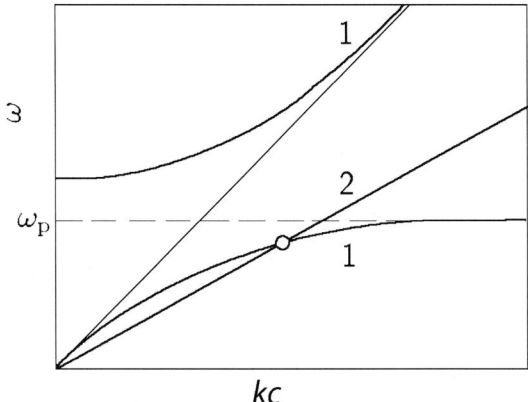

Fig. 2.3. Dispersion for cold magnetized plasma. Curve 1: electromagnetic wave. Curve 2: $\omega = kv$. The circle corresponds to Cherenkov radiation

When the plasma density is small this wave transforms to the well–known axially symmetric fast ($v_{\mathrm{ph}} > c$) TM–wave of a circular cylindrical waveguide with conductive walls.

The lower pass-band ($0 < \omega < \omega_{\mathrm{p}}$) exists only under finite values of the plasma density. The phase velocities of TM–waves within this band are limited from above by a value smaller than the velocity of light:

$$\frac{\omega}{k} < \frac{\omega_{\mathrm{p}} c}{\sqrt{\omega_{\mathrm{p}}^2 + \lambda_n^2 c^2/b^2}} < c \; . \qquad (2.30)$$

It is interesting to note that this determines the maximal number $n_{\max}$ of the radial mode, which still can be emitted by the particle. In particular, if the particle velocity is high enough or the plasma density is low, the Cherenkov radiation in the system considered cannot be emitted at all. In this case, the excited field decreases exponentially on both sides of the emitter and is antisymmetric with respect to the point $z - vt$. Therefore, it can provide neither the radiation emission nor the emitter deceleration and should be classified as a distorted Coulomb field.

In addition, the anisotropic magnetized plasma considered in this subsection has no longitudinal polarization oscillations, while in isotropic plasma such oscillations do exist (e.g., they can be excited by a charge moving uniformly).

Concluding this section we regret that many interesting features of Cherenkov radiation remained out of its scope. More details and information about its discovery and unduly forgotten preceding works the reader can find in the monographs [6, 7].

2.2 Transition Radiation

As it has been mentioned above, emission of transition radiation takes place when a charged particle, moving uniformly, transverses an inhomogeneity of its electrodynamic environment, be it a spatial change of the medium dielectric properties or a discontinuity of the boundary conditions. While the particle is moving in vicinity of the inhomogeneity, some deformation of the proper field pattern has to occur. During this process, the derivatives of the electric and magnetic field components with respect to time and space are nonzero. Consequently, the electromagnetic fields under excitation are moving away from the charged particle in the form of the medium proper waves. The set of such waves defines the transition radiation emission.

To demonstrate the quantitative characteristics of transition radiation, we choose the simplest models of medium electrodynamic inhomogeneities. These models, chosen to illustrate the applied problems of microwave electronics and particle accelerators, also permit presenting the final results in the explicit analytical form.

We now suppose that an origin of the transition radiation is a thin current filament with the current density in the form of a wave propagating with a given phase velocity v (on the basis of the reasoning given above, one can also clearly see that the result to be obtained will permit restoring the field structure of the radiation emitted by a single particle as well).

2.2.1 Medium Step–Like Inhomogeneity

Thus, we now consider an infinite circular waveguide ($-\infty < z < +\infty$) with perfectly conductive walls of radius b filled with some nondispersive dielectric with a step–like profile of the dielectric constant $\varepsilon(z)$:

$$\varepsilon(z) = \begin{cases} \varepsilon_- & \text{if } z < 0 \,; \\ \varepsilon_+ & \text{if } z > 0 \,. \end{cases}$$

For simplicity, we also assume that the conditions for emission of the Cherenkov radiation are not met in either of the semi–infinite homogeneous dielectric cylindrical waveguides under consideration (i.e., $\beta^2 \varepsilon_\pm < 1$).

As in the case of the Cherenkov radiation examined above, an emitter excites only axially symmetric TM–waves, their longitudinal electric fields taking away kinetic energy from the particle. For each of the half–spaces, the corresponding system of Maxwell equations has the form:

$$\frac{\partial}{\partial z} B_\varphi(r, z, \omega) - i k_0 \varepsilon_\pm E_r(r, z, \omega) = 0 \,, \tag{2.31a}$$

$$\frac{\partial}{\partial z} E_r(r, z, \omega) - \frac{\partial}{\partial r} E_z(r, z, \omega) - i k_0 B_\varphi(r, z, \omega) = 0 \,, \text{ and} \tag{2.31b}$$

$$\frac{1}{r} \frac{\partial}{\partial r} [r B_\varphi(r, z, \omega)] + i k_0 \varepsilon_\pm E_z(r, z, \omega) = \frac{q \delta(r)}{\pi c r} \exp\left(i k_0 z / \beta\right) \,, \tag{2.31c}$$

where zero boundary conditions for E_z on the cylindrical surface $r = b$ must be taken into account.

In both the half–spaces, the solutions of the homogeneous system of equations (2.31a)–(2.31c) are the $\exp(ik_{\pm}z)$–type free waves with the propagation constants $k_{\pm}$; their field radial distributions are determined by the corresponding boundary problem.

Elimination of B_φ and E_r from the system (2.31a)–(2.31c) yields the homogeneous Bessel equation for E_z:

$$\frac{1}{r}\frac{d}{dr} r \frac{d}{dr} E_z + \left(k_0^2 \varepsilon_{\pm} - k_{\pm}^2\right) E_z = 0 \quad (2.32)$$

$$E_z(r=b) = 0 \,.$$

So, the solution can be presented as:

$$E_z(r, z, \omega) = \sum_n C_n^{\pm} J_0\left(\frac{\lambda_n r}{b}\right) \exp(\pm ik_{\pm}z)$$

$$B_\varphi(r, z, \omega) = \sum_n \frac{ik_0 \varepsilon_{\pm} b}{\lambda_n} C_n^{\pm} J_1\left(\frac{\lambda_n r}{b}\right) \exp(\pm ik_{\pm}z) \quad (2.33)$$

$$E_r(r, z, \omega) = \sum_n \frac{\pm ik_{\pm} b}{\lambda_n} C_n^{\pm} J_1\left(\frac{\lambda_n r}{b}\right) \exp(\pm ik_{\pm}z)$$

with

$$k_{\pm} = \sqrt{k_0^2 \varepsilon_{\pm} - \lambda_n^2/b^2} \,.$$

Here λ_n are the roots of the Bessel function: $J_0(\lambda_n) = 0$, $\lambda_1 = 2.405$.

The amplitudes $C_n^{\pm}$ are unknown yet and the frequency ω is considered to be higher than the cutoff frequency of the corresponding mode of the "cold" waveguide. These are the fast free waves, leaving the discontinuity surface with the phase velocity exceeding that of light, which represent transition radiation. The low-frequency field component ($k_0^2 \varepsilon_{\pm} < \lambda_1^2/b^2$) decays exponentially as a distance from the discontinuity surface increases.

The solution driven by the current is characterized by the z–dependence $\propto \exp(ik_0 z/\beta)$. Using the well–known expansion of the δ–function over Bessel functions

$$\frac{\delta(r)}{r} = \sum_{n=1}^{\infty} \frac{2 J_0(\lambda_n r/b)}{b^2 J_1^2(\lambda_n)}, \quad (2.34)$$

one can also develop the forced solution as an expansion of the type of (2.33):

$$E_z(r, z, \omega) = \sum_n \frac{-2iqk_0 \left(1 - \varepsilon_{\pm} \beta^2\right)}{\pi c \varepsilon_{\pm} \beta^2 b^2 J_1^2(\lambda_n) \Delta_{\pm}} J_0\left(\frac{\lambda_n r}{b}\right) \exp(ik_0 z/\beta) \quad (2.35a)$$

$$B_\varphi(r, z, \omega) = \frac{2q}{\pi c b^3} \sum_n \frac{\lambda_n J_1(\lambda_n r/b)}{\Delta_{\pm} J_1^2(\lambda_n)} \exp(ik_0 z/\beta) \quad (2.35b)$$

$$E_{\rm r}(r, z, \omega) = \frac{B_\phi(r, z\omega)}{\beta_0 \varepsilon_\pm} \,. \tag{2.35c}$$

Here

$$\Delta_\pm \equiv \frac{\omega^2}{v^2} + \frac{\lambda_n^2}{b^2} - k_0^2 \varepsilon_\pm \,.$$

The physical meaning of the expressions obtained is rather clear. Indeed, on the right of (2.31c), the propagation constant of the driving current wave does not correspond to the proper wave dispersion. Consequently, the forced solution depicts a field "tied up" to the charged particle and incapable of propagating by itself over any large distances from the emitter.

As the total field components B_φ and E_r have to be continuous at $z = 0$, it is easy to find the amplitudes $C_n^\pm$ determining the mode composition of the transition radiation emitted along the emitter trajectory $(+)$ and in the opposite direction $(-)$:

$$C_n^\pm = \frac{\pm 2 i q (\varepsilon_+ - \varepsilon_-) \lambda_n^2}{\pi c b^4 \Delta_+ \Delta_- \varepsilon_\pm (\varepsilon_- k_+ + \varepsilon_+ k_-) J_1^2(\lambda_n)}$$
$$\times \left\{ k_0 k_\mp \varepsilon_\pm \mp \frac{1}{\beta} \left[\frac{\omega^2}{v^2} + \frac{\lambda_n^2}{b^2} - k_0^2 (\varepsilon_+ + \varepsilon_-) \right] \right\} \,. \tag{2.36}$$

In the general case (under arbitrary values of the constant $\varepsilon_\pm$) the expressions for the transition radiation fields are rather complicated. However, they assume a relatively simple form if we address to the case of a practical interest, regarding one of the media (e.g., input one) as a perfectly conductive metal. Correspondingly, considering the limit $\varepsilon_+ \to \infty$, we get the total field in the metal $(z > 0)$ equal to zero, whereas in the backward wave of the transition radiation $(z < 0)$ it takes the form:

$$B_\varphi = \sum_n \frac{k_0}{k_-} \frac{2q\lambda_n}{\pi b^3 c J_1^2(\lambda_n) \Delta_-} J_1\left(\lambda_n \frac{r}{b}\right) \exp(-i k_- z) \tag{2.37a}$$

$$E_{\rm r} = \sum_n \frac{-2q\lambda_n}{\pi b^3 c \beta \varepsilon_- J_1^2(\lambda_n) \Delta_-} J_1\left(\lambda_n \frac{r}{b}\right) \exp(-i k_- z) \tag{2.37b}$$

$$E_z = \sum_n \frac{-2 i q \lambda_n^2}{\pi b^4 c k_- \beta J_1^2(\lambda_n) \varepsilon_- \Delta_-} J_0\left(\lambda_n \frac{r}{b}\right) \exp(-i k_- z) \,. \tag{2.37c}$$

In the majority of cases of a single discontinuity (detectors of charged particles, etc.) one is interested not in the field distributions but in a magnitude of the particle total energy losses ΔW by the emission into the left half–space. To calculate these losses for the given model, it is necessary to consider the longitudinal electric field at the point $r = 0$, $z = vt$ and find its work on the particle all over its trajectory. It is unnecessary to take into account the Coulomb field because the total work performed by the latter on the particle is evidently equal to zero. As (2.33) describes the Fourier transforms,[3] the

---

[3] Because of the reasons already discussed in the previous subsection, all the operations must be performed under the assumption that $\operatorname{Im}\omega > 0$.

calculation of the energy losses also yields the mode content and frequency spectrum of radiation:

$$\Delta W_-^{(n)} \equiv q \int_{-\infty}^{0} v\, E_{zn}(r=0, z=vt, t) dt \qquad (2.38)$$

$$= -\frac{4q^2 \lambda_n^2}{\pi v b^4 J_1^2(\lambda_n)} \int_0^\infty d\omega \left(\frac{\omega^2}{v^2 \gamma^2} + \frac{\lambda_n^2}{b^2} \right)^{-2} = -\frac{q^2 \gamma}{b \lambda_n J_1^2(\lambda_n)}.$$

Here, for simplicity, it is supposed that $\varepsilon_- = 1$, which conditions that (2.38) yields the radiation losses at the boundary between a conductor and a vacuum.

As the right–hand side of (2.38) indicates, these losses increase with the relativistic factor γ. From the viewpoint of physics, this effect is explicable in the following way. The characteristic longitudinal size of the region of localization of the emitter Coulomb field diminishes while γ grows. Consequently, the rate of the temporal rearrangement of the field structure has to increase. As a result, the intensity of transition radiation, the emission of which is driven by this rearrangement, rises up as well.

In addition, from (2.38), one can also judge that the emitter energy losses in question consist of the energy spent on the emission of the transverse waves moving away from the conductor ($\omega^2 > c^2 \lambda_n^2/b^2$; $\mathrm{Re}\, k_-^2 > 0$). The emitter total energy losses also include the energy spent on exciting those electromagnetic fields, the amplitudes of which decay exponentially when $z \to -\infty$ ($\omega^2 < c^2 \lambda_n^2/b^2$; $\mathrm{Re}\, k_-^2 < 0$).

2.2.2 Smooth Inhomogeneity of Medium

The results obtained above relate to the models where the dielectric constant undergoes a discontinuity, that is, it alters within the distances, considerably shorter than the radiation wavelength. Surely, this approximation overestimates the contribution of the radiation emitted at high frequencies that could be of importance to the applications. Seemingly, in the presence of a smooth inhomogeneity, the radiation spectrum should be cut off effectively at the wavelengths of order of the inhomogeneity size. Nevertheless, this general rule is not fulfilled in the case.

If the dielectric constant varies along z arbitrarily, in Eqs. (2.31a)–(2.31c) one should substitute $\varepsilon(z)$ for $\varepsilon_\pm$. Then the harmonic functions $\exp(\pm i k_\pm z)$ cannot be the proper waves. However, the separation of variables is still admissible in the system of equations. Expanding δ–function in eigenfunctions of the boundary problem (see (2.34)), one obtains two coupled equations for Bessel transforms of the transverse fields:

$$\frac{d}{dz} B_\varphi(z,\omega) - ik_0 \varepsilon(z) E_r(z,\omega) = 0 \qquad (2.39a)$$

$$ik_0 \varepsilon(z) \frac{d}{dz} E_r(z,\omega) + \left(k_0^2 \varepsilon - \frac{\lambda_n^2}{b^2} \right) B_\varphi(z,\omega)$$

$$= -\frac{2q\lambda_n}{\pi c b^3 J_1^2(\lambda_n)} \exp(ik_0 z/\beta_0) \ . \tag{2.39b}$$

If $\epsilon(z)$ varies smoothly enough over the wavelength–order distance, it is convenient to present a solution in the form of waves propagating along the emitter trajectory and in the opposite direction with their propagation constants, amplitudes, and phases varying smoothly as well (so-called WKB–approximation):

$$B_\varphi(z,\omega) \tag{2.40}$$
$$= \sqrt{\frac{\varepsilon}{k}} \left[D^+(z) \exp\left(i\int k(z)\,dz\right) + D^-(z) \exp\left(-i\int k(z)\,dz\right) \right]$$

where

$$k(z) = \sqrt{k_0^2 \varepsilon - \lambda_n^2/b^2} \ . \tag{2.41}$$

Omitting simple calculations, we get for the complex amplitudes $D^\pm$:

$$\frac{D^\pm}{dz} = \pm \frac{iq\lambda_n}{\pi c b^3 J_1^2(\lambda_n)\sqrt{k\varepsilon}} \exp\left(i\int (k_0/\beta \mp k)\,dz\right)$$
$$\pm \frac{i}{2}\sqrt{\frac{\varepsilon}{k}} \left[D^\pm + D^\mp \exp\left(\mp 2i\int k\,dz\right) \right] \frac{d}{dz}\left(\frac{1}{\varepsilon}\frac{d}{dz}\sqrt{\frac{\varepsilon}{k}}\right) \ . \tag{2.42}$$

Neglecting the second term on the right (it contains the derivatives of the slowly varying function $\varepsilon(z)$) and taking into account the natural conditions at $z \to \pm\infty$, we obtain[4]

$$D^\pm = \pm \frac{iq\lambda_n}{\pi c b^3 J_1^2(\lambda_n)} \int_{\mp\infty}^{z} \exp\left(i\int (k_0/\beta \mp k)\,dz\right) (k\varepsilon)^{-1/2}\,dz \ . \tag{2.43}$$

The structure of the latter expression permits explaining the process of radiation wave emergence all over the $\varepsilon(z)$ inhomogeneity. If this parameter varies smoothly, the integral on the right of (2.43) is small because of the exponential factor fast oscillations. However, if within a certain section of the waveguide the wave "instantaneous" phase velocity $\omega_0/k(z)$ is close to the emitter velocity v, the integrand substantially contributes to the excited wave amplitude. The physical sense of this phenomenon is as follows. In a weakly inhomogeneous medium, an emitter with its velocity below the Cherenkov radiation threshold ($v < c/\sqrt{\varepsilon}$) cannot transfer any substantial energy to the transition radiation field. Indeed, the particle proper Coulomb field structure varies too slowly under such conditions. Thus, in the case of a weak inhomogeneity, the emitter energy transferred to the transition radiation is exponentially small.

---

[4] Generally speaking, the omitted terms change only the phase of the wave emitted, but in some cases their inclusion reveals the possibility of the wave parametric excitation, related to the "resonant" transition radiation considered below.

The situation changes essentially if the emitter for a rather long time locks in the Cherenkov synchronism with the wave generated on the inhomogeneity. Really, let us suppose that at a certain region of a weakly inhomogeneous medium the latter condition is satisfied. Consequently, the "length of radiation formation" (defined as the distance where the wave phase shift with respect to the emitter is of the order of 2π) essentially exceeds the wavelength (sometimes, by several orders of magnitude). From the viewpoint of applications, it means the following: in a detector of transition radiation, emitted by relativistic particles, it is utterly unnecessary to match the dielectric boundary with an optic accuracy. As regards the fundamental characteristics of radiation, the conclusion once more emphasizes an important role of the condition of the synchronism between an emitter and the excited wave. Note that for the backward wave this effect does not take place.

Generally speaking, the existence of an inhomogeneity of the medium dielectric constant does not make the necessary condition for emitting the transition radiation. An inhomogeneity in the boundary conditions would be enough for realizing this effect. A good example is the radiation emission by a charged particle passing over an edge of a semi–infinite waveguide with conductive walls or through a hole in a conducting plane, etc. We will not discuss these cases here because they are of importance mainly for single particle detection purposes rather than for generation of radiation. The theory of transition radiation can be found in numerous works starting from the pioneer papers by I. Frank and V. Ginsburg [18] and in contemporary reviews [19, 20]. Those interested in the used mathematical methods and in their applications for more complicated geometry can be addressed, for example, to [21, 22].

2.3 Cherenkov Radiation in Periodic Structures

Transition radiation examined above relates to the case of a single isolated inhomogeneity passed by a moving charged particle. Naturally, radiation of this type should be characterized by the total energy emitted (or by its spectral–angular distribution) rather than by the power of losses as it takes place in distributed systems with the quasi–resonant interaction.

At the same time, it is easy to imagine identical inhomogeneities that a particle periodically encounters during its motion. The transition radiation emitted in the latter case is being summed up coherently. The amplification conditioned by coherent summation is maximal for those waves the phase shifts of which are equal to a multiple of 2π during the time of the particle passage over one period of the structure D:

$$(\omega - kv)\frac{D}{v} = 2\pi n; \qquad n = 0, \pm 1, \pm 2, \dots \qquad (2.44)$$

Hence, at a given frequency, such resonant (or parametric) transition radiation will be characterized by a petal–like angular distribution with the characteristic angles ϑ for which

$$\cos\vartheta = \frac{kc}{\omega} = \frac{1}{\beta} - \frac{2\pi c}{D\omega}n; \qquad n = 1;\ 2;\ \ldots \qquad (2.45)$$

For instance, a particle moving parallel to the surface of a diffraction grating should emit radiation of that pattern.[5] Such relatively simple arguments are inapplicable if the radiation occurs in a periodic waveguide structure typical for microwave devices and linear accelerators. Such structures are known as retarding electromagnetic waves, meaning that a certain component of the wave propagates with a phase velocity $v_{ph} \approx v$. Then the resonant transition radiation can be interpreted as Cherenkov radiation regarding this quasi-synchronous harmonic. Obviously, this interpretation is more adequate because the term "transition" implies more or less localized inhomogeneities.

2.3.1 Proper Waves in Periodic Structures

According to fundamental principles of physics, there exists a close link between the symmetry properties of a system and the inherent conservation laws (Noether's theorem). In our case, it is a correlation between characteristics of the structure proper electromagnetic waves and the system symmetry. For example, as it follows from the system steadiness (i.e., physical equivalence of all moments of time), a wave is characterized by a constant value of its frequency. Translational symmetry with respect to any of the generalized coordinates provides the conservation of the corresponding generalized momentum; that is, a definite wave propagation constant does exist being a vector in an isotropic system or a scalar in systems uniform only along z. In the first case, the proper waves are plane waves of the $\exp(\mathbf{ikr})$–type with a fixed photon momentum $\hbar\mathbf{k}$, while in the second case they are cylindrical waves with the projection of a momentum on z–axis equal to $\hbar k$. The proper waves with a fixed angular momentum correspond to axially symmetric systems; the waves with a given left–(right–) hand circular polarization correspond to a parity symmetry, etc.

As it is clear from the above, any of definite k values cannot be prescribed to proper waves of inhomogeneous structures. Strictly speaking, these waves are not even periodic in space, being monochromatic in time. Correspondingly, the notion of a proper wave phase velocity makes no direct physical sense. Nevertheless, because of periodicity of all the characteristics along z–axis, there must exist a more general conservation law (or a wave characteristic, which would be analogous with k to some extent).

Indeed, let us consider two points located on the same radius and separated longitudinally exactly by the structure period D along the z–axis. The fields at these points on a fixed moment of time can differ only in their phases.

---

[5] In the particular case of a nonrelativistic particle moving over a comb–like conductive structure, this radiation can be regarded as the radiation emitted by the charge electrostatic image changing its position periodically (Smith–Purcell effect [23]).

2.3 Cherenkov Radiation in Periodic Structures

This phase shift per period D denoted by μ is independent of the choice of the points pair. Furthermore, restricting ourselves just to the case of axially symmetric proper waves and writing their potentials as

$$\mathbf{A}(r,z) = \mathbf{a}^{(\mu)}(r,z) \exp(i\mu z/D)$$
$$\Phi(r,z) = \phi^{(\mu)}(r,z) \exp(i\mu z/D) .$$

We see that the functions $\mathbf{a}^{(\mu)}(r,z)$ and $\phi^{(\mu)}(r,z)$ are periodic in z and meet the gauge condition

$$k_0 \phi^{(\mu)} = \mu a_z^{(\mu)}/D - i \operatorname{div} \mathbf{a}^{(\mu)} .$$

In a certain respect, there exists essential difference between the proper wave characteristic μ/D and the propagation constant k in the case of a homogeneous system. According to its physical meaning, the phase shift μ is determined only within an interval, equal to 2π, the choice $-\pi \leq \mu \leq \pi$ being handy. Within this interval a discrete set of branches of the dispersion curve $\omega(\mu)$ exists, each of them being determined by its proper cutoff frequency ω_i. To avoid misunderstanding, we note that the cutoff frequency spectrum does not coincide with the spectrum of a regular (homogeneous) waveguide, being considerably more complicated. However, the lowest cutoff frequencies are still of the order c/b, where b is the system transverse size.

A proper wave is completely determined by the cutoff frequency and by the value of μ. Its field configuration depends on the periodicity element geometry and on boundary conditions. Nevertheless, an important feature remains the same: as μ is varying from $-\pi$ up to $+\pi$, the proper wave frequency remains within a finite frequency band. There exist certain frequency values that any of real μ do not correspond to. As they say, the waves with such frequencies belong to a structure stop band. Experiencing a total internal reflection in a periodic structure, these waves cannot propagate there. The corresponding stop bands are of the same physical nature as the analogous bands in the case of Bragg diffraction of X–rays in crystals (there exists a lattice period, corresponding to certain incidence angles, for which a given frequency belongs to a stop band). The physical nature of the total reflection consists in coherent addition of the waves reflected from a periodic sequence of disturbances. On both the boundaries of the transmission bands (i.e., for $\mu = 0; \pm\pi$), the notion of the propagation direction loses its sense, and the proper wave ought to be a standing wave here. Note that if $\mathbf{a}^{(\mu)} \exp(i\mu z/D)$ is a solution of Maxwell equations, the complex conjugate function $\mathbf{a}^{(\mu)*} \exp(-i\mu z/D)$ (the opposite wave) is the solution as well because the coefficients of these equations are real. In the systems satisfying the condition of reciprocity (i.e., both the opposite directions of the wave propagation are equivalent to one another), the following equality is true[6]:

---

[6] Although this condition would seem obvious, it is not always satisfied. The propagation of clockwise and counterclockwise polarized waves in helically symmetric systems contrasts with the previous example.

$$\mathbf{a}^{(\mu)}(r,-z) = \mathbf{a}^{(\mu)*}(r,z) \ .$$

As in regular systems, the proper waves of a periodic structure may be classified according to supplementary physical features. In particular, in axially symmetric systems, all axially symmetric waves may be divided into the two independent classes: M or TE–waves with transverse component of the electric field and E or TM–waves having an axial electric component and a purely azimuthal magnetic field. In what follows, we are interested mainly in the waves of the second type because only they can be excited by a charged particle moving along the waveguide axis. Such classification is impossible for waves nonsymmetric azimuthally.

As the functions $\mathbf{a}^{(\mu)}(r,z)$ and $\phi^{(\mu)}(r,z)$ are periodic in z with the period D, they can be presented as a sum of Fourier harmonics having the form $\exp(i2\pi s z/D)$ within the interval $0 \le z \le D$. Thus, in a periodic structure, each of its proper waves is a sum of the $\exp[i(\mu+2\pi s)z/D - i\omega(\mu)t]$–type harmonic waves traveling along z–axis with the wave numbers $k = (\mu+2\pi s)/D$ and phase velocities $v_{\mathrm{ph}} = \omega(\mu)D/(\mu+2\pi s)$. In contrast to a homogeneous system, these traveling harmonics cannot exist independently because each of them cannot meet alone the necessary boundary conditions. In other words, their amplitudes and phases are closely correlated.

In this representation, it is important that a proper wave of a periodic structure contains harmonics propagating with different phase velocities, including small (and even negative) ones. Moving along the waveguide axis, a charged particle is always in synchronism with some of these harmonics. Consequently, the particle participates in the energy exchange with the proper waves in all pass bands (as regards the interchange intensity, that is a quite another question). Investigating this phenomenon, it is easy to see that in a periodic system emission of the Cherenkov radiation is possible in its "pure" form (noncomplicated by polarization losses).

For better understanding of all the possibilities available, it is expedient to plot the dispersion curves in the plane (k_0, k). As a set of harmonics with all the integral numbers s corresponds to each μ, these graphs can be obtained by repeating the dependencies $k_0(\mu)$ periodically along the k–axis with the period $2\pi/D$. In the units of the velocity of light, the phase velocity of every harmonic is equal to the tangent of the straight line $k_0/k = \mathrm{const}$ drawn from the origin of the coordinates up to the point in the dispersion curve corresponding to a given harmonic (see Fig. 2.4). Note that all harmonics have the same group velocity $\partial \omega/\partial k$.

To use a standard presentation of the charge field as a superposition of the proper waves, we have to prove their orthogonality and completeness. The proper wave potentials $\mathbf{A}_n(\mathbf{r},\mu)$, $\Phi_n(\mathbf{r},\mu)$ obey the homogeneous wave equations

$$\left(\Delta + \frac{\omega_n^2(\mu)}{c^2}\right)\mathbf{A}_n(\mathbf{r},\mu) = 0 \text{ and} \tag{2.46a}$$

2.3 Cherenkov Radiation in Periodic Structures

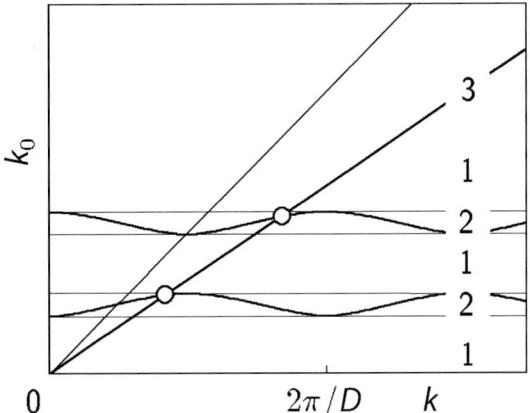

Fig. 2.4. Dispersion in a periodic structure near the cutoff. 1: stop-bands. 2: transparency bands. Curve 3: $k_0 = k\beta$. The circles correspond to Cherenkov radiation into the first two bands

$$\left(\Delta + \frac{\omega_n^2(\mu)}{c^2}\right)\Phi_n(\mathbf{r},\mu) = 0\;, \qquad (2.46b)$$

where $\omega_n(\mu)$ is the eigenfrequency of the mode corresponding to the transmission band number n and the given phase shift μ per structure period D. As it follows from the general theory of wave equations [4], each solution $\Phi_n(\mu)$ of the scalar wave Eq. (2.46b) generates three noncolinear vector wave functions being the solutions of Eq. (2.46a) (this statement can be immediately verified by direct substitution). One of these functions may be called potential one:

$$\mathbf{A}_{n0}(\mathbf{r},\mu) = -\nabla\Phi_n(\mathbf{r},\mu), \qquad \mathrm{rot}\,\mathbf{A}_{n0}(\mathbf{r},\mu) = 0\;. \qquad (2.47)$$

The two others will be called "solenoidal functions":

$$\mathbf{A}_{n1}(\mathbf{r},\mu) = \mathrm{rot}\,[\mathbf{e}\times\Phi_n(\mathbf{r},\mu)] \quad \text{and} \qquad (2.48)$$
$$\mathbf{A}_{n2}(\mathbf{r},\mu) = \frac{c}{\omega_n}\mathrm{rot}^2[\mathbf{e}\times\Phi_n(\mathbf{r},\mu)]\;,$$

where $\mathbf{e}$ is a unit constant vector (the unit vector in the z direction is chosen here). The correlation between these functions is as follows:

$$\mathbf{A}_{n1} = [\mathbf{e}\times\mathbf{A}_{n0}]\;; \quad \mathbf{A}_{n2} = \frac{c}{\omega_n}\mathrm{rot}\,\mathbf{A}_{n1}\;; \quad \mathbf{A}_{n1} = \frac{c}{\omega_n}\mathrm{rot}\,\mathbf{A}_{n2}\;. \qquad (2.49)$$

Under the supposition that the particle is moving along the z-axis, one may treat $\mathbf{A}_{n0}(\mathbf{r},\mu)$ and $\mathbf{A}_{n2}(\mathbf{r},\mu)$ as a complete orthogonal set of the system proper waves because $\mathbf{A}_{n1}$ has no longitudinal component and, consequently, it corresponds to the proper waves of the TE–type.

As regards a periodic system, the orthogonality of the vector wave functions with different values of μ follows from the fact that any bilinear form composed of $\mathbf{A}_n(\mathbf{r},\mu)$ and $\mathbf{A}_n^*(\mathbf{r},\mu')$ goes to zero when integrated over the system period. This is conditioned by the periodicity of the functions $\mathbf{a}(r,z)$ and $\mathbf{a}^*(r,z)$ in z and by the presence of the oscillating factor $\exp\left(\mathrm{i}(\mu-\mu')z/D\right)$. If $\mu=\mu'$, the corresponding integral is finite.

However, for the further consideration, orthogonality of the eigenfunctions with the same μ but with different subscripts n still ought to be proved as well as that of $\mathbf{A}_{n0}$ to $\mathbf{A}_{n2}$. For simplicity, we limit ourselves to the periodic waveguide system with a perfectly conducting surface. If the vector identity

$$\mathrm{div}[\mathbf{A}\times\mathbf{B}] = \mathbf{B}\cdot\mathrm{rot}\mathbf{A} - \mathbf{A}\cdot\mathrm{rot}\mathbf{B},$$

and the wave equation (2.46a) are used, the scalar product of two solenoidal functions can be presented as

$$\mathbf{A}_{n2}(\mathbf{r},\mu)\mathbf{A}_{m2}^*(\mathbf{r},\mu')$$
$$=\frac{c^2}{\omega_m^2(\mu')-\omega_n^2(\mu)}\mathrm{div}\left\{[\mathbf{A}_{n2}\times\mathrm{rot}\mathbf{A}_{m2}^*]-[\mathbf{A}_{m2}^*\times\mathrm{rot}\mathbf{A}_{n2}]\right\}. \quad (2.50)$$

Integrating this equality over a volume V of an element of periodicity and taking into account that tangential components of the vector potential vanish on conductive surfaces, one gets

$$\int_V \mathbf{A}_{n2}\cdot\mathbf{A}_{m2}^* dV = \frac{c^2}{\omega_m^2(\mu')-\omega_n^2(\mu)}$$
$$\times\int\left\{[\mathbf{A}_{n2}\times\mathrm{rot}\mathbf{A}_{m2}^*]-[\mathbf{A}_{m2}^*\cdot\mathrm{rot}\mathbf{A}_{n2}]\right\}\mathrm{d}\mathbf{s}. \quad (2.51)$$

On the right-hand side, the integral is taken over two arbitrary surfaces limiting the element of periodicity from both sides. An arbitrary choice of the element of periodicity and the application of (2.51) yield

$$\int_V \mathbf{A}_{n2}\mathbf{A}_{m2}^* dV = \frac{c\left[\exp(\mathrm{i}(\mu-\mu'))-1\right]}{\omega_m^2(\mu')-\omega_n^2(\mu)}$$
$$\times\int_S\left\{\omega_m(\mu')[\mathbf{a}_{n2}\times\mathbf{a}_{m1}^*]-\omega_n(\mu)[\mathbf{a}_{n1}\times\mathbf{a}_{m2}^*]\right\}\cdot\mathrm{d}\mathbf{s}.$$
$$(2.52)$$

Taken over an arbitrary cross–section S, the integral on the right-hand side is independent of its choice. Now one can see that if $\mu=\mu'$, the functions $\mathbf{A}_{n2}(\mathbf{r},\mu)$ and $\mathbf{A}_{m2}(\mathbf{r},\mu)$ are orthogonal in the volume of one element of periodicity with the norm that follows from (2.51) via the limiting transition $\mu'\to\mu$:

$$N_n = \int_V |\mathbf{A}_{n2}|^2 dV \quad (2.53)$$
$$= \frac{\mathrm{i}c}{2(\partial\omega_n/\partial\mu)}\int_S\left([\mathbf{a}_{n2}\times\mathbf{a}_{n1}^*]-[\mathbf{a}_{n1}\times\mathbf{a}_{n2}^*]\right)\cdot\mathrm{d}\mathbf{s}.$$

2.3 Cherenkov Radiation in Periodic Structures

The terms $i(\omega_n/c)\mathbf{a}_{n2}\exp(i\mu z/D)$ and $(\omega_n/c)\mathbf{a}_{n1}\exp(i\mu z/D)$ represent the electric and magnetic field components of the proper wave, correspondingly. Therefore, on the right-hand side of (2.53), where the integral is taken over the waveguide cross–section, the integrand coincides up to a constant factor with Poynting's vector of the solenoidal mode. The volume integral on the left is proportional to the time–average energy of the electric field in one element of periodicity. Hence, the value

$$v_g = D\frac{\partial \omega_n}{\partial \mu} \tag{2.54}$$

describes the group velocity, defined as the ratio of the transmitted power P_n to the average electromagnetic energy per unit of the system length. The norm of the solenoidal functions may be written as

$$N_n = \frac{8\pi c^2 D}{\omega_n^2 v_g} P_n \, . \tag{2.55}$$

Orthogonality of the potential–type eigenfunctions to all the solenoidal functions can be proved in the same way. As the scalar functions Φ_n with different n are mutually orthogonal as well, the equality

$$\int_V \mathbf{A}_{n0} \cdot \mathbf{A}_{m0}^* dV = \frac{\omega_n^2}{c^2} \int_V \Phi_n \Phi_m^* dV = \text{const} \times \delta_{nm} \tag{2.56}$$

is satisfied.

It is worth while mentioning that these modes do not transfer energy and correspond to Coulomb field components distorted by the presence of conducting walls. At the same time, the set of solenoidal modes propagating freely should be identified with the radiation field emitted.

2.3.2 Excitation by Moving Charge

Our goal is now to present the resonant transition radiation as Cherenkov radiation emitted in the form of a set of the structure proper waves. So, we begin with an inhomogeneous wave equation for the vector potential of the field of a charged particle moving along the z–axis. After the Fourier transformation with respect to time, the equation takes the form

$$\left(\triangle + k_0^2\right)\mathbf{A} = -\frac{q}{\pi c r}\mathbf{e}\delta(r)\exp\left(ik_0 z/\beta\right) \, . \tag{2.57}$$

Applying the standard procedure of the vector potential expansion over the solenoidal modes

$$\mathbf{A}(\mathbf{r},\omega) = \sum_n \int \{C_{n1}(\mu)\mathbf{a}_{n1}(\mathbf{r},\mu) + C_{n2}(\mu)\mathbf{a}_{n2}(\mathbf{r},\mu)\}$$
$$\times \exp\left(i\mu z/D\right) d\mu \tag{2.58}$$

to (2.57), one gets the following expression for the coefficients of the radiation field mode distribution:

$$C_{n2}(\mu) = \frac{2qca_n^*(s)}{N_n(\omega_n^2 - \omega^2)} \delta\left(\mu - \frac{\omega D}{\beta c} + 2\pi s\right). \tag{2.59}$$

Here the symbol

$$a_n(s) = \frac{1}{D}\int_{-\infty}^{+\infty}(\mathbf{e}\cdot\mathbf{a}_{n2})|_{r=0}\exp(-2\pi i s z/D)\,dz \tag{2.60}$$

designates the longitudinal field spatial harmonic in the n-th transmission band, which is determined by the synchronism condition.

Thus, the vector potential solenoidal component has the form:

$$\mathbf{A}_n(\mathbf{r},\omega) = \sum_n \frac{2qca_n^*(s)}{N_n(\omega_n^2 - \omega^2)}\mathbf{a}_{n2}(\mathbf{r},\mu)\exp(i\mu z/D) \tag{2.61}$$

$$\mu = \frac{\omega D}{\beta c} - 2\pi s(n).$$

Note that we do not consider here the potential part of the field described by the scalar potential and by expansion over $\mathbf{a}_{n0}$, which is not of importance for radiation.

The radiation frequency spectrum is determined by the poles of this expression; that is, the conditions

$$\omega = \pm\omega_n(\mu); \qquad \mu = \frac{\omega D}{\beta c} - 2\pi s(n) \tag{2.62}$$

must be satisfied, which is in full agreement with Fig. 2.4, plotted on the basis of the synchronism condition. Moreover, it is evident that the emitter–field interaction is realized via the eigenfunction synchronous harmonic, the phase velocity of which coincides with the particle velocity. These harmonics are distinct from each other in different transmission bands. Therefore, generally speaking, the radiation spectrum must be a line spectrum. All of these prove the total physical identity of the effect with Cherenkov radiation. The only difference is due to waves retardation and dispersion provided by the specific boundary conditions rather than by dielectric properties of the medium.

To find the microwave power emitted into different transmission bands, one should calculate the work performed on the particle per unit time by the proper solenoidal electric field longitudinal component[7]:

$$\mathbf{E}(\mathbf{r},t) = \frac{i}{c}\int_{-\infty}^{+\infty}\exp(-i\omega t)\mathbf{A}(\mathbf{r},\omega)\omega\,d\omega. \tag{2.63}$$

---

[7] Considerations based on the field asymptotic behavior fail in this case because the radiation field remains in the system.

2.3 Cherenkov Radiation in Periodic Structures

On the analogy of the calculations in Sect. 2.1, the integral over frequencies is taken along the path going around the poles from above. Hence, the solenoidal field in front of the particle is zero. Behind the emitter, this field, determined by the sum of the residues at the points $w = \pm w_n$

$$E_z(r=0; z < \beta ct)$$
$$= -\sum_n \frac{4\pi a_n^*(s)}{N_n} (\mathbf{e}\,\mathbf{a}_{n2})|_{r=0} \exp\left(-\frac{2\pi i s z}{D}\right) \cos\left[w_n \left(\frac{z}{\beta c} - t\right)\right] \quad (2.64)$$

has the form of the structure proper waves "dragging" behind the particle. The acting effective field is a half-sum of these fields.

Multiplying the latter expression by $-\beta cq$, averaging it over z within the system period and substituting $z = \beta ct$, one gets the power emitted:

$$P = \sum_n \frac{\pi q^2 \beta c}{N_n} |a_n(s)|^2 \,, \quad (2.65)$$

where (2.62) is taken into account. This power naturally splits into the partial constituents, which are being noncoherently emitted into different transmission bands because the corresponding microwave fields are mutually orthogonal.

This expression may be presented in a form more suitable for the microwave applications by introducing the notion of a "coupling impedance" R_n. The latter is defined as the ratio of the squared amplitude of the longitudinal component of the field harmonic synchronous with the emitter system axis ($E_{zn}(r=0)$) to the power flow transferred by the corresponding proper wave. Now, making use of the norm (2.55), one can rewrite the expression in the form containing only the system integral characteristics:

$$P = \sum_n \frac{q^2 \beta c v_{gn} R_n}{8D} \,. \quad (2.66)$$

This formula presents the "resonant" transition radiation as a particular case of the Cherenkov radiation emitted in the form of proper slow waves. This representation is especially handy if the wavelength is comparable with the structure period or exceeds the latter. For instance, this occurs under small values of μ. Then all high harmonics can be neglected considering the wave as a slow one with the definite propagation constant $k = \mu/D$. Below we will refer to this so-called impedance approximation just talking about slow waves in periodic systems characterized by a definite frequency and wave number. So far as a uniformly moving particle "feels" the synchronous harmonic only there would not be essential losses in physics of the process but radiation going into higher passbands of the waveguide.

However, this approach is not productive for nonrelativistic particles interacting with very slow harmonics of the proper wave. The above–given physical considerations and quantitative estimations, derived formally, also remain justified in the case. Nevertheless, sometimes it would be more appropriate to consider separate acts of transition radiation emitting at single inhomogeneities and sum coherently the fields generated in the process. Such an approach is preferable also if a real number of the system periods is not too large.

The radiation emission in a helical slow–wave structure, widely applied in microwave electronics (e.g., traveling–wave tubes and backward–wave oscillators), makes a classical example of the emission process in a periodic system. Strictly speaking, the azimuthal symmetry, assumed above for simplicity, is not characteristic of proper waves of such a system. Notwithstanding this fact, the effect of the Cherenkov interaction between a charged particle moving uniformly and a slow wave is realized in the latter case as well.

The same mechanisms work in waveguide structures of other types, applicable in microwave electronics and for charged particle accelerators. In particular, the following systems belong to this class:

- waveguides with corrugated walls, applicable in relativistic microwave electronics;
- chains of weakly coupled cavities in short–pulse electron linear accelerators operating in the stored energy regime;
- cavities with drift tubes (Alvarez systems), applicable for ion linacs; and
- "comb–like" slow–wave structures and diffractive gratings used in millimeter–band devices.

As regards the last item, charged particles moving along the system in a vacuum interact with a periodic structure only through peripheral fringes of their proper fields. The radiation emitted under such conditions should be considered as Cherenkov radiation in the limiting case of a relatively short length of the structure period. For nonrelativistic particles, this can be interpreted as the Smith–Purcell effect (see [23]).

For the case of a stratified medium, the effects considered above had been first examined in [25] under the name "parametric Cherenkov radiation." Later on, the phenomenon of transition radiation emission in a rather wide wavelength range was intensively investigated in analytical and experimental works from the viewpoint of the possibility of its applications in microwave electronics, detectors of charged particles, and plasma physics.

3

Microwave Bremsstrahlung

Existence of the charged particle acceleration (i.e, temporal variations in either a magnitude or a direction of its velocity) always causes the emission of a specific radiation called bremsstrahlung.

Below we will describe qualitative and quantitative characteristics of emission of the microwave bremsstrahlung of several types, which make the key points to microwave electronics and charged particle accelerators. They are as follows:

- dipole radiation;
- the radiation emitted by a particle moving along a wave-like trajectory with a relativistic velocity (the undulator radiation);
- the magnetic bremsstrahlung emitted by a charged particle uniformly rotating in an external magnetic field (including cyclotron and synchrotron radiation);
- the radiation emitted by a charged particle oscillating in the field of an external electromagnetic wave (wave scattering by a free charged particle); and
- the radiation emitted by a bound charged particle in the field of an external electromagnetic wave.

3.1 Radiation Field as Superposition of Proper Waves

For determining the electromagnetic field of radiation emitted by a point charged particle moving in an arbitrary manner, they usually apply the retarded potentials, which enable presenting the unknown field characteristics at a given moment via the coordinate of the charged particle $\mathbf{r}_0(t)$, its velocity $\mathbf{v} \equiv \dot{\mathbf{r}}_0$ and acceleration $\mathbf{a} \equiv \ddot{\mathbf{r}}_0$ at the previous moments [1]. The corresponding electric field at a point $\mathbf{r}$ at a moment t may be expressed as

$$\mathbf{E} = \frac{q}{(Rc - \mathbf{R}\mathbf{v})^3} \left\{ c^2 \left(1 - \beta^2\right) (\mathbf{R}c - R\mathbf{v}) + [\mathbf{R} \times [(\mathbf{R}c - R\mathbf{v}) \times \dot{\mathbf{v}}]] \right\}. \quad (3.1)$$

3 Microwave Bremsstrahlung

Here all values in the r.h.s. are to be taken at the moment t' determined by the finite speed of signal propagation:

$$t' + R(t')/c = t \; ; \quad \mathbf{R}(t') = \mathbf{r} - \mathbf{r}_0(t') \, . \tag{3.2}$$

As it has already been mentioned above, the second addendum on the right of (3.1), proportional to the particle acceleration $\mathbf{a} \equiv \dot{\mathbf{v}}$, is decreasing as R^{-1}. This very addendum is known to be responsible for microwave radiation emitted by the charged particle. Another addendum, depending on the emitter velocity, describes the quasi–stationary field "attached" to the particle. Note that this field can take a rather specific form, conditioned by the dependence $\mathbf{v}(t)$.

Notwithstanding the undoubted elegance of the formulae (3.1), (3.2) they are not rather handy for the analytical calculations corresponding to practical physical conditions. The Lienard–Wiechert potentials actually represent not the solutions of Maxwell equations but just another form of these equations and need further complicated calculations. In addition, generalizations of these potentials either on the case of the boundary conditions prescribed at a finite distance from the emitter or on that of a dispersion medium are, at least, nontrivial. Moreover, there exist other problems. For instance, let the microwave field be emitted under channelling (e.g., if a charged particle is moving in a waveguide with conducting walls). Certainly, it would be rather senseless to expect the radiation field decreasing strictly in proportion to R^{-1}. Moreover, for certain problems discussed below the field in the near zone is of the same importance as radiation in the far zone. For all these reasons, a more formal approach would be preferable.

Thus, in accordance with the reasoning given above, the general procedure for determining the radiation field emitted by a charged particle consists of its presentation as a superposition of vector eigenfunctions of the wave equation with the further restoration of the explicit temporal and coordinate dependencies. Simultaneously, the principle of radiation emission has to be taken into account. According to the latter, in the expansion in eigenfunctions, only the waves corresponding to the choice of the retarded potentials must be preserved. At large distances from the emitter, this field component ought to be identified with the radiation field.[1]

3.1.1 Proper Waves of Free Space

It is notorious that the eigenfunctions of Maxwell equations in free space are plane waves. The coordinate dependence of these eigenfunctions is determined by the factor $\exp(i\mathbf{kr})$. Therefore, Maxwell equations, describing the temporal dependence of the complex amplitudes of the electric and magnetic field vectors, take the form:

---

[1] However, the microwave radiation also contributes to the field component localized in vicinity of the particle.

$$\frac{d}{dt}\mathbf{E}(\mathbf{k},t) - ic\left[\mathbf{k}\times\mathbf{B}\right] = -\frac{q\mathbf{v}(t)}{2\pi^2}\exp[-i\mathbf{k}\mathbf{r}_0(t)] \qquad (3.3a)$$

$$\frac{d}{dt}\mathbf{B}(\mathbf{k},t) + ic\left[\mathbf{k}\times\mathbf{E}\right] = 0 \qquad (3.3b)$$

$$i\mathbf{k}\mathbf{E} = \frac{q}{2\pi^2}\exp[-i\mathbf{k}\mathbf{r}_0(t)] \qquad (3.3c)$$

In calculating the right–hand sides of the system (3.3a), the use is made of the fact that the charged particle of a charge q, moving along the trajectory $\mathbf{r}_0(t)$, induces the charge and current densities:

$$\rho(\mathbf{r},t) = q\delta(\mathbf{r}-\mathbf{r}_0(t))\;;\quad \mathbf{j}(\mathbf{r},t) = q\mathbf{v}(t)\delta(\mathbf{r}-\mathbf{r}_0(t))\;;\quad \mathbf{v}\equiv\frac{d\mathbf{r}_0}{dt}, \qquad (3.4)$$

The radiation field has to be perpendicular to the direction of the wave propagation. Hence, excluding the magnetic field $\mathbf{B}$ and the longitudinal electric field $E_l \equiv \mathbf{k}\mathbf{E}/k$ from (3.3a) with the help of (3.3b), one gets the following equation:

$$\left(k^2c^2 + \frac{d^2}{dt^2}\right)\mathbf{E}_{\mathrm{tr}}(\mathbf{k},t) = -\frac{q}{2\pi^2 k^2}\frac{d}{dt}[\mathbf{k}\times[\mathbf{v}\times\mathbf{k}]]\exp[-i\mathbf{k}\mathbf{r}_0(t)]\,, \qquad (3.5)$$

for the transverse electric field $\mathbf{E}_{\mathrm{tr}} \equiv [\mathbf{k}\times[\mathbf{E}\times\mathbf{k}]]/k^2$.

To solve (3.5), initial conditions, prescribed by a statement of the problem, must be set up. Seemingly, acceptance of the zeroth initial conditions be natural for singling out the radiation field vector. However, this supposition would imply as though the field existed at the initial moment had been compatible only with the charged particle at rest. Therefore, the field calculated would contain an additional splash conditioned by the emitter instantaneous acceleration up to the velocity $\mathbf{v}(0)$. This splash hampers the physical explanation of the results.

To avoid such misunderstandings, the emitter initial acceleration must be related to the remote past (when $t\to -\infty$). For each of the unknown field spectral harmonics, proportional to $\exp(-i\omega t)$, we find a stationary solution by the Fourier method. This very approach, formally reducible to certain rules of encompassing poles of the integrand in the Fourier transform, has been applied above in calculating the field emitted by a charged particle uniformly moving in a material medium (see Sect. 2.1). In this case, (3.5) affords:

$$\mathbf{E}_{\mathrm{tr}}(\mathbf{k},\omega) = \frac{iq\omega\left[\mathbf{k}\times[\mathbf{U}\times\mathbf{k}]\right]}{2\pi^2 k^2 (k^2 c^2 - \omega^2)}. \qquad (3.6)$$

Here we designate:

$$\mathbf{U}(\mathbf{k},\omega) = \frac{1}{2\pi}\int_{-\infty}^{+\infty}\mathbf{v}(t)\exp\left[i\left(\omega t - \mathbf{k}\mathbf{r}_0(t)\right)\right]dt. \qquad (3.7)$$

As it has to be expected, if the emitter dynamics is described by a continuous spectrum (the particle nonperiodic motion), the field excited is characterized by a continuous spectrum as well.

For many applications, it is appropriate to suppose that the particle motion is periodic (or, to be more precise, there exists an inertial coordinate system where the motion is periodic[2] with the period T). The periodicity means that the particle velocity and corresponding field consist of discrete sets of harmonics with the frequencies $\omega_s = 2\pi s/T$; $s = \pm 1, \pm 2 \ldots$.

The radiation field harmonics propagate away from the emitter location, their wave numbers corresponding to the frequencies mentioned. The results of the calculations may be Lorentz–transformed to the laboratory system. In this respect, the phenomena to be examined below differ from the examples of the radiation emission by charged particles moving uniformly (see Chap. 2), where the frequency spectra of the radiation emitted are determined by an electrodynamic environment of the emitter only.

As regards the conditions for synchronism, they are nominally reducible to (1.5) when $s \neq 0$. Physically, this means that, in a reference frame where the particle is on an average at rest, the frequencies of the radiation emitted are multiples of the emitter periodic motion frequency. A more detailed study of the conditions for the approximate relatively short–time synchronism can yield a more detailed information–for instance, about the radiation angular distribution.

To restore the spatial characteristics of the radiation field, the inverse Fourier transform (i.e., an integral over $\mathbf{k}$) has to be applied under the supposition that $\mathrm{Im}\,\omega > 0$. We will take for granted the following statement, almost evident from the viewpoint of physics: at a sufficiently large distance from the emitter location, essentially exceeding the wavelength of the radiation emitted, the wave vector $\mathbf{k}$ is directed along the vector $\mathbf{r}$ (formally, it is provable by expressing the plane wave as a series in spherical ones). We now ascribe a unit vector $\mathbf{e}$ to this direction. Hence, in the inverse Fourier transform, one may suppose that $\mathbf{k} = k\mathbf{e}$ in the integrand except the exponent $(\exp(i\mathbf{k}\mathbf{r}))$, where the coefficient multiplying $\mathbf{k}$ is large $(r \to \infty)$. Taking into account that the following equalities hold

$$\int \frac{\exp(i\mathbf{k}\mathbf{r})\,d\mathbf{k}}{k^2 - \omega^2/c^2} = 2\pi^2 \frac{\exp(\pm i\omega r/c)}{r} \quad \text{for} \quad \mathrm{Im}\,\omega = \pm i0, \qquad (3.8)$$

one gets

$$\mathbf{E}_{\mathrm{tr}}(\mathbf{r}, \omega) = \frac{iq\omega\,[\mathbf{e} \times [\mathbf{U} \times \mathbf{e}]]}{c^2 r} \exp(i\omega r/c)\,; \qquad (3.9a)$$

$$\mathbf{U} = \frac{1}{2\pi} \int_{-\infty}^{+\infty} \mathbf{v}(t)\exp\left[i\omega\left(t - \mathbf{e}\mathbf{r}_0(t)/c\right)\right]dt\,. \qquad (3.9b)$$

This is a well–known result, describing an asymptotically spherical wave with its amplitude decreasing as r^{-1} when $r \to \infty$.

---

[2] To avoid a physically senseless notions, it is preferable to imply a quasi–periodic motion because any process is characterized by its natural duration, which is considered to be much longer than the period T.

3.1.2 Proper Waves of Waveguide Systems

Structure and Spectra of Waveguide Proper Waves

The quantitative description of microwave bremsstrahlung emitted by charged particles in waveguide structures is of interest for many applications. In general, to classify proper waves of such structures as well as to describe their dispersion and the field configuration, the vector eigenfunctions of Maxwell equations must be used. The corresponding mathematical procedure, necessary for the further consideration, is detailed below.

In general, under the boundary conditions prescribed on surfaces located at a finite distance from the emitter, presentation of the total system of vector eigenfunctions of Helmholtz equation

$$\triangle \mathbf{A} + k_0^2 \mathbf{A} = 0 \qquad (3.10)$$

is much more complicated than in the analogous case of a scalar field. It is known that the three noncolinear solutions of (3.10)

$$\mathbf{A}_0 = -\nabla \Phi \ ; \quad \mathbf{A}_1 = \frac{1}{k_0}[\nabla \Phi \times \mathbf{e}] \ ; \quad \mathbf{A}_2 = \frac{1}{k_0^2} \nabla(\mathbf{e}\nabla \Phi) + \Phi \mathbf{e} \qquad (3.11)$$

(**e** is an arbitrary constant unit vector) can be generated by an arbitrary solution of the scalar equation

$$\triangle \Phi + k_0^2 \Phi = 0 \ . \qquad (3.12)$$

If the boundary condition on Φ is correctly specified, these solutions form a discrete set of the vector solutions (or the proper waves of the vector field). The parameter $k_0 c$ plays the role of the proper wave frequency. In particular, in the regular waveguide, the proper waves with the frequency lower than the cutoff frequency of the principal mode cannot propagate moving away from their origin, and decay exponentially moving away from their origin. In spatially periodic waveguide systems, such waves exist as well (see also Sect. 2.3). The values of k_0, corresponding to these waves, are located between the transmission bands (i.e., in the stop bands).

It is natural to suppose that the radiation field may be presented as a superposition of proper waves freely propagating in the system. Such waves are the solenoidal eigenfunctions $\mathbf{A}_1$ and $\mathbf{A}_2$, correlated as

$$\mathbf{A}_1 = \frac{1}{k_0} \text{rot}\, \mathbf{A}_2 \ ; \quad \mathbf{A}_2 = \frac{1}{k_0} \text{rot}\, \mathbf{A}_1 \ . \qquad (3.13)$$

They also have to satisfy the vector boundary conditions, following from the physical considerations. However, determination of these eigenfunctions in the explicit (analytical) form makes a rather difficult mathematical problem. The latter is solvable if the boundary conditions are prescribed on the coordinate surfaces of one of the six systems of coordinates.[3] So far as we know, the most detailed description of this question is given in [24].

For the sake of simplicity, below we will limit ourselves to the case of the boundary conditions prescribed by a system of perfect conductors, homogeneous along the z–axis. It is appropriate to choose the vector **e** in the same direction. For instance, determination of the radiation field in homogeneous waveguides of an arbitrary cross section, in transmission lines, etc., belongs to this set of problems.

Because of the translational symmetry, all the eigenfunctions of the waveguide systems in question depend on z as $\exp(ikz)$, and the operator ∇ is divisible into the transverse and longitudinal components:

$$\nabla = \nabla_\perp + ik\mathbf{e}$$

Now, the functions $\mathbf{A}_j$ can be expressed via the scalar Φ:

$$\mathbf{A}_0 = -\nabla_\perp \Phi - ik\Phi \mathbf{e} ,$$
$$\mathbf{A}_1 = \frac{1}{k_0} [\nabla_\perp \Phi \times \mathbf{e}] , \qquad (3.14)$$
$$\mathbf{A}_2 = \frac{ik}{k_0^2} \nabla_\perp \Phi + \mathbf{e}\left(1 - \frac{k^2}{k_0^2}\right) \Phi ,$$

where Φ satisfies the two–dimensional Helmholtz equation:

$$\triangle_\perp \Phi + \left(k_0^2 - k^2\right) \Phi = 0 . \qquad (3.15)$$

The boundary–value problem for this equation has solutions under discrete positive values of the parameter

$$\omega_j^2 = c^2 \left(k_0^2 - k^2\right) . \qquad (3.16)$$

From the viewpoint of physics, the latter is the square of the cutoff frequencies of the proper waves $\propto \exp\left[i\left(kz - k_0 ct\right)\right]$. These waves propagate along the z–axis, obeying the hyperbolic law of dispersion (3.16), characteristic of regular waveguides of an arbitrary cross section.

In the case of Dirichlet boundary condition ($\Phi = 0$ on the boundary), the vectors $\mathbf{A}_0$ and $\mathbf{A}_1$ on the boundary are directed along the vector **n** normal to the boundary surface. Consequently, $\mathbf{A}_0$ and $\mathbf{A}_1$ must be identified with the electric field, whereas the magnetic field must be presented as a superposition of the vectors $\mathbf{A}_{1j}$.

As regards the problem with $\mathbf{n}\nabla_\perp \Phi = 0$ on the boundary, the electric field takes the form of a superposition of the vectors $\mathbf{A}_{1j}$, normal to the boundary. The magnetic field may be presented as a superposition of the vectors $\mathbf{A}_{0j}$

---

[3] They are Cartesian, the three cylindrical, spherical, and conical coordinates. As it is known, the variables in the scalar equation (3.12) are divisible in 11 coordinate systems.

and $\mathbf{A}_{2j}$ (in fact only as $\mathbf{A}_{2j}$ because the magnetic field is purely solenoidal). It is easy to see that, in practice, one deals with presentation of the solenoidal field component as subsystems of TM– and TE–waves, which compose a set of the regular waveguide proper waves.

In both the boundary–value problems, the vector $\mathbf{A}_0$ is orthogonal to $\mathbf{A}_2$ within the volume limited by lateral walls of the waveguide and its two cross sections. Really, the corresponding scalar product of these vectors is

$$\int_V \mathbf{A}_0 \mathbf{A}_2^* \, dV = -\frac{1}{k_0} \int_V \nabla \Phi \operatorname{rot} \mathbf{A}_1^* \, dV$$
$$= -\frac{1}{k_0} \int_V \operatorname{div} [\nabla \Phi \times \mathbf{A}_1^*] \, dV = -\frac{1}{k_0} \int_S [\nabla \Phi \times \mathbf{A}_1^*] \, d\mathbf{s} \,, \tag{3.17}$$

where $d\mathbf{s}$ is the vector of the limiting surface element. In the integrand, the vector product is equal to zero on the lateral surface. As the vector $d\mathbf{s}$ is contrary–directed at the waveguide cross sections, the total integral is really equal to zero.

We will not prove orthogonality and completeness of the system of the vector functions $\mathbf{A}_{0j}$, $\mathbf{A}_{1j}$ and $\mathbf{A}_{2j}$ under the same value of k_0. Both the properties follow just from orthogonality and completeness of the set of the scalar functions Φ_j and $\Phi_{j'}$, which correspond to different values of k and j. If these functions are normalized as

$$\int_V \Phi_j(\mathbf{r}, k) \Phi_{j'}^*(\mathbf{r}, k') \, dV = 2\pi \delta_{jj'} \delta(k - k') \,, \tag{3.18}$$

one gets for the potential functions:

$$\int \mathbf{A}_{0j} \mathbf{A}_{0j'}^{'*} \, dV = \int_V \nabla \Phi_j \nabla \Phi_{j'}^{'*} \, dV$$
$$= \int_V \nabla \left((\nabla \Phi_j \nabla \Phi_{j'}^{'*}) \right) dV + k_0^2 \int_V \Phi_j \Phi_{j'}^{'*} \, dV$$
$$= 2\pi k_0^2 \delta_{jj'} \delta(k - k') \,. \tag{3.19}$$

Normalization of the solenoidal functions can be proved in the same way[4]:

$$\int_V \mathbf{A}_{1j} \mathbf{A}_{1j'}^{'*} \, dV = \int_V \mathbf{A}_{2j} \mathbf{A}_{2j'}^{'*} \, dV = 2\pi \left(1 - \frac{k^2}{k_0^2}\right) \delta_{jj'} \delta(k - k') \,. \tag{3.20}$$

The point to be made here is that the relations may be interpreted as the equality of average energies stored per a unit length of the waveguide in the electric and magnetic fields of the solenoidal mode.

---
[4] If the system is periodic in z, it is more handy to normalize the solenoidal functions to the transferred power flow (see Sect. 2.3).

Naturally, by presenting the unknown field as a superposition of the waveguide proper waves, one makes the problem of determining the power angular distribution physically meaningless. Instead, the question of the radiation mode composition has to be regarded. Because $\mathbf{A}_1$ and $\mathbf{A}_2$ are related to the electric and magnetic fields, the Poynting vector is determined by the product $[\mathbf{A}_{1j} \times \mathbf{A}_{2j'}^*]$. If the factors in the latter correspond to the same value of k_0 while $j \neq j'$, the vector flow through an arbitrary cross section of the waveguide is equal to zero by virtue of orthogonality of the proper functions Φ_j and $\Phi_{j'}$. If the values of j are identical ($j = j'$), whereas the corresponding frequencies are different ($\omega_j \neq \omega_{j'}$), the time–average power flow goes to zero. From the point of view of physics, this fact means that the total power of the radiation emitted is equal to the sum of partial powers of the radiation emitted as individual modes. Under a prescribed value of k_0 (and, correspondingly, a given k), the integral

$$\int_S \left[\mathbf{A}_{1j} \times \mathbf{A}_{2j}^*\right] \mathrm{d}s = -\frac{ik}{k_0^3} \int_S |\nabla_\perp \Phi_j|^2 \, \mathbf{e} \, \mathrm{d}s = -\frac{ik}{k_0^3}\left(k_0^2 - k^2\right), \qquad (3.21)$$

is independent of the choice of the waveguide cross-section S. Here the use is made of (3.14) and normalization

$$\int_S |\Phi_j|^2 \, \mathrm{d}s = 1, \qquad (3.22)$$

which follows from (3.18).

Excitation of Waveguide Proper Waves

We now expand the excited electromagnetic field over vector eigenfunctions of Maxwell equations with taking into account purely solenoidal nature of the magnetic field:

$$\mathbf{E}(\mathbf{r},t) = \int \sum_j \left[\mathcal{E}_{0_j}(t)\mathbf{A}_{0_j}(\mathbf{r},k) + \mathcal{E}_{1_j}(t)\mathbf{A}_{1_j}(\mathbf{r},k)\right.$$
$$\left. + \mathcal{E}_{2_j}(t)\mathbf{A}_{2_j}(\mathbf{r},k)\right] \mathrm{d}k \qquad (3.23\mathrm{a})$$

$$\mathbf{B}(\mathbf{r},t) = \int \sum_j \left[\mathcal{B}_{1_j}(t)\mathbf{A}_{1_j}(\mathbf{r},k) + \mathcal{B}_{2_j}(t)\mathbf{A}_{2_j}(\mathbf{r},k)\right] \mathrm{d}k . \qquad (3.23\mathrm{b})$$

The integration is over all the values of the proper wave propagation constants.

We now substitute these rows into Maxwell equations

$$\mathrm{rot}\, \mathbf{B} = \frac{4\pi}{c}\mathbf{j} + \frac{1}{c}\frac{\partial \mathbf{E}}{\partial t}; \qquad \mathrm{rot}\, \mathbf{E} = -\frac{1}{c}\frac{\partial \mathbf{B}}{\partial t} \qquad (3.24)$$

multiply (3.24) by $\mathbf{A}'_{nj'}$ ($n = 0, 1, 2, \ldots$) and integrate over $\mathbf{r}$. Because of orthogonality of the proper waves, we get a system of the equations for the amplitudes $\mathcal{E}_{nj}(t)$ and $\mathcal{B}_{nj}(t)$:

3.1 Radiation Field

$$\dot{\mathcal{E}}_{0_j} = -\frac{2}{k_0^2}\int \mathbf{j}(\mathbf{r},t)\mathbf{A}_{0_j}^*\,d\mathbf{r} = -\frac{2q}{k_0^2}\mathbf{v}(t)\mathbf{A}_{0_j}^*(\mathbf{r}_0(t))$$

$$\dot{\mathcal{E}}_{1_j} - ck_0\mathcal{B}_{2_j} = -\frac{2c^2k_0^2}{\omega_j^2}\int \mathbf{j}(\mathbf{r},t)\mathbf{A}_{1_j}^*\,d\mathbf{r}$$

$$= -\frac{2q\left(k^2c^2+\omega_j^2\right)}{\omega_j^2}\mathbf{v}(t)\mathbf{A}_{1_j}^*(\mathbf{r}_0(t))$$

$$\dot{\mathcal{E}}_{2_j} - ck_0\mathcal{B}_{1_j} = -\frac{2k_0^2c^2}{\omega_j^2}\int \mathbf{j}(\mathbf{r},t)\mathbf{A}_{2_j}^*\,d\mathbf{r}$$

$$= -\frac{2q\left(k^2c^2+\omega_j^2\right)}{\omega_j^2}\mathbf{v}(t)\mathbf{A}_{2_j}^*(\mathbf{r}_0(t)) \qquad (3.25)$$

$$\dot{\mathcal{B}}_{1_j} + ck_0\mathcal{E}_{2_j} = 0$$

$$\dot{\mathcal{B}}_{2_j} + ck_0\mathcal{E}_{1_j} = 0\,.$$

Here the dots imply the total derivatives with respect to t.

It should be emphasized that the solution of the system (3.25) represents the solenoidal component of the driven electric field. The latter contains not only the radiation fields (the waves freely propagating away from their source), but also the time–dependent near–zone field of the moving charged particle. The corresponding field vector is not proportional to the emitter acceleration.

We now recur to the problem with the initial conditions determined when $t \to -\infty$. In the general case of the emitter finite motion, the right–hand sides of (3.25) can be expressed via Fourier transforms:

$$\mathbf{v}(t)\mathbf{A}_{n_j}^*(\mathbf{r}_0(t)) = \int_{-\infty}^{+\infty} U_{n_j}(\omega,k)\exp(-i\omega t)\,d\omega \qquad (3.26a)$$

$$U_{n_j}(\omega,k) = \frac{1}{2\pi}\int_{-\infty}^{+\infty}\mathbf{v}(t)\mathbf{A}_{n_j}^*(\mathbf{r}_0(t),k)\exp(i\omega t)\,dt\,. \qquad (3.26b)$$

In accordance with the above–mentioned causality principle, (3.26b) determines the amplitudes $U_{nj}(\omega,k)$ only for $\operatorname{Im}\omega > 0$. It has to be expanded analytically into the lower half–plane of the argument ω, regarded as a complex variable. Taking notice of this, one gets the following expression for the Fourier components of the amplitudes of the electric and magnetic field from (3.25):

$$\mathcal{E}_{0_j}(\omega,k) = -\frac{2iqc^2}{\omega\left(\omega_j^2+k^2c^2\right)}U_{0_j}(\omega,k)$$

$$\mathcal{E}_{1,2_j}(\omega,k) = -\frac{2iq\omega\left(k^2c^2+\omega_j^2\right)}{\omega_j^2\left(k^2c^2+\omega_j^2-\omega^2\right)}U_{1,2_j}(\omega,k) \qquad (3.27)$$

$$\mathcal{B}_{1,2_j}(\omega,k) = -\frac{2q\left(k^2c^2+\omega_j^2\right)^{3/2}}{\omega_j^2\left(k^2c^2+\omega_j^2-\omega^2\right)}U_{2,1_j}(\omega,k)\,.$$

We now integrate the expansion (3.23a) over k, which restores the unknown fields as the coordinate functions. Here the distances along z are considered to be large, for sure exceeding the sizes of the emitter trajectory localization. Thus, if $z \to +\infty$, the contour of integrating in (3.23a), going in the real axis, has to be enclosed with Jordan arc into the upper half–plane of the complex variable k. It encircles only the poles $\mathcal{E}_{1,2j}$, located at the points $k = k_j(\omega) = (\omega^2 - \omega_j^2)^{1/2}$. As it should be expected, the parameter k_j coincides with the propagation constant for the j–th mode, traveling with the frequency ω. The poles $\mathcal{E}_{0j}$ are located on the imaginary axis. In fact, it means that the field potential component decays exponentially with the damping constant of the order of c/ω_j when the emitter is far from the point of observation. The analogous physical considerations are true for the wave frequencies $\omega < \omega_j$. If $z \to -\infty$, the field pattern is the same on the supposition that one disregards the phase of the field traveling away from the emitter, which is of no interest at present.

So, the spectral components of the electric and magnetic fields of the radiation emitted into the j–th mode are as follows:

$$\mathbf{E}_{1,2_j}(\mathbf{r}, \omega) = \frac{2\pi q \beta_{\text{ph}j}^3}{c\left(\beta_{\text{ph}j}^2 - 1\right)} U_{1,2_j}(\omega, k_j(\omega)) \mathbf{A}_{1,2_j}(\mathbf{r}, k_j(\omega)) \tag{3.28a}$$

$$\mathbf{B}_{1,2_j}(\mathbf{r}, \omega) = -\frac{2\pi i q \beta_{\text{ph}j}^3}{c\left(\beta_{\text{ph}j}^2 - 1\right)} U_{2,1_j}(\omega, k_j(\omega)) \mathbf{A}_{2,1_j}(\mathbf{r}, k_j(\omega)) \;. \tag{3.28b}$$

Here $\beta_{\text{ph}j} = \omega/ck_j(\omega)$ is the j–th mode phase velocity (in the units of the velocity of light).

The point to be made is that the formal symmetry of these expressions is somewhat deceptive because the TM– and TE–waves differ in their cutoff frequencies ω_j and dispersion.

Anyway, the given reasoning relates to the inner boundary–value problem for (3.15) and (3.10), respectively. The outer boundary–value problem (e.g., the radiation emitted in open transmission lines) requires separate consideration–at least, because of the two reasons. First, there exists the possibility of propagation of TEM–modes, characterized by the zero cutoff frequency. Second, which is of no less importance, as regards the outer problem, the condition for the radiation at infinity has to be used. In particular, because of the second reason, the formulae obtained for the radiation emission in the waveguide of a sufficiently large cross section, derived with taking into account the field reflection from the waveguide walls, cannot be just transformed into the corresponding expressions for free space.

However, outer boundary–value problems are rare in applications. Therefore, as regards the problems of this class, we will limit our considerations to the radiation emission into free space.

3.2 Dipole Radiation

The simplest example of the radiation emitted into free space takes place under conditions of the particle motion localized within a region the sizes of which are small in comparison with the characteristic wavelength of the radiation c/ω. As it is evident, this condition should be realized in the frame where the motion remains finite, providing such a reference frame exists at all. In this case, the magnitude $\mathbf{kr}_0$ in (3.7) is small and $\mathbf{U}$ coincides with the Fourier harmonic of the particle velocity. Correspondingly, the electromagnetic radiation field at large distances from the emitter may be presented as

$$\mathbf{E}(\mathbf{r},t) = \frac{q}{c^2 r} \left[[\mathbf{a} \times \mathbf{e}] \times \mathbf{e}\right] \exp(ikr) \tag{3.29a}$$

$$\mathbf{B}(\mathbf{r},t) = \frac{q}{c^2 r} [\mathbf{a} \times \mathbf{e}] \exp(ikr) . \tag{3.29b}$$

The field spectral composition coincides with that of the particle acceleration $\mathbf{a} = \dot{\mathbf{v}}$. This radiation is known as dipole radiation.

The formulae (3.29a) permit calculating the radiation power emitted into a unit solid angle $d\Omega = \sin\vartheta\, d\vartheta\, d\phi$ (θ is the angle between the vectors $\mathbf{a}$ and $\mathbf{e}$; ϕ is the polar angle of $\mathbf{e}$):

$$\frac{dP_\infty}{d\Omega} = q \frac{|\mathbf{a}|^2}{4\pi c^3} \sin^2 \vartheta . \tag{3.30}$$

The subscript "∞" denotes that the angular distribution of the radiation power and the total radiation flow are calculated using asymptotics of the proper field of the emitting dipole in the far zone of the latter.

The corresponding expression for the total power of the radiation has the form [1]:

$$P_\infty = \frac{2q^2 |\mathbf{a}|^2}{3c^3} . \tag{3.31}$$

As (3.31) indicates, the harmonic oscillator emits the energy per unit time proportional to the fourth power of the oscillation frequency ω and to the amplitude squared.

Another important feature peculiar to the dipole radiation should be mentioned: no momentum is carried away by the radiation itself. Really, direct substitution of $-\mathbf{e}$ for $\mathbf{e}$ changes just the sign of the Poynting vector but not its magnitude. Hence, the flows of momentum carried away by the radiation field in any of two opposite directions are identical.

As it is known, higher–order terms in expansion of the function $\exp(ikr_0)$ in powers of the argument $\mathbf{kr}_0$ describe the radiation emitted by higher multipoles (electric quadrupole, magnetic dipole, etc.). Within the framework of applicability of expansion in multipoles, these terms contribute essentially to the total radiation power only for zero intensity of the electric dipole radiation. For instance, this takes place if the radiation is emitted by an isolated system of identical particles characterized by zero total charge. In the case of the radiation emission by a single particle, such a situation cannot be realized. Therefore, we will omit higher order multipoles. It is worth mentioning that even in the case of the purely harmonic oscillator, higher harmonics of its frequency contribute to the multipole radiation.

The pattern radically alters for relativistic particles. Generally speaking, the expansion in multipoles in itself loses its sense. The radiation emission by an infinitely small oscillating dipole that as a whole is moving with the relativistic velocity radically differs from the case examined above. First of all, this radiation is nonmonochromatic. Really, because of Doppler effect

$$\omega = \omega'\gamma(1+\beta\cos\vartheta')\ ;\quad \gamma = (1-\beta)^{-1/2}\ . \tag{3.32}$$

So, the frequency of radiation in the laboratory frame of reference depends on the angle ϑ' between the propagation vector of radiation and the emitter velocity vector (the prime denotes that the corresponding parameters relate to a reference frame where the radiation preserves its dipole character). The angle ϑ' itself obeys the laws:

$$\cos\vartheta = \frac{\cos\vartheta'+\beta}{1+\beta\cos\vartheta'}\ ;\quad \cos\vartheta' = \frac{\cos\vartheta-\beta}{1-\beta\cos\vartheta}\ . \tag{3.33}$$

We now take into account that in the primed system the radiation frequency ω', coinciding with the oscillator frequency, is independent of the radiation propagation vector. Thus, one gets the following presentation of ω as a function of ϑ:

$$\omega = \frac{\omega'/\gamma}{1-\beta\cos\vartheta} = \frac{\Omega}{1-\beta\cos\vartheta}\ , \tag{3.34}$$

where $\Omega = \omega'/\gamma$ is the oscillator frequency in the laboratory frame. The point to be made here is that (3.34) has been already derived above from the condition of phase synchronism between the particle and the radiation field emitted.

Thus, under conditions of relativistic motion ($\gamma \gg 1$), the particle emits waves of a high frequency, which approximately $2\gamma^2$ times exceeds the emitter oscillation frequency, if the wave is emitted along the particle velocity. The frequency of waves emitted in opposite direction is two times smaller than that of the emitter oscillations. The angle $\vartheta' = \pi/2$ in the emitter frame of reference (or the angle $\vartheta \approx \gamma^{-1}$ in the laboratory system) makes a conventional boundary between the high–frequency and low–frequency microwave radiation. In this connection, it may be worth recalling peculiarities of the microwave transition radiation emitted by an ultra-relativistic particle moving uniformly (see Sect. 2.3).

It is easy to see that these features not only are characteristic of the spectral distribution but are also appropriate to the total power angular distribution. Really, a number of quanta emitted at the angles $\vartheta' > \pi/2$ ($\vartheta > \gamma^{-1}$) and $\vartheta' < \pi/2$ ($\vartheta < \gamma^{-1}$) is the same. However, the quantum energy in the second case is about γ^2 times higher by the order of magnitude than in the first one. Therefore, the major flow of the radiation power is really concentrated in a narrow space angle on the order of γ^{-2} in vicinity to the direction of the emitter velocity vector. Here it is appropriate to mention that the radiation carries away a momentum from the emitter, while the emitter itself is subjected to the recoil (decelerating) force acting in the direction opposite to its velocity.

It is somewhat more difficult to do with the analogous simple considerations in calculating the dependence of the radiation total intensity versus the particle energy. According to [1], let us present (3.31) in the four–dimensional form, which is valid for an arbitrary reference frame. In its structure essence, it is the law of variation of energy (i.e., that of the zeroth component of the particle 4-momentum p_i) in the concomitant system, where the emitter momentum spatial components are equal to zero. As it is known, the three–dimensional acceleration squared in the concomitant system coincides with the four–dimensional acceleration squared with the coefficient c^4. Hence, the radiation energy total flow (3.31) has to be written as

$$\mathrm{d}p_i = \frac{2q^2}{3c}\left(\frac{\mathrm{d}u_k}{\mathrm{d}s}\right)^2 \mathrm{d}x_i = \frac{2q^2}{3c}\left(\frac{\mathrm{d}u_k}{\mathrm{d}s}\right)^2 u_i \mathrm{d}s \ . \tag{3.35}$$

Here u_k is the particle 4–velocity and $\mathrm{d}s$ is the interval differential.

If one proceeds to the system where the velocity of the particle is $\mathbf{v}$ and its momentum $\mathbf{p} = m\gamma\mathbf{v}$ is the zeroth component in the four–dimensional expression (3.35) yields for P_∞:

$$P_\infty = \frac{2q^2\gamma^2}{3c}\left[\left(\frac{\mathrm{d}\gamma\mathbf{v}}{c\mathrm{d}t}\right)^2 - \left(\frac{\mathrm{d}\gamma}{\mathrm{d}t}\right)^2\right] \tag{3.36}$$

(the three spatial components of (3.35) afford the above–mentioned recoil momentum).

In (3.36), it is worth while to present the derivatives with respect to time as the functions of the forces acting upon the particle in the directions transverse and parallel to the vector of its velocity:

$$P_\infty = \frac{2q^2}{3m^2c^3}\left[\gamma^2(\mathbf{F}_\perp)^2 + F_\parallel^2\right] \ . \tag{3.37}$$

Thus, the total intensity of the radiation emitted by an ultra-relativistic particle in vacuum increases in the direct proportion to the emitter energy squared. The cases of the particle motion in parallel to the electric field, when the emitter radiation energy losses are independent of γ, makes the only exception.

3.3 Undulator Radiation

The above–given presentation of radiation characteristics of an oscillating dipole moving with the relativistic velocity along z–axis may be qualitatively

3 Microwave Bremsstrahlung

extended onto the case of an undulator. The latter is one of the basic devices used in the modern relativistic microwave electronics. In the broad sense, it is a system where the electron moves (or, to be more precise, can move) along a periodic trajectory that on an average coincides with z–axis and has a large number of periods. In the narrow sense of the word, this term is applicable to magnetostatic systems, where the electron kinetic energy remains constant with an accuracy to the radiation losses. The relativistic point dipole above can be considered as an undulator with small transverse deviations of the emitter from the straight line and small transverse velocities.

Applications of the undulator radiation are conditioned by the three factors. First, if the Lorentz force induced by the external magnetic field is purely transverse, the radiation intensity can be essentially heightened by making use of relativistic particles. Second, the length of the wave emitted forward under the same conditions can be substantially shorter than the trajectory period. Third, on the assumption that the emitter trajectory is strictly periodic, the radiation emitted must be characterized by a line spectrum with the line width decreasing with the growth of the number of periods. This fact is important not only for heightening the radiation spectral density: as it will be demonstrated below, it also plays an important role in obtaining coherent radiation emitted by a large number of electrons collectively.

The radiation emitted in a magnetostatic undulator differs from the radiation emitted by the moving dipole, which has been considered above. In the undulator, the transverse oscillations are forced and their harmonic composition is determined by the dependence of the magnetic field versus z. In addition, if the particle total energy γ is prescribed, the longitudinal velocity depends not only on this parameter but also on the transverse oscillation amplitude. This circumstance somewhat changes the condition of synchronism and, consequently, the correlation between the length of the wave emitted and the length of the undulator period.

On a periodic trajectory, the shift of the vector $\mathbf{r}_0$ during one period D is equal to $D\mathbf{e}_z$. We now present the integral in the general expression (3.9b) as the sum of the integrals over one period with the corresponding phasers:

$$\mathbf{U} = \frac{1}{2\pi} \int_0^{S_0} \mathbf{t} \exp\left[i\omega\left(\frac{s}{v} - \frac{\mathbf{e}\mathbf{r}_0(s)}{c}\right)\right] ds$$

$$\times \sum_{n=-\infty}^{+\infty} \exp\left[i\omega n \left(\frac{S_0}{v} - \frac{D}{c}\cos\vartheta\right)\right]. \tag{3.38}$$

Here the arc length $s = vt$, varying from zero to S_0 during one period, serves as the integration variable; $\mathbf{t}(s)$ is a unit vector tangent to the trajectory; $\cos\vartheta \equiv \mathbf{e}\mathbf{e}_z$ is the cosine of the angle with respect to z–axis.

Taking into account the relation

$$\sum_{-\infty}^{+\infty} \exp(i\alpha\mu n) = 2\pi \sum_{n=-\infty}^{\infty} \delta(\alpha\mu - 2\pi n) = \frac{2\pi}{\alpha} \sum_{n=-\infty}^{\infty} \delta(\mu - 2\pi n/\alpha)$$

3.3 Undulator Radiation

and applying (3.9a) and (3.38), we obtain the expression for the electric radiation field:

$$\mathbf{E}_{tr}(\mathbf{r},\omega) = \frac{iq\omega\omega^*}{2\pi c^2 r} \exp(i\omega r/c) \sum_{n=1}^{\infty} \int_0^{S_0} [\mathbf{e} \times [\mathbf{t} \times \mathbf{e}]]$$
$$\times \exp\left[i\omega^* n \left(\frac{s}{v} - \frac{\mathbf{e}\mathbf{r}_0}{c}\right)\right] \, \mathrm{d}s \times \delta(\omega - n\omega^*). \quad (3.39)$$

This equation confirms that under the given conditions radiation is really characterized by the line spectrum concentrated at harmonics of the basic frequency

$$\omega^* = \frac{2\pi v/D}{1 - \beta \cos\vartheta + K^2/2\gamma^2}. \quad (3.40)$$

This formula contains the so–called undulator coefficient K, here determined by the somewhat artificial relation

$$\frac{S_0}{D} = 1 + \frac{K^2}{2\gamma^2}. \quad (3.41)$$

However, in the cases of practical importance it has a simple physical sense. Generally speaking, the definition (3.41) itself indicates that this coefficient is a measure of curvature of the particle trajectory in the undulator (or, to be more precise, it depicts the ratio of the transverse oscillation amplitude to the period length D). If this ratio is not large, the approximate equality

$$\frac{S_0}{D} = \int_0^D \sqrt{1 + (\mathrm{d}\mathbf{r}_\perp/\mathrm{d}z)^2} \, \mathrm{d}z \approx 1 + \langle (\mathrm{d}\mathbf{r}\perp/\mathrm{d}z)^2 \rangle/2 \quad (3.42)$$

is justified (here $\mathbf{r}_\perp = [\mathbf{e}_z \times [\mathbf{r}_0 \times \mathbf{e}_z]]$ is the vector of the particle transverse displacement).

By comparing (3.41) with (3.42), one can see that the parameter K/γ has the sense of a small mean–square (effective) angle between the emitter trajectory and z–axis. In the relativistic case, this angle should be inversely proportional to γ. Therefore, the parameter K is the undulator characteristic independent of particles energy. The point to be made that for $K < 1$ the emitter transverse motion is nonrelativistic. In this case, for the small ϑ, (3.40) yields the dependence of the basic frequency ω^* versus the system external parameters:

$$\omega^* \approx \frac{2\gamma^2}{1 + \vartheta^2\gamma^2 + K^2} \frac{2\pi v}{D}. \quad (3.43)$$

As it follows from this expression, the radiation frequency ω^* for small K essentially exceeds that of the emitter forced transverse oscillations $2\pi v/D$. Naturally, this is accompanied by smallness the emitter transverse velocity and by decreasing the total radiation intensity. Large values of K, useful in some applications, diminish the basic frequency ω^*. Therefore, the radiation emission at a high frequency becomes obtainable only due to generating higher harmonics. Effectiveness of this process substantially depends on a particular design of the undulator.

3.4 Cyclotron Radiation

We now consider radiation produced by an emitter plane motion in an uniform magnetic field, the field vector being orthogonal to the plane of motion. The same procedure can be used as that for calculation of undulator radiation. The difference is that the vector $\mathbf{r_0}$ in the given case keeps on returning to its initial value after each cycle of motion. Hence, the fundamental frequency is equal to the cyclotron one: $\omega^* \equiv \omega_c$. As the charged particle is uniformly moving along a circumference, it is natural to relate the origin of coordinates to its center and to choose the z-axis perpendicular to the orbital plane. In addition, let us introduce the transverse unit vectors $\mathbf{e_x}$ and $\mathbf{e_y}$ orienting the first one in the plane of the vectors $\mathbf{e}$ and $\mathbf{e_z}$ so that $\mathbf{e}\mathbf{e_x} = \sin\vartheta$; $\mathbf{e}\mathbf{e_y} = 0$.
Then

$$\mathbf{r_0} = R\left(\mathbf{e_x}\sin(s/R) + \mathbf{e_y}\cos(s/R)\right) \tag{3.44}$$

$$\mathbf{t} = \mathbf{e_x}\cos(s/R) - \mathbf{e_y}\sin(s/R).$$

The integral in (3.39) can be expressed via Bessel functions:

$$\int_0^{S_0} \mathbf{t}\exp\left[\frac{i\omega_c n}{v}(s-\beta\mathbf{e}\mathbf{r_0})\right]ds \tag{3.45}$$

$$= 2\pi R\left[\mathbf{e_x}\frac{J_n(n\beta\sin\vartheta)}{\beta\sin\vartheta} - i\mathbf{e_y}J'_n(n\beta\sin\vartheta)\right]$$

(the prime denotes derivatives with respect to the argument).

It is worth mentioning that the factor i at $\mathbf{e_y}$ corresponding to the phase shift $\pi/2$ physically implies that, generally speaking, the radiation is ellipticaly polarized.

Making use of relations

$$[\mathbf{e}\times[\mathbf{e_y}\times\mathbf{e}]] = \mathbf{e_y},$$

$$[\mathbf{e}\times[\mathbf{e_x}\times\mathbf{e}]] = \mathbf{e_x} - \mathbf{e}\sin\vartheta,$$

$$(\mathbf{e_x} - \mathbf{e}\sin\vartheta)^2 = \cos^2\vartheta,$$

one gets the well-known Shott relation for intensity of the n-th harmonic of the rotational frequency emitted into a solid angle $d\Omega$ [1]:

$$\frac{dI_n}{d\Omega} = q^2\frac{n^2\omega_c^2}{2\pi c}\left[\cot^2\vartheta\, J_n^2(n\beta\sin\vartheta) + \beta^2 J'^2_n(n\beta\sin\vartheta)\right]. \tag{3.46}$$

The terms in the square brackets depict the polarization contributions in the plane $\mathbf{e}, \mathbf{e_z}$ and orthogonal to the latter.

Naturally, for a low velocity emitter all the formulae give characteristics of the dipole radiation. The problem of generating high harmonics in the ultra-relativistic case will be studied separately in Sect. 3.4.2.

3.4.1 Cyclotron Radiation Emitted in Waveguide

The cyclotron radiation emitted by moderately relativistic particles is very important for applications, and the work of many microwave devices is based on it. However, as regards these applications, the classical formulae given above are applicable only under essential limitations because real boundary conditions have not been taken into account. For instance, the frequency in gyrotrons [26] is on purpose made close to the cutoff frequency of the waveguide where the radiation emission takes place. Consequently, the waveguide is not to be ignored. We will estimate the waveguide influence on the radiation characteristics by an example of a charged particle rotating around the cylindrical waveguide axis when the helix radius and step are constant. The motion of this type can occur in a longitudinal homogeneous magnetic field as well as in a periodic transverse magnetic field, which enables qualifying the radiation emitted as that of cyclotron and undulator type simultaneously. And what is more important, the transverse motion frequency is comparable with the waveguide cutoff frequency while the emitter constant longitudinal velocity can be close to c.

The eigenfunctions of the scalar wave equation (3.15) in a perfectly conducting cylinder of radius b may be written in the cylindrical coordinates (r, φ, z) in the form:

$$\Phi_j(r, \varphi, z) = N_j J_n\left(\frac{\omega_j r}{c}\right) \exp\left[i\left(n\varphi + kz\right)\right] . \tag{3.47}$$

Here the norm N_j and the spectrum of eigenvalues ω_j are determined by a type of the boundary-value problem. In particular, the following equalities are valid for TM waves:

$$N_j^{-2} = \frac{b^2}{4\pi} J_n'^2\left(\frac{\omega_j b}{c}\right) ; \qquad J_n\left(\frac{\omega_j b}{c}\right) = 0 . \tag{3.48}$$

For TE waves the corresponding relations take the form:

$$N_j^{-2} = \frac{b^2}{4\pi}\left(1 - \frac{n^2 c^2}{\omega_j^2 b^2}\right) J_n^2\left(\frac{\omega_j b}{c}\right) ; \qquad J'_n\left(\frac{\omega_j b}{c}\right) = 0 . \tag{3.49}$$

Respectively, according to (3.14),

$$\mathbf{A}_{1_j} = \frac{N_j c}{\sqrt{\omega_j^2 + k^2 c^2}} \left[i\frac{n}{r}\mathbf{e}_r J_n - \frac{\omega_j}{c}\mathbf{e}_\varphi J'_n\right] \exp\left[i\left(n\varphi + kz\right)\right] ; \tag{3.50}$$

$$\mathbf{A}_{2_j} = \frac{N_j k^2 c^2}{\omega_j^2 + k^2 c^2}\left[i\frac{\omega_j}{kc}\mathbf{e}_r J'_n - \frac{n}{kr}\mathbf{e}_\varphi J_n + \frac{\omega_j^2}{k^2 c^2}\mathbf{e}_z J_n\right] \tag{3.51}$$
$$\times \exp\left[i\left(n\varphi + kz\right)\right] .$$

3 Microwave Bremsstrahlung

To calculate the components of **U**, determined by (3.38), one has to substitute the particle coordinates $r = R$; $\varphi = \omega_c t$; $z = \beta c t$ into the right-hand sides of these relations and integrate them over time from $-\infty$ to $+\infty$ with the multiplier $\exp(i\omega t)$. One should also take into account the equality

$$\frac{1}{2\pi}\int_{-\infty}^{+\infty} \exp\left[it\left(\omega - n\omega_c - \beta\sqrt{\omega^2 - \omega_j^2}\right)\right] dt$$
$$= \sum \frac{\delta(\omega - \omega_j^*)}{\left|1 - \beta\beta_{\mathrm{ph}_j}\right|}, \qquad (3.52)$$

where summation is extended over all roots ω_j^* of the equation

$$\omega - n\omega_c - \beta\sqrt{\omega^2 - \omega_j^2} = 0 . \qquad (3.53)$$

In addition, one may now substitute ω_j^* for ω and $\omega^*/\beta_{\mathrm{ph}_j}$ for kc, where β_{ph_j} is the phase velocity (in units of c) of the j-th mode at the frequency ω_j^*.

It is appropriate to note that (3.53), derived in a formal way, coincides with the system of equations

$$\omega = \beta k c + n\omega_c ; \qquad \omega = \sqrt{\omega_j^2 + k^2 c^2} . \qquad (3.54)$$

It is easy to see that the first one represents the condition of the oscillator synchronism with the wave while the second describes the dispersion law a the system homogeneous along the z-axis. As it follows from (3.53), the j-th mode either is not emitted at all or is emitted at two frequencies, which are the roots of (3.54).

Presenting the particle velocity in the form

$$\mathbf{v} = \omega_c R \mathbf{e}_\varphi + \beta c \mathbf{e}_z ,$$

one may rewrite the expressions for $U_{\varphi,z}$ in (3.38) as

$$U_\varphi = -\sum N_j \frac{R\omega_c \sqrt{\beta_{\mathrm{ph}_j}^2 - 1}}{\beta_{\mathrm{ph}_j} \left|\beta\beta_{\mathrm{ph}_j} - 1\right|} J'_n\left(\frac{\omega_j R}{c}\right) \delta(\omega - \omega_j^*)$$

$$U_z = \sum N_j \frac{c\left(\beta\beta_{\mathrm{ph}_j} - 1\right)}{\beta_{\mathrm{ph}_j}\left|\beta\beta_{\mathrm{ph}_j} - 1\right|} J_n\left(\frac{\omega_j R}{c}\right) \delta(\omega - \omega_j^*) . \qquad (3.55)$$

As it has been mentioned above the cyclotron radiation total power may be presented as a sum of powers of the spectral components P_j. Each of them is equal to the Poynting vector flux through the waveguide cross section.

After some algebra with making use of (3.21),(3.28a), and (3.28b), it may be presented in the form

$$P_j = N_j^2 \frac{q^2 \pi R \omega_c \beta_{\text{ph}_j}}{4c\sqrt{\beta_{\text{ph}_j}^2 - 1}} \left| \frac{J_n^{2'}(\omega_j R/c)}{\beta \beta_{\text{ph}_j} - 1} \right| . \quad (3.56)$$

As it had to be expected, this expression bears little resemblance to Shott formula, or, to be more precise, to its generalization on the case of a nonzero translation velocity. In place of distribution in continuous parameter (the angle or the corresponding frequency), the emitter in a waveguide excites a discrete set of modes with fixed frequencies. Only those of the modes are emitted, the phase velocities of which satisfy the condition for synchronism. The latter is especially important from the viewpoint of mode selection, which aims to maximization of the radiation spectral brightness. It is easy to prove that (3.54) possess a real solution only under the condition

$$n^2 \omega_c^2 > \omega_j^2 \left(1 - \beta^2\right) . \quad (3.57)$$

For the cyclotron radiation in the direct meaning of the word (i.e., radiation emitted in a homogeneous magnetic field B_0) the inequality determines the condition for the radiation emission of the j-th mode—the magnetic field has to be high enough:

$$\frac{qB_0}{mc} > \frac{\omega_j}{n} \gamma \sqrt{1 - \beta^2} . \quad (3.58)$$

If the particle helical trajectory step $D = 2\pi/k_w$ is prescribed (as it occurs in a helical undulator) the analogous condition reads:

$$k_w > \frac{\omega_j}{nc\gamma\beta} . \quad (3.59)$$

For large frequencies, it is easier to satisfy than (3.58).

It is worth to mention that at the threshold (3.56) forecasts infinitely large power which is, of course, meaningless. In vicinity of this point where the particle velocity is equal to the mode group velocity, the both frequencies ω_j^* merge together and corresponding fields are not independent (i.e., are coherent).

3.4.2 Synchrotron Radiation

Although (3.46) (Shott formula) is applicable under arbitrary values of the emitter velocity, the case of particle ultra-relativistic motion is so specific that corresponding radiation has a special name—synchrotron radiation.[5]

---

[5] The roots of this term are historical: in forties, the notions "cyclotron" and "synchrotron" were associated with cyclic accelerators of nonrelativistic and relativistic particles, respectively.

The practical value of synchrotron radiation is conditioned by two factors. First, it limits the energy obtainable in cyclic accelerators of light particles. Second, it provides a unique possibility of generating electromagnetic waves in a small angle with a well-determined spectrum in the ultraviolet and soft x-ray regions.

As regards the energy limitations, the radiation spectrum characteristics are of minor importance, whereas the angular distribution and total intensity are of much greater interest. Extending summation over all harmonic numbers in (3.46) and making use of the sums known in the theory of Bessel functions [27]:

$$\sum_{n=1}^{\infty} n^2 J_n^2(nx) = \frac{x^2(4+x^2)}{16(1-x^2)^{7/2}} \ ; \quad \sum_{n=1}^{\infty} n^2 J_n'^2(nx) = \frac{4+3x^2}{16(1-x^2)^{5/2}} \quad (3.60)$$

one can get the angular distribution of the radiation intensity integrated over the frequency spectrum:

$$\frac{dP}{d\Omega} = \frac{q^2 c \beta^4}{32\pi R^2} \left[\frac{\cos^2\vartheta \left(4 + \beta^2 \sin^2\vartheta\right)}{\left(1 - \beta^2 \sin^2\vartheta\right)^{7/2}} + \frac{4 + 3\beta^2 \sin^2\vartheta}{\left(1 - \beta^2 \sin^2\vartheta\right)^{5/2}} \right] . \quad (3.61)$$

In the square brackets, we have on purpose singled out the two addenda corresponding to different linear polarizations. In the ultra-relativistic case ($\beta \to 1$), this angular distribution is characterized by a sharp maximum in the plane of the particle rotation (i.e., at $\vartheta \to \pi/2$) because of the presence of the small factor $(1 - \beta^2 \sin^2\vartheta)$ in the denominator of (3.61). As this feature is not characteristic of low-number harmonics, one may conclude that the major part of the radiation power is concentrated in high harmonics of the cyclotron frequency.

To avoid a tiresome procedure of integrating over the angle in order to determine the total radiated power, let us straightaway consider the ultra-relativistic case $\gamma \gg 1$. In addition, because of the feature mentioned above we will limit ourselves with small angles $\psi = \pi/2 - \vartheta$ supposing that

$$1 - \beta^2 \sin^2\vartheta \approx \gamma^{-2} + \psi^2 \ll 1 . \quad (3.62)$$

Nevertheless, the integration over ψ may be extended to infinite limits because of the sharp decrease of the integrand when $\psi > \gamma^{-1}$. Thus, the total radiation power can be written down as

$$P = \frac{q^2 c \gamma^4}{12 R^2} [1 + 7] \quad (3.63)$$

showing that the major part of the radiation emitted (87.5%) is polarized in the plane of the particle rotation.

As it follows from (3.63), under a prescribed value of the guiding magnetic field, the radiation losses increase in direct proportion to the square of the particle energy. For light particles, the radiation power can be so high that, in fact, the losses place an upper limit of the electron energy in accelerators. Even modern accelerating systems are unable to compensate the losses at energies exceeding 100 GeV. At the same time the radiation losses for protons start to indicate themselves only at the energies of the order of scores of TeV.

If one is mostly interested in characteristics of synchrotron radiation itself, its spectral–angular distribution at high harmonics is of a special importance. To describe it quantitatively, it is handy to use the asymptotic representation of Bessel functions with large values of subscripts and arguments [27]:

$$J_n(nx) \approx \frac{1}{\pi\sqrt{3}} \left(1 - x^2\right)^{1/2} K_{1/3}\left(\frac{n}{3}(1-x^2)^{3/2}\right)$$

$$J_n'(nx) \approx \frac{1}{\pi\sqrt{3}} \left(1 - x^2\right) K_{2/3}\left(\frac{n}{3}(1-x^2)^{3/2}\right), \qquad (3.64)$$

which is valid for[6] $n \gg 1 \quad x \to 1 - 0$. Putting

$$x = \beta \sin\vartheta \,; \quad 1 - x^2 = 1 - \beta^2 \sin^2\vartheta \approx \frac{1 + \gamma^2\psi^2}{\gamma^2}$$

the substitution of (3.64) into (3.46) affords

$$\frac{dP_n}{d\Omega} \approx \frac{q^2 n^2 \omega_c^2}{6\pi^3 c n_c^{2/3}} \left[\psi^2 K_{1/3}^2\left(\frac{n}{3n_c}\right) + n_c^{-2/3} K_{2/3}^2\left(\frac{n}{3n_c}\right)\right], \qquad (3.65)$$

where

$$n_c = \left(1 - \beta^2 \sin^2\vartheta\right)^{-3/2} \approx \left(\gamma^{-2} + \psi^2\right)^{-3/2} \gg 1\,.$$

Thus, for harmonics with relatively small numbers $1 \ll n \ll n_c$ the radiation intensity increases with n according to the fractional power law. If $n > 3n_c$ the intensity exponentially decays. The critical harmonic number n_c reaches its maximum of order of γ^3 inside the characteristic cone $\psi < \gamma^{-1}$.

The synchrotron radiation characteristics not only bear resemblance to those of high harmonics of the undulator radiation with large values of the parameter K, but their patterns are also conditioned by the same physical reasons. Really, in both the cases, the frequency spectrum of radiation emitted forward by a relativistic particle is shifted toward short-wave regions. This fact stipulates the factor γ^2, multiplying the basic frequency. The "twinkling" effect taking place at a fixed position of an observer provides one more factor γ.

As regards the asymptotic formulae for the spectrum and intensity of the synchrotron radiation, one should take care applying them for moderate harmonic numbers. Rather than purely mathematical complications, one can miscalculate because of the role of realistic orbit surroundings principally different

---

[6] Unfortunately, the residual terms of this asymptotic are poorly investigated. Probably, the condition $x \to 1$ would be just the sufficient one.

from the vacuum case. In particular, harmonics of the wavelength comparable with the channel transverse dimensions are influenced by the boundary conditions and cannot be described by (3.46).

The peculiarities of short-wave synchrotron radiation considered above permit determining the radiation characteristics under arbitrary trajectory if its properties vary smoothly enough. So far as the length of the radiation emission is much smaller than the instantaneous radius of curvature:

$$l_{rad} \approx \frac{2\pi}{k}(1-\beta) \ll \rho,$$

the latter can be considered as constant. Hence, the field characteristics have to be similar to those for the motion along a tangent circumference of radius ρ. In other words, the radiation emitted in the given direction at high frequencies is determined by the trajectory local characteristics in the zone of the radiation emission. Only those sections of the trajectory contribute where at least an approximate particle–wave synchronism exists, that is, where the phase slip of the particle in the forward directed wave is minimal. If the trajectory is smooth enough, it may be presented in the emission zone as the expansion:

$$\mathbf{r}_0(t) \approx \mathbf{r}_0^* + s\mathbf{t} - \frac{s^2}{2\rho}\mathbf{n} - \frac{s^3}{6\rho^2}\left(\mathbf{t} - \frac{d\rho}{ds}\mathbf{n} - \rho\kappa\mathbf{b}\right) + \cdots; \quad (3.66)$$

$$\mathbf{v}(t) \approx \beta c\mathbf{t} - s\frac{\mathbf{n}}{\rho} + \cdots, \quad (3.67)$$

Here $\mathbf{r}_0^*$ denotes the point of the trajectory where the vector of the principal normal $\mathbf{n}$ is perpendicular to the wave vector $\mathbf{k}$; $\mathbf{t}$ and $\mathbf{b}$ are the vectors of the tangent and binormal to the trajectory at the point $\mathbf{r}_0^*$, respectively; s is the arc length referred to this point; κ is the trajectory torsion at $\mathbf{r}_0^*$. Thus, the phase factor is approximately equal to

$$i\mathbf{k}\mathbf{r}_0 \approx i\mathbf{k}\mathbf{r}_0^* + is\mathbf{k}\mathbf{t} + i\frac{s^3}{6\rho^2}(\mathbf{t}\mathbf{k} - \rho\kappa\mathbf{b}\mathbf{k}) + \cdots \quad (3.68)$$

Substitution of (3.68) into (3.9b) yields:

$$\mathbf{U} = \frac{1}{2\pi}\int_{-\infty}^{+\infty}\left(\mathbf{t} - \mathbf{n}\frac{s}{\rho}\right) \quad (3.69)$$

$$\times \exp\left\{i\frac{\omega}{\beta c}\left[s(1-\beta\mathbf{e}\mathbf{t}) + s^3\frac{\beta}{6\rho^2}(\mathbf{t}\mathbf{e} - \mathbf{b}\mathbf{e}\rho\kappa)\right]\right\}ds.$$

It is legitimate to approximate the particle trajectory as (3.68) because only vicinity of the point of a quasi-stationary phase $\mathbf{r}_0^*$ contributes to the integral in (3.69). This integral has already been calculated for undulator radiation. It may be presented as

$$\mathbf{U} = \frac{\rho}{\pi\sqrt{3\beta}} \left(\frac{\beta c}{\rho\omega_{\max}(\mathbf{te} - \rho\kappa\mathbf{be})} \right)^{1/3} \quad (3.70)$$

$$\times \left[\mathbf{t} K_{\frac{1}{3}}\left(\frac{\omega}{3\omega_{\max}}\right) - \mathbf{in}\left(\frac{2c}{\beta^{1/2}\rho\omega_{\max}}\right)^{1/3} K_{\frac{2}{3}}\left(\frac{\omega}{3\omega_c}\right) \right],$$

where

$$\omega_{\max} = \frac{\beta c}{\rho}\left(\frac{\mathbf{te}-\rho\kappa\mathbf{be}}{8(1-\beta \mathbf{et})^3}\right)^{1/2}. \quad (3.71)$$

It is easy to see that for a plane trajectory this expression affords the asymptotic formula (3.65) if one equals the basic rotation frequency ω^* to $c\beta/\rho$. However, to provide absolute identity of these expressions, an additional supposition is required: the particle must appear in the zone of the radiation emission with a strict periodicity. Naturally, there is no reason for realizing this condition. Considering a single passage of the particle through the trajectory section under observation, (3.70) affords just the energy emitted into the corresponding solid angle. In the case of the particle quasi-periodic motion (i.e., if, on an average, the particle periodically reappears in the zone of the radiation emission), the average radiation power is equal to the radiation energy divided by the time interval between the emitter returns. However, if the emitter motion is not maintained within the "optical" accuracy, the radiation spectrum is continuous rather than a discrete one. In other words, radiation emitted during a sequence of the particle turns is characterized by an arbitrary phase (i.e., it is incoherent). A discrete character of the synchrotron radiation spectrum follows from the analysis of the idealized emitter trajectory. Surely, finiteness of the beam emittance is neglected in the idealized model. A series of randomizing factors, which disturb a strict periodicity of rotation of an electron bunch, is not taken into account within the framework of this model as well. In particular, there are the following ones are relevant here:

- collisions between the emitters themselves as well as with atoms of the residual gas and their ions; and
- betatron and synchrotron oscillations of the emitting particles; and
- quantum fluctuations of emission.

By virtue of the reasoning given above, one can hardly expect that each electron would periodically return to the same point with an accuracy of tenths of a micron. Surely, with this accuracy, it would be more realistic to consider the electron location in two consequent turns utterly noncorrelated. Consequently, under such conditions, synchrotron radiation is characterized by a continuous spectrum.

3.5 Scattering by Free Charged Particle

Contents of the previous subsections as well as conclusions drawn there are based on the supposition that the particle motion is prescribed, regardless

3 Microwave Bremsstrahlung

of the physical fields conditioning this motion. However, there exists a wide range of specific problems where the particle acceleration is driven by electromagnetic fields of waves excited by external sources. Generally speaking, the electromagnetic radiation field emitted by the charged particle under such conditions is characterized by frequencies and wave vectors different from the corresponding characteristics of the incident wave. The field has to be regarded as a result of the incident wave scattering. The processes of scattering are not only interesting and important by themselves but they also make physical grounds for a series of applications, where either single particles or rather dense bunches are used. In the given subsection, we will limit ourselves with discussion of scattering by a single charged particle.

Within the framework of the problem without prescribed initial conditions (or, to be more precise, with the initial conditions when $t \to -\infty$), we will suppose the scattering to be a stationary process independent of the incident wave phase. As it is known, the effective cross section serves as a quantitative characteristic of the scattered wave intensity. It is defined as a ratio of the total radiation power flux to the incident power flux density. The corresponding characteristic, related to a unit frequency interval and/or to a unit angle, is called the scattering differential cross section.

Important peculiarities of the process of scattering can be established even on the basis of simple kinematic considerations. The latter, in many respects analogous with the condition of synchronism for the particle moving uniformly, directly follows from the conservation laws. Semiquantum description is rather handy here. Of course, it is not difficult to "translate" it to the language of wave mechanics.

So, we now investigate the process of scattering of a plane wave (its frequency is ω_0; the wave vector is $\mathbf{k} = \mathbf{n}\omega_0/c$) by a free charged particle with the velocity $\mathbf{v}$ when the incident wave intensity is low. This process may be presented as absorption of a quantum of energy $\hbar\omega_0$, and momentum $\hbar\mathbf{k} = \mathbf{n}\hbar\omega_0/c$ and consequent emission of a quantum of energy $\hbar\omega$ and momentum $\mathbf{e}\hbar\omega/c$. So we start with energy conservation law

$$\gamma' - \gamma = (\hbar/mc^2)(\omega_0 - \omega)$$

and conservation of momentum (in units of mc):

$$\mathbf{p}' - \mathbf{p} = (\hbar/mc^2)(\omega_0\mathbf{n} - \omega\mathbf{e}) \; ; \quad \gamma'^2 = p'^2 + 1 \; ; \quad \gamma^2 = p^2 + 1 \; .$$

Solving these equations with respect to ω to the first order in $\hbar$, one gets[7]

$$\omega = \omega_0 \frac{1 - \mathbf{v}\mathbf{n}/c}{1 - \mathbf{v}\mathbf{e}/c} \; . \tag{3.72}$$

As (3.72) does not contain Planck constant $\hbar$, this formula is of the purely classical nature. It describes a relation of the radiation frequency ω to the

---

[7] Higher orders in $\hbar$ correspond to the quantum Compton effect, essential only for very hard quanta.

angles of the wave incidence and scattering. Surely, this equation does not determine the process intensity or the effective cross section. Nevertheless, it permits us to make some important conclusions. First, scattering of a low–intensity wave by a particle that is at rest initially ($\mathbf{v} = 0$) occurs without transformation of the scattered wave frequency. This process is called Thomson scattering [1]. Second, the wave emitted in the direction of motion of a relativistic mirror $(1 - \mathbf{ve}/c \ll 1)$ has a frequency essentially exceeding that of the incident wave. This becomes especially clear if the particle moves in the direction opposite to the incident wave vector. Under such conditions, the frequencies are correlated as

$$\omega = \omega_0 \frac{1+\beta}{1-\beta} \approx 4\gamma^2 \omega_0 \, . \qquad (3.73)$$

In this case, the process may be interpreted as the incident wave reflection from a relativistic mirror moving to meet the incident wave. The coefficient of the order of γ^2 is a general characteristic of the radiation emitted by an oscillator moving with the relativistic velocity. Here it is appropriate to note that this coefficient is twice as large as in the case of the undulator radiation. Such a substantial alteration in the wave frequency during the scattering permits calling the latter Compton scattering (note that it is not quite correct). Another point to be made is that the energy spent on emitting a hard quantum of radiation, naturally, is taken off the energy of the emitter onward motion. Respectively, the emitter, obtaining a recoil momentum, is decelerated.

To determine the scattering effective cross section, it is necessary to examine the emitter steady motion in the incident wave field and calculate the radiation field by the general expression (3.9b). As it is clear from general considerations, in the wave of the amplitude E_0 and of the wavelength $\lambda = 2\pi c/\omega_0$, the emitter can absorb energy only of the order of $qE_0\lambda$ (in that reference frame where the particle rests at an average). As a rule, in practice, this energy is much less than mc^2. Consequently, in the reference frame under consideration, the emitter keeps on being a nonrelativistic particle, while the radiation emitted is almost of the purely dipole type. However, first of all, this statement is not quite evident with regard to the laboratory system. Second, concerning essential progress in laser technology, the problem of large–amplitude wave scattering becomes more and more urgent. Therefore, below we will investigate the emitter relativistic motion in the wave field, not restricting ourselves to the case of smallness of the electric field amplitude of the incident plane wave.

We now introduce the dimensionless variables:

$$\mathbf{g}(\varphi) = \frac{q\mathbf{E}(\varphi)}{mc\omega_0} \, ; \quad \varphi = \mathbf{kr} - \omega_0 t \, .$$

The point to be made here is that the dependence $\mathbf{g}(\varphi)$ is not specified yet, which enables examining waves of arbitrary polarization. The only suppositions necessary: the wave is to be regarded as transverse and harmonic one:

$$\mathbf{Ek} = 0\,; \quad k = \frac{\omega_0}{c}\,; \quad \frac{d^2\mathbf{g}}{d\varphi^2} = -\mathbf{g}\,. \tag{3.74}$$

Under these conditions, the wave dimensionless magnetic field is $[\mathbf{n} \times \mathbf{g}]$. As $\mathbf{gn} = 0$, the relativistic equation of the emitter motion

$$\frac{1}{\omega_0} \frac{d\gamma \mathbf{v}}{c\,dt} = \tag{3.75}$$

$$\mathbf{g}(\varphi) + \left[\frac{\mathbf{v}}{c} \times [\mathbf{n} \times \mathbf{g}]\right] = \mathbf{g}(1 - \mathbf{vn}/c) + \mathbf{n}(\mathbf{vg}/c)$$

and the corresponding equation for the particle energy variation

$$\frac{1}{\omega_0} \frac{d\gamma}{dt} = \mathbf{vg}/c \tag{3.76}$$

yields the relation:

$$J = \gamma\,(1 - \mathbf{vn}/c) = \text{const}. \tag{3.77}$$

It is the exact integral of motion, very important for the further reasoning.

Taking into account the relation

$$d\varphi/dt = -\omega_0(1 - \mathbf{vn}/c)\,,$$

we will regard φ as an independent variable. Introduction of a new unknown function

$$\mathbf{u} = \frac{\mathbf{v}}{c - \mathbf{vn}}\,, \quad \mathbf{v} = c\frac{\mathbf{u}}{1 + \mathbf{un}}\,, \tag{3.78}$$

reduces the initial equation (3.75) to the linear vector equation of the first order for $\mathbf{u}$:

$$J\dot{\mathbf{u}} = -\mathbf{g} - \mathbf{n}\,(\mathbf{ug}) \tag{3.79}$$

(the dotted notation implies differentiating with respect to φ).

A general solution of (3.79)

$$\mathbf{u} = \mathbf{u}_0 + \frac{\dot{\mathbf{g}}}{J} + \mathbf{n}\frac{(\dot{\mathbf{g}}\mathbf{u}_0)}{J} + \mathbf{n}\frac{\dot{g}^2}{2J^2} \tag{3.80}$$

depends on a single arbitrary vector constant $\mathbf{u}_0$. The particle velocity at a conventional moment of the scattering initiation (or the particle initial velocity under the conditions of the adiabatically slow increase of the scattered wave amplitude) may be expressed via this constant:

$$\mathbf{v}_0 = c\frac{\mathbf{u}_0}{1 + \mathbf{u}_0\mathbf{n}}\,; \quad 1 - \mathbf{v}_0\mathbf{n}/c = \frac{1}{1 + \mathbf{u}_0\mathbf{n}}\,.$$

Surely, the integral of motion obtained above (3.77) can be expressed via the same parameter as well:

$$J = \gamma_0\,(1 - \mathbf{v}_0\mathbf{n}/c)\,; \quad \gamma_0 = \left(1 - \beta_0^2\right)^{-1/2}\,. \tag{3.81}$$

3.5 Scattering by Free Charged Particle

Not only the velocity of the particle but also its trajectory in the parametric form is easily obtainable from this equation. However, this is unnecessary because the integral of interaction (3.10) we are looking for takes the form:

$$\mathbf{U} = -\frac{c}{2\pi\omega_0} \int_{-\infty}^{+\infty} \mathbf{u} \exp\left[-i\frac{\omega}{\omega_0}\left(\varphi - \int (\mathbf{n}-\mathbf{e})\mathbf{u}\,d\varphi\right)\right] d\varphi. \quad (3.82)$$

The emitter motion under investigation may be presented as a superposition of the uniform drift and oscillations with a unit frequency (with respect to φ). The direction and magnitude of the drift depend on the phase of the wave under scattering.

As well as in the case of undulator radiation, we again make the use of the fact that the integrand in (3.82) contains a product of two periodic functions. One of them has the period 2π, while the other one is the exponent of the form

$$\exp\left[-i\frac{\omega}{\omega_0}\left(1 + (\mathbf{n}-\mathbf{e})\bar{\mathbf{u}}\right)\varphi\right].$$

Here the vector $\mathbf{u}(\varphi)$ is presented as a sum of the average value and the oscillating component: $\mathbf{u}(\varphi) = \bar{\mathbf{u}} + \tilde{\mathbf{u}}(\varphi)$.

Hence, on the analogy of (3.38), one gets that the scattered radiation field is a sum of harmonics (neglecting phase factors):

$$\mathbf{U} = -\frac{\omega^* c}{2\pi\omega_0} \sum_{s=-\infty}^{\infty} \int_0^{2\pi} \tilde{\mathbf{u}} \exp\left[is\left(\varphi + \frac{\omega^*}{\omega_0}\int (\mathbf{n}-\mathbf{e})\tilde{\mathbf{u}}\,d\varphi\right)\right] d\varphi$$
$$\times \delta(\omega - s\omega^*). \quad (3.83)$$

The basic frequency ω^* is given by

$$\omega^* = \frac{\omega_0}{1 + (\mathbf{n}-\mathbf{e})\bar{\mathbf{u}}}. \quad (3.84)$$

Because of the Doppler effect, it somewhat differs from the incident wave frequency. This difference exists even if the wave is scattered by a charged particle that initially was at rest because the particle obtains finally the drift velocity $\bar{\mathbf{u}}$. The latter is negligible for small wave amplitudes which correspond to the dipole approximation in the particle rest frame. In our case, $\bar{\mathbf{u}}$ and ω^* take the form:

$$\bar{\mathbf{u}} = \frac{\mathbf{v}_0}{c - \mathbf{v}_0 \mathbf{u}}; \quad \omega^* = \omega_0 \frac{1 - \mathbf{v}_0 \mathbf{n}/c}{1 - \mathbf{v}_0 \mathbf{e}/c}. \quad (3.85)$$

These relations completely correspond to the above-given kinematic considerations.

Within the framework of the dipole approximation, the contribution of $\mathbf{u}$ to the exponent power in (3.83) is negligible as well. Calculations of the energy flow per unit of the solid angle yield:

$$P = \frac{c}{2\pi} \left(\frac{q^2}{mc^2}\right)^2 \frac{(1 - \mathbf{v}_0\mathbf{n}/c)^2}{(1 - \mathbf{v}_0\mathbf{e}/c)^6 \gamma_0^2}$$
$$\times \left|\left[\mathbf{e} \times \left(\mathbf{n}\left(\mathbf{v}_0\bar{\mathbf{E}}/c\right) + \bar{\mathbf{E}}\left(1 - \mathbf{v}_0\mathbf{n}/c\right)\right)\right]\right|^2 . \qquad (3.86)$$

Here $\bar{\mathbf{E}}$ denotes the incident wave electric field vector averaged with the multiplier $\exp(-i\varphi)$:

$\bar{\mathbf{E}} = E_0 \mathbf{e}_x / 2$ for the wave with linear polarization along $\mathbf{e}_x$;

$\bar{\mathbf{E}} = E_0 \left(\mathbf{e}_x \pm i\mathbf{e}_y\right)/2$ for the wave with circular polarization.

Division of both the sides of (3.86) by the incident wave energy flux density (i.e., by $cE_0^2/8\pi$ or $cE_0^2/4\pi$, respectively) affords the desirable expression for the differential cross section $d\mathcal{G}/d\Omega$. It should be noted that the latter is independent of the incident wave frequency.

In a particular case of a linearly polarized incident wave scattered by an immobile charged particle, the well–known Thomson formula follows from (3.86):

$$\frac{d\mathcal{G}}{d\Omega} = r_c^2 \sin^2 \vartheta ; \qquad \mathcal{G} = \frac{8\pi}{3} r_c^2 . \qquad (3.87)$$

Here ϑ is the angle between the direction of scattering (the vector $\mathbf{e}$) and the incident wave electric field vector ($\mathbf{g}$). This scattering is not accompanied by any alterations in the wave frequency.

Among other particular cases of the process under examination, the wave scattering by an ultra-relativistic emitter moving in the direction opposite to that of the wave propagation should be mentioned. The scattering of this type provides the maximal increase of the radiation frequency. Under these conditions, $\mathbf{n}\mathbf{v}_0 \approx -c$ and

$$\frac{d\mathcal{G}}{d\Omega} = r_c^2 \frac{16}{\gamma_0^2 (1 - \beta_0 \cos \vartheta)^6} , \qquad (3.88)$$

where ϑ is the angle between the vectors $\mathbf{v}_0$ and $\mathbf{e}$.

If the scattered wave field amplitude is large, the radiation field pattern is much more complicated. Nevertheless, the principal characteristics are traceable by the example of the particular case of the wave with circular polarization. Here the emitter motion is a superposition of a direct linear drift and a circular motion. Correspondingly, characteristics of the radiation field under such conditions are similar to those of the cyclotron radiation. One has to take into account the Doppler effect, conditioned by the emitter drift velocity. Apropos, as (3.83) indicates, criterion of the dipole approximation applicability in the laboratory system (i.e., the weak–field criterion) can be reduced now to the inequality $|g| \ll J$, or

$$\frac{qE_0\lambda}{mc^2} \ll \gamma_0 \left(1 - \mathbf{v}_0\mathbf{n}/c\right) .$$

However, calculation of the radiation spectral–angular distribution for the strong field (when the last inequality is not justified) is rather pointless not only because of bulkiness of the analytical expressions. Really, then the particle drift velocity reaches relativistic values being dependent on the incident wave phase. Therefore, correct definition of the scattering transverse cross section as a characteristic of the scattering process implies averaging over all the possible initial phases of the incident wave. Doppler effect, corresponding to the emitter drift motion, heavily indicates itself when $J \gg |g|$. Hence, as a result of averaging, substantial smearing of the radiation spectral–angular characteristics takes place. In this case, the dependence of the scattered wave frequency versus the direction of propagation discussed above becomes disguised.

3.6 Scattering and Absorption by Bound Particle

The scattering by a bound charge differs from the case of a free particle, discussed above, in the two aspects of principal importance:

- First, in the scattered radiation spectrum there have to exist those frequencies which differ from that of the incident wave by multiples of the proper frequencies. This phenomenon is conditioned by the emitter motion corresponding to the internal degrees of freedom. The scattering of this type is called nonelastic or Raman one.
- Second, if the wave frequency coincides with any of the frequencies, the possibility of a wave–particle resonance is rather evident.

As regards the single particle motion, to the examination of which we limit ourselves here, the internal degrees of freedom of the scattering system can be determined only by external fields. The principal attention should be paid to an external homogeneous magnetic field, where the system as a whole is moving with the velocity $\mathbf{v}(t)$. The latter is a superposition of the particle uniform displacement in parallel to the magnetic field vector $\mathbf{e}_z B_0$ and the uniform rotation with the cyclotron frequency $qB_0/mc\gamma$. To exclude the contribution of the proper cyclotron radiation emitted by the particle to the wave scattering characteristics, we will consider only those wave fields which are driven by alterations in the particle motion in the incident wave fields.

The corresponding generalized equation of motion (3.77) in the variables $\mathbf{u}, \varphi$ takes the form:

$$\frac{\mathrm{d}\mathcal{J}\mathbf{u}}{\mathrm{d}\varphi} = -\mathbf{g} - \mathbf{n}(\mathbf{u}\mathbf{g}) - [\mathbf{u} \times \mathbf{e}_z]\,\Omega\,; \quad \Omega = \frac{qB_0}{\omega_0 mc}\,. \tag{3.89}$$

Here $\mathcal{J} = \gamma\,(1 - \mathbf{v}\mathbf{n}/c)$ satisfies the equation:

$$\frac{\mathrm{d}\mathcal{J}}{\mathrm{d}\varphi} = \Omega \mathbf{n}\,[\mathbf{u} \times \mathbf{e}_z]\,. \tag{3.90}$$

Generally speaking, now it is not an integral of motion. This fact considerably complicates the calculations and leads to the bulkiness of the final expressions. Here we limit ourselves to the case of a wave propagating strictly in parallel to the external magnetic field. This situation is of the main interest for practical applications.

In this particular case $\mathcal{J} = $ const, and Eq. (3.89) becomes a linear one. We may look for the general solution of (3.89) in the form of a sum of the two addenda. The first one is a forced solution characterized by a unit frequency (related to the variable φ):

$$\mathbf{u}_1 = \frac{1}{\Omega^2 - \mathcal{J}^2} \left\{ -\mathcal{J}\dot{\mathbf{g}} + \Omega\left[\mathbf{g} \times \mathbf{e}_z\right] - \mathbf{n}\frac{\dot{g}^2}{2} \right\}. \tag{3.91}$$

The second addendum is a general solution of the corresponding homogeneous equation for transverse and longitudinal components:

$$\dot{\mathbf{u}}_\perp = -\frac{\Omega}{\mathcal{J}} [\mathbf{u}_\perp \times \mathbf{e}_z] \; ; \qquad \dot{u}e_z = -\frac{1}{\mathcal{J}}(\mathbf{g}\mathbf{u}_\perp) \;.$$

The transverse component of the general solution of (3.89) is a constant length vector uniformly rotating around $\mathbf{e}_z$ with the frequency $\Omega/\mathcal{J}$. The longitudinal component amplitude is modulated by a product of the two harmonic functions of frequencies $\Omega/\mathcal{J}$ and unity. It is easy to see that the corresponding longitudinal part of $\tilde{\mathbf{u}}$ from (3.83) is characterized by the frequencies $1-\Omega/\mathcal{J}$ (the so-called Stokes component) and $1+\Omega/\mathcal{J}$ (anti–Stokes component). Surely, amplitudes and phases of these beats depend on the way of the system exciting. Rotation of $\mathbf{u}_\perp$ plays the role of an internal degree of freedom in the given case.

The wave scattering without alteration in the incident wave frequency, sometimes called the "coherent" scattering, is of a special interest for our purposes (microwave electronics). The corresponding effective cross section is determined by averaging the vector $\mathbf{u}_\perp(\varphi)$. Naturally, the process of coherent scattering transforms to Thomson scattering in the limiting case of high frequencies of the wave (i.e., when the inequality $\Omega \ll \mathcal{J}$ holds). Here the magnetic field influence is negligible. In the opposite limiting case of low frequencies $\Omega \gg \mathcal{J}$, only the addendum proportional to Ω^{-1} must be taken into account in (3.91). Hence, the scattering cross section becomes proportional to the wave frequency squared:

$$\mathcal{G} \approx \frac{8\pi}{3}\left(\frac{q^2}{mc^2}\right)^2 \frac{1}{\Omega^2} = \frac{8\pi}{3}\left(\frac{q\omega_0}{B_0 c}\right)^2 \tag{3.92}$$

(one should keep in mind that we consider the wave under scattering to be propagating in parallel to the magnetic field vector).

It is interesting to note that, as it is known [8], the cross section of the scattering by the bound oscillator is proportional to the 4th power of the

3.6 Scattering and Absorption by Bound Particle

incident wave frequency (the so–called Rayligh scattering, responsible for the blue color of the sky).

Finally, as should be expected, the scattering cross section increases in resonant manner if $\Omega \to \mathcal{J}$ (see also (3.91)). Naturally, this unlimited growth, being physically meaningless, requires separate discussion. It is well known that the analogous situation arises in scattering by an oscillator. The maximum amplitude of the stimulated oscillations is limited by two reasons. They are the forces of the radiation friction (see Chap. 4) and RF nonlinearity. As regards scattering in an external magnetic field, finiteness of the time of the particle motion in this field most often plays the role of the restricting factor. This time can be less than the time of establishment of the stimulated oscillations, which is by the order of magnitude equal to $|\Omega/\mathcal{J} - 1|^{-1}$ (with respect to φ). Under such conditions, it would be more appropriate to dwell not upon the wave scattering, but upon the wave energy absorption which is spent on exciting the degrees of the particle freedom.

From this viewpoint, the conditions of the precise resonance $\Omega = \mathcal{J}$ are of particular interest. This case lately attracts attention as a method of accelerating charged particles and (the inverse situation) that of generating the microwave field in cyclotron autoresonance masers (CARM). In its essence, the effect in question represents the well–known cyclotron resonance generalized for the particle three–dimensional motion in the traveling plane wave.

All the necessary relations to discuss the autoresonance have been already obtained above. Here one has to find not the stationary solution, which is inappropriate for the precise resonance, but the solution of (3.86) with taking into account initial conditions. The latter, for simplicity, can be chosen in the form of the particle purely longitudinal motion with the velocity $\mathbf{v}_0$ and, correspondingly, with the integral of motion

$$\mathcal{J} = \gamma_0 \left(1 - \beta_0\right) = \sqrt{\left(1 - \beta_0\right)/\left(1 + \beta_0\right)} \,.$$

Here we address a particular case: the incident wave is characterized by circular polarization with the direction prescribed by the external magnetic field. One can prove that the existence of the wave field component rotating in the opposite direction does not change the results qualitatively. Therefore, the equation of the charged particle motion in the incident wave field is:

$$\dot{\mathbf{g}} = -[\mathbf{g} \times \mathbf{e}_z] \;;\quad \mathbf{g} = [\dot{\mathbf{g}} \times \mathbf{e}_z] \;;\quad g^2 = \text{const} \,. \tag{3.93}$$

The solution of (3.89) under the chosen initial conditions takes the form:

$$\mathbf{u} = -\varphi \mathbf{g} + \mathbf{n}\left(\frac{\beta_0}{1 - \beta_0} + \frac{g^2 \varphi^2}{2}\right) \tag{3.94}$$

or

$$\mathbf{v}/c = \frac{\mathbf{n}\left(\beta_0 + g^2\varphi^2\left(1 - \beta_0\right)/2\right) - \mathbf{g}\varphi\left(1 - \beta_0\right)}{1 + \varphi^2 g^2 \left(1 - \beta_0\right)/2} \;; \tag{3.95}$$

$$\gamma = \gamma_0 \left(1 + g^2\varphi^2 \left(1 - \beta_0\right)/2\right) . \qquad (3.96)$$

The dependence $\varphi(t)$ has to be determined as well. Taking into account that $d\varphi/dt = -\omega_0 \left(1 - \mathbf{v}\mathbf{n}/c\right)$, one may write this dependence as

$$\omega_0 t = -\frac{\varphi}{1-\beta_0}\left(1 + g^2\varphi^2 \left(1 - \beta_0\right)/6\right) . \qquad (3.97)$$

Equations (3.95), (3.96), and (3.97) yield the parametric temporal dependencies of the velocity and energy of the particle. They indicate that, under the resonance condition

$$B_0 = (\omega_0 mc/q)\sqrt{(1-\beta_0)/(1+\beta_0)} ,$$

the particle experiences a monotonous acceleration as a result of the wave energy resonant absorption in the magnetic field. The particle trajectory has the form of a helix unwinding with an increasing step and the diminishing pitch angle.

In this connection, the two interesting facts are worth mentioning. First, the resonance under consideration in its essence is of the three–dimensional cyclotron type. This resonance is not limited by the relativistic increase in the particle energy. The absence of such restrictions is a result of existence of the exact integral of motion $\mathcal{J}$. From the viewpoint of physics, this integral indicates that Larmor frequency of the particle rotation in the magnetic field decreases exactly in the same way as the frequency of the incident wave field acting upon the particle (with taking into account the heightening of the longitudinal velocity). This circumstance gave the name to the phenomenon [28, 29].

Second, in the relativistic case, the autoresonance is possible at rather high frequencies the magnetic field strength being relatively moderate:

$$\frac{qB_0}{mc} = \omega_0\sqrt{\frac{1-\beta_0}{1+\beta_0}} \approx \frac{\omega_0}{2\gamma_0} .$$

Such a combination of the system parameters is of interest for practical applications. The coefficient of frequency multiplication is lower than under the conditions of the undulator radiation. Nevertheless, the autoresonance may be prospective for generating the short–wave radiation as well as for the schemes of particle acceleration where high–frequency fields are used.

Naturally, there exist discrepancies between the idealized case under discussion and the real conditions (e.g., deviation of the magnetic field from the resonance value or the field inhomogeneity, inequality of the scattered wave phase velocity to c, convergence between the directions of the incident wave propagation and the magnetic field vector, etc.). Such deviations result both in limiting the obtainable energy and in a periodic energy interchange between the wave and the particle. Really, this energy interchange is accompanied by

3.6 Scattering and Absorption by Bound Particle

the radiation emission. Correspondingly, if the time of the wave–particle interaction is much longer than the interchange period, the wave scattering is implicated. Otherwise, the process principally consists in the wave energy absorption by the charged particle.

In conclusion, certain limits of description above should be mentioned. The matter is that the processes of emitting the microwave bremsstrahlung has been described above in terms of classical electrodynamics. It is evident that this approach is justified if there is a sufficiently large number of the electromagnetic field quanta ΔN_ω emitted by the charged particle during the characteristic time interval ΔT_ω. This number can be estimated by the known intensity of the radiation emitted P_ω:

$$\Delta N_\omega \approx \left(\frac{P_\omega}{\hbar\omega}\right) \Delta T_\omega . \qquad (3.98)$$

The inequality $\Delta N_\omega \gg 1$ determines the range of applicability of classical electrodynamics. In particular, one may regard the radiation field phase φ as a physical parameter under such conditions. Its uncertainty $\Delta\varphi$ may be considered to be in inverse proportion to the number of the photons ΔN_ω:

$$\Delta\varphi \Delta N_\omega \approx 1 .$$

Otherwise, if the radiation intensity and the number of photons determined by the latter are small, the field phase is not defined. In this case, the first multiplier on the right-hand side of (3.98) means just the probability of the energy quantum emission per unit time.

The microwave bremsstrahlung plays an important role not only in microwave electronics but also in plasma physics, cyclic accelerators, and storage rings. Therefore, at present, various theoretical aspects of this subject are in the complete way elucidated in the numerous monographs (e.g., see [1, 5]). However, lesser attention was paid to the undulator radiation characteristics. Nowadays the main interest to the subject is due to so-called free electron lasers. The idea to use the effect for the microwave radiation emission belongs to V. Ginzburg [11], the magnetostatic undulator has been proposed, designed, and applied by H. Motz [30].

4

Radiation Reaction

In the previous sections, almost all the results are obtained under the supposition that a character of the emitter motion has been prescribed. The physical factors providing this motion are deliberately ignored. Evidently, such statement of the problem is not quite correct. Really, the electromagnetic radiation always carries away certain amounts of the emitter energy and momentum. These losses either alter the corresponding characteristics of the emitter motion or have to be compensated by an external field providing the given motion.

Alterations in dynamic characteristics of a charged particle conditioned by the radiation field are negligible only within a limited time interval when these alterations are sufficiently small. For instance, the absence of noticeable violations of the conditions for the wave–particle synchronism, which considerably influence the radiation parameters, may serve as a criterion of the neglect admissibility. However, being justified in many cases, this approximation evidently contradicts the problem of transferring an essential amount of the particle energy to the wave emitted. Under experimental conditions, this problem can be partly solved by the corresponding alterations in the medium parameters. In particular, an adiabatic diminution of the wave phase velocity in amplifiers of the TWT and free electron laser types serves this purpose as well as the phase velocity increase along the accelerating structure in ion liner accelerators.

The exact compensation for the alterations in the emitter parameters with the help of external fields is impossible. The matter is that the characteristics of the electromagnetic field "prescribing" the emitter motion (and there exist no other means of influencing a charged particle) essentialy differ from those of the microwave radiation field. Consequently, external field cannot provide fulfillment of the corresponding conservation laws. The simplest case of the emitter uniform and rectilinear motion in a material medium makes the only exception. Under such conditions, the emitter energy losses by Cherenkov radiation can rather precisely be compensated by an external electric field, spatially homogeneous, constant in time and directed along the emitter trajectory.

Below we describe the physical processes responsible for the radiation reaction force as well as the quantitative characteristics of the force influence on the emitter dynamics.

4.1 Conservation Laws

In general, the problem of compensating alterations in the emitter trajectory parameters by external fields has no solution. The basic physical reason of this consists in the principal difference between the mechanical conservation laws for a point charged particle and for the radiation field. The electrodynamic conservation laws are integral ones (i.e., they are not localized either in space or in time). In particular, the law of conservation of the field energy is reducible to the following statement: within a closed volume, temporal alterations in the total energy of the field and emitting particles are equal to the time average flux of Poynting vector through the surface bounding this volume. Were the radiation field strengths decreasing in a sufficiently quick manner with the growth of a distance from the emitter, it would be possible to neglect Poynting vector flux by choosing the bounding surface at a distance large enough. Then the supposition about preservation of the "field + emitter" system total energy within the given volume would be justified. However, it is clear that the radiation field does not satisfy this supposition because of the very definition of this field. Therefore, the time–average total flux of Poynting vector has to be equal to the rate of the emitter energy variation (naturally, time averaged as well).[1] The analogous considerations are also true for the particle linear and angular momenta.

The reasoning presented indicates that to preserve the equation of motion in the standard form, it must be complemented with a force providing fulfillment of the conservation laws – if only on an average. This force is called the radiation reaction force. The averaging should be included because variations in the particle energy and of the power flux of the electromagnetic radiation field do not coincide either in time or in space. Not containing the degrees of the radiation field freedom, this equation (with taking into account the radiation reaction force) is irreducible to the canonical (Hamiltonian) form. Therefore, sometimes they define the radiation reaction force as the radiation friction or the force of deceleration by radiation. The necessity of introducing the radiation reaction force becomes especially clear by using as an example the plane-wave scattering by a free charged particle (see Sect. 3.6). There we have considered not the emitter motion to be prescribed but the electromagnetic field stimulating this motion. Surely, in the problem examined, an

---

[1] In a condensed medium, a part of the emitter kinetic energy is spent on exciting the medium polarization oscillations (see Sect. 2.1) and also on ionization of the medium atoms [10].

intensity of the transmitted wave has to be less than that of the incident one due to emergence of a scattered wave. However, this simultaneously implies the incident wave momentum decrease. At the same time, the scattered wave does not carry away any momentum (at least, in the dipole approximation). Consequently, an extra momentum, taken away from the incident wave and not carried away by the scattered one, has to be absorbed by the particle. The momentum absorbtion may be interpreted as the incident wave pressure. As a rule, this pressure may be presented as Lorentz force, described by the vector product of the magnetic field strength of the wave under emission and the particle velocity driven by the wave electric field. However, this explanation lacks in an important detail. That is, as the particle velocity is $\pi/2$-shifted in phase with respect to the wave electric field (and with the corresponding magnetic field as well), the average value of Lorentz force is precisely equal to zero. The only possible way out of this contradiction could consist in supposing that the particle acceleration conditioned by the electric field of the wave be phase–shifted with respect to the latter. In its turn, this implies the necessity to introduce an additional friction force – the radiation reaction – into the equation of motion.

Surely, the radiation reaction force may (and must) be interpreted as a result of the particle proper field action. During the emitter uniform rectilinear motion in vacuum, such an interaction does not occur. This follows even from the condition of symmetry of a vacuum environment with respect to the emitter.[2] In general, the calculations of the radiation backward influence on the emitter under the conditions of its acceleration is complicated by some difficulties of the fundamental nature, the discussion of which is beyond the scope of this monograph. Such calculations are reliable solely in the case of the emitter acceleration low enough. Below we will restrict ourselves to the simplifying considerations following from the conservation laws.

As regards a reference frame where the particle velocity is low, one may consider the radiation to be of the dipole nature and the energy flux carried away with the radiation to be proportional to the average square of the emitter acceleration. Consequently, the radiation reaction force $\mathbf{F}_{\mathrm{rad}}$ must meet the condition

$$\left\langle \mathbf{F}_{\mathrm{rad}} \mathbf{v} + \frac{2q^2}{3c^3}\left(\frac{d\mathbf{v}}{dt}\right)^2 \right\rangle = 0 , \qquad (4.1)$$

where $\langle \ \rangle$ signifies the time average.

As the right–hand side of (4.1) is equal to zero, the expression in the angular brackets must represent the total derivative with respect to time of a certain function of velocity and its derivatives. Presenting the acceleration squared as

$$\left(\frac{d\mathbf{v}}{dt}\right)^2 = \frac{d}{dt}\left(\mathbf{v}\frac{d\mathbf{v}}{dt}\right) - \mathbf{v}\frac{d^2\mathbf{v}}{dt^2} \qquad (4.2)$$

---

[2] In the meantime, we are not going to consider here the problem of the radiation reaction during the emitter motion in a material medium.

we now rewrite (4.1) in the following form:

$$\left\langle \left(\mathbf{F}_{rad} - \frac{2q^2}{3c^3} \frac{d^2\mathbf{v}}{dt^2} \right) \mathbf{v} \right\rangle = 0 . \tag{4.3}$$

Judging by this formula, one may put in a rather general case

$$\mathbf{F}_{rad} = -\mu \frac{d\mathbf{v}}{dt} + \frac{2q^2}{3c^3} \frac{d^2\mathbf{v}}{dt^2} . \tag{4.4}$$

The first addendum, containing an undetermined factor μ, has the form of a nonrelativistic inertial force. It indicates that at least a part of the emitter inertial mass has to be conditioned by the electromagnetic field surrounding the particle (but again on an average). However, this is rather the problem of interpretation because no other mass except for the mass directly measured in the experiment and already included into the equation of motion can be ascribed to the emitter.[3]

Thus, we have demonstrated that in the system under consideration the equation of motion must be complemented with the radiation reaction force, proportional to the time derivative of the particle acceleration:

$$\mathbf{F}_{rad} = \frac{2q^2}{3c^3} \frac{d^2\mathbf{v}}{dt^2} . \tag{4.5}$$

The procedure of the time averaging used in the presented "derivation" of (4.5), based on the demand of fulfilling the energy conservation law. This can produce an impression that (4.5) is valid only in the cases of the emitter periodic (or quasi–periodic) motion. However, the same expression could be obtained by examining the acceleration enduring for a sufficiently short-time interval, in which the rate of the emitter kinetic energy diminution remains approximately constant. The necessity of averaging is conditioned by a nonlocalized character of the energy conservation law. That is, the radiation friction work on the charged particle per unit time is equal not only to the radiation power flux through a closed surface (the radiation power), but it also includes the field energy variation within the volume enclosed in this surface. By taking into account this variation, one can avoid many contradictions caused by applying (4.5). Anyway, the radiation friction concept is justified for sufficiently slow variations in dynamic parameters of the "radiation + emitter" system when, at least in one reference frame, the radiation reaction force is small in comparison with the forces of external fields acting upon the particle. The universal criterion of such smallness looks like

---

[3] Any model of the emitter, where the mechanical and electromagnetic components of the particle mass are singled out, contradicts its presentation as an elementary particle.

$$2\pi r_c \ll \lambda, \tag{4.6}$$

where $r_c = q^2/mc^2$ is the particle classical radius and λ is the characteristic wavelength of the field acting upon the particle or its path curvature radius. As for an electron $r_c = 2.8 \times 10^{-15}$ m, the condition (4.6) is surely fulfilled for all parameters of motion interesting applications.

As a real physical meaning is appropriate to $\mathbf{F}_{\text{rad}}$ (at least, within the framework of the perturbation theory), this force has to be interpreted as the spatial component of a 4-vector f_i, orthogonal to the 4-velocity u_i and reducing to

$$\left(\frac{2q^2}{3c}\right)\left(\frac{d^2 u_i}{ds^2}\right), \quad \text{if } u_{1,2,3} = 0.$$

As $u_i^2 \equiv -1$ and $u_i(du_i/ds) \equiv 0$, this vector has the form [1]:

$$f_i = \frac{2q^2}{3c}\left[\frac{d^2 u_i}{ds^2} - u_i\left(\frac{du_k}{ds}\right)^2\right]. \tag{4.7}$$

The expression obtained permits us to write the radiation reaction force in the three-dimensional form in an arbitrary inertial reference frame where $\mathbf{v} \neq 0$:

$$\mathbf{F}_{\text{rad}} = -P_0 \mathbf{v} + \frac{2q^2}{3c^3\gamma}\frac{d}{dt}\gamma^3\frac{d\mathbf{v}}{dt}. \tag{4.8}$$

Here P_0 is the radiation power (or, to be more precise, it is the work performed by the particle against the radiation reaction force per unit time):

$$P_0 = \frac{2q^2}{3c}\left[\gamma^4\left(\frac{d\mathbf{v}}{cdt}\right)^2 - \gamma\frac{d^2\gamma}{dt^2}\right]. \tag{4.9}$$

As the factor γ is raised to the high power on the right-hand side of (4.8), we can limit ourselves in the relativistic case to the first addendum. This, in the literal sense, attaches to $\mathbf{F}_{\text{rad}}$ the meaning of the friction directed against the emitter velocity. This force is conditioned by the recoil momentum of the radiation that is directed mainly along $\mathbf{v}$. Consequently, the expression for the radiation losses is simplified also because the relativistic particle acceleration consists in variations in the velocity direction but not in its magnitude. Therefore, the emitter acceleration and the radiation power may be presented as

$$\frac{d\mathbf{v}}{cdt} \approx \frac{\mathbf{F}_\perp}{m\gamma c} \quad \text{and} \quad P_0 \approx \frac{2q^2 F_\perp^2}{3m^2 c^3}\gamma^2. \tag{4.10}$$

Here $\mathbf{F}_\perp$ is the force transverse to the instantaneous velocity $\mathbf{v}$.

As this force is almost independent of the emitter energy in the relativistic case, the relativistic particle power losses are proportional to γ^2 for fixed curvature of the trajectory.

4.2 Radiation Reaction and Emitter Proper Field

The considerations above clearly indicate that the radiation reaction force is not a force in the literal sense of the word because it is conditioned not only by variations in the mechanical characteristics of the emitter motion but also by structural changes in the electromagnetic field surrounding the charged particle. As (4.8) and (4.10) indicate, a simple physical interpretation of this force as a recoil stimulated by the radiation is justifiable only in the ultra-relativistic case. In this connection, it would be appropriate to note that introducing the radiation friction on the basis of the conservation laws, we have not specified any concrete model of the emitter. At the same time, we have ignored the possibilities of variations in the energy of the particle internal degrees of freedom (for instance, the rotational motion). Both these simplifying suppositions are justifiable in the case of an elementary particle because, within the framework of the relativistic principles, it must be point one. At the same time, the force acting upon the emitter is the Lorentz force, coinciding with the electric field in the rest frame. The emitter proper electric field only can make the physical reason for the radiation friction (to be more precise, the proper field deviation from the Coulomb field, caused by the particle acceleration and finiteness of the velocity of light).

Calculation of the emitter proper field, which again yields (4.4), is usually based on applying either the retarded potentials ([1]) or the retarded field.

It is also supposed that the particle motion varies slowly enough. In other words, the change of the particle acceleration during the time required for light to pass through the charge distribution is small in comparison with the acceleration itself. Further, at the limit of the zero particle "radius," one gets grounds for calling (4.4) "an exact expression in a certain sense" [1]. All the same, it remains not quite satisfactory. Really, taking into account $\mathbf{F}_{\text{rad}}$ in the equation of motion for a free particle results in

$$m\frac{d\mathbf{v}}{dt} = \frac{2q^2}{3c^3}\frac{d^2\mathbf{v}}{dt^2} . \qquad (4.11)$$

The latter, in addition to the trivial solution $\mathbf{v} = \text{const}$, possesses another one, physically senseless: both the acceleration and velocity of the particle are exponentially increasing with the characteristic time $\tau = 2q^2/3mc^2$ [1].

As we avoid using the retarded potentials because of the above–given reasons, our starting point in calculations of the emitter proper field is the system of Maxwell equations for the electric field Fourier–amplitudes

$$\mathbf{E}(\mathbf{k},t) = \frac{1}{8\pi^3}\int \exp(-i\mathbf{k}\mathbf{r})\mathbf{E}(\mathbf{r},t)\,d\mathbf{r} .$$

The current and charge densities for a point charged particle, moving along the trajectory $\mathbf{r}_0(t)$ with the velocity $\mathbf{v}(t) \equiv d\mathbf{r}_0/dt$, can be written as

$$\rho(\mathbf{r},t) = q\delta(\mathbf{r}-\mathbf{r}_0) \quad \rho(\mathbf{k},t) = \frac{1}{8\pi^3}\exp(-i\mathbf{k}\mathbf{r}_0) \qquad (4.12a)$$

$$\mathbf{j}(\mathbf{r},t) = q\mathbf{v}\delta(\mathbf{r}-\mathbf{r}_0) \qquad \mathbf{j}(\mathbf{k},t) = \frac{\mathbf{v}}{8\pi^3}\exp(-i\mathbf{k}\mathbf{r}_0) \ . \tag{4.12b}$$

Making use of the latter relations, one gets the following equation for $\mathbf{E}(\mathbf{k},t)$:

$$\left(\frac{d^2}{dt^2} + k^2c^2\right)\mathbf{E}(\mathbf{k},t) \tag{4.13}$$

$$= -\frac{q}{2\pi^2}\left(\frac{d\mathbf{v}}{dt} - i(\mathbf{k}\mathbf{v})\mathbf{v} + ikc^2\right)\exp(-i\mathbf{k}\mathbf{r}_0(t)) \ .$$

It should be noted that (4.13) is written down for the total electric field without singling out its longitudinal and transverse components.

According to the method applied in [2], let us present the total field as a sum of the static Coulomb field of the charged particle located at the point $\mathbf{r}_0(t)$ and an additional field $\mathbf{E}'$:

$$\mathbf{E}(\mathbf{k},t) = -\frac{iq\mathbf{k}}{2\pi^2 k^2}\exp(-i\mathbf{k}\mathbf{r}_0) + \mathbf{E}'(\mathbf{k},t) \ . \tag{4.14}$$

For $\mathbf{E}'(\mathbf{k},t)$ one gets the equation:

$$\left(\frac{d^2}{dt^2} + k^2c^2\right)\mathbf{E}'(\mathbf{k},t) = -\frac{q}{2\pi^2 k^2}\frac{d}{dt}[\mathbf{k}\times[\mathbf{U}\times\mathbf{k}]] \ , \tag{4.15}$$

where $\mathbf{U} = \mathbf{v}\exp(-i\mathbf{k}\mathbf{r}_0(t))$.

By the direct check it is easy to see that (4.15) admits the solution satisfying zero initial conditions:

$$\mathbf{E}'(\mathbf{k},t) = \frac{q^2}{2\pi^2 c^2 k^4}\left\{\frac{d}{dt}[\mathbf{k}\times[\mathbf{k}\times\mathbf{U}]]\right.$$

$$-\cos(kct)\left(\frac{d}{dt}[\mathbf{k}\times[\mathbf{k}\times\mathbf{U}]]\right)_{t=0}$$

$$\left. -\int_0^t \cos(kc(t-t'))\frac{d^2}{dt'^2}[\mathbf{k}\times[\mathbf{k}\times\mathbf{U}]]\,dt'\right\} \ . \tag{4.16}$$

To restore the time dependence of the emitter proper field at the point of the particle location, the right-hand side of (4.16) must be integrated over space of $\mathbf{k}$ values with the multiplier $\exp(i\mathbf{k}\mathbf{r}_0(t))$. The integral is to be taken over the spherical volume $|\mathbf{k}| < k_m$, where the condition $\mathbf{k}(\mathbf{r}_0(t) - \mathbf{r}_0(t')) \ll 1$ holds, so that the exponent may be considered equal to unit. Integration all over the directions of $\mathbf{k}$ yields:

$$\int \mathbf{k}\,d\Omega = 0 ; \qquad \int [\mathbf{k}\times[\mathbf{k}\times\mathbf{U}]]\,d\Omega = -\frac{8\pi}{3}\mathbf{U} \ .$$

These formulae indicate that Coulomb field component vanishes. Hence, the emitter proper field may be written as

$$\mathbf{E}\left(\mathbf{r}_0(t),t\right) = -\frac{4qk_m}{3\pi c^2}\frac{\mathrm{d}\mathbf{U}}{\mathrm{d}t} + \frac{4q}{3\pi c^2}\left(\frac{\mathrm{d}\mathbf{U}}{\mathrm{d}t}\right)_0 \frac{\sin k_m ct}{ct}$$

$$+\frac{4q}{3\pi c^2}\int_0^t \mathrm{d}t' \frac{\mathrm{d}^2\mathbf{U}}{\mathrm{d}t'^2}\int_0^{k_m}\cos\left(kc(t-t')\right)\mathrm{d}k. \quad (4.17)$$

In the particle rest system, $\mathrm{d}\mathbf{U}/\mathrm{d}t$ coincides with $\mathrm{d}\mathbf{v}/\mathrm{d}t$. Therefore, after multiplying by q, the first addendum represents the inertial force, corresponding to the emitter electromagnetic mass. In accordance with the above–given considerations, it has to be included into the total inertial force. However, if $k_m \to \infty$, this term is formally divergent. [4]

In calculations of the third addendum, the equality

$$\lim_{k_m \to \infty}\int_0^{k_m}\cos\left(kc(t-t')\right)\mathrm{d}k = \frac{\pi}{c}\delta(t-t')$$

is to be used. This results in the following expression for the electric field acting upon the emitter:

$$q\mathbf{E}\left(\mathbf{r}_0,t\right) = \frac{2q^2}{3c^3}\frac{\mathrm{d}^2\mathbf{v}}{\mathrm{d}t^2} + \frac{4q^2}{3\pi c^2}\left(\frac{\mathrm{d}\mathbf{v}}{\mathrm{d}t}\right)_0 \frac{\sin k_m ct}{ct}$$
$$+ \text{ the terms going to zero when } k_m \to \infty. \quad (4.18)$$

If the time is long enough, the second term in (4.18) vanishes so that one again obtains (4.4) describing the radiation friction force.

In its essence, the above–given derivation does not differ from the traditional one based on the smallness of the electromagnetic signal retardation [6]. Nevertheless, in our opinion, the limitations on applicability of the notion of radiation friction more distinctly indicate themselves during derivation presented. First, as a result of a not quite correct (but necessary) procedure of the emitter mass renormalization, there arises an additional addendum in the form of $\sin k_m ct/ct$, which has no definite limit when $k_m \to \infty$, but vanishes for large values of t. This makes it meaningless to solve the equation of motion of the (4.11) type under any possible initial conditions. Thus, the above–mentioned paradox concerning the charged particle "self–acceleration" during the time τ can be avoided. Second, the limiting transition $k_m \to \infty$ is correct only under a sufficiently smooth behavior of the function $\mathrm{d}^2\mathbf{v}/\mathrm{d}t^2$. The latter supposition deprives the vector $\mathbf{F}_{\mathrm{rad}}$ of the sense of an instantaneous force acting upon the emitter. In the long run, the inequality of the (4.6) type serves as a criterion of applicability of the "radiation friction" notion.

The methodological difficulties presented are conditioned by the enforced introduction of the concept of a "point" charged particle into electrodynamics, where the intrinsic fundamental length (the particle classical radius r_c exists). For solving these problems, one should address to quantum electrodynamics,

---

[4] The necessity of the particle mass renormalization arises from infiniteness of the point particle proper field energy.

where the notion of the Compton wavelength $\hbar/mc$ arises (the latter is by two orders of magnitude larger r_c). Discussion of these problems is beyond the scope of this book. We just would like to emphasize that the similar difficulties still remain within the framework of the quantum theory as well.

4.3 Radiation Friction and Charged Particle Dynamics Radiation Cooling

The radiation friction forces are supposed to be much weaker than the external ones. Nevertheless, they can influence essentially the motion and dynamics of charged particles. This is clear even by intuition. Really, the radiation friction work upon the emitter monotonously, even if slowly, diminishes the particle energy. This causes violation of the conditions for synchronism, which determine the spectral–angular composition of the radiation emitted. For instance, as a result of such violation, infinitely narrow spectral lines, characteristic of the emitter periodic motion, in reality have a finite "natural" width. Under certain conditions, this effect can be of principal importance. At least, it limits the radiation source spectral brightness (or the scattering resonant cross section). However, as it will be demonstrated below, other effects usually determine the spectral line width in the processes of the radiation emission by intense beams. Such factors are either finiteness of the radiation emission time and/or the coherency of individual emitters, which are not quite independent.

Nevertheless, in many cases the radiation reaction forces are of a principal importance. They, for example, should be taken into account to explain the nature of a force acting upon the particle and directed along the incident wave Poynting vector (the light pressure). Really, for a small enough amplitude of the incident plane–polarized wave, one may suppose that the charged particle is moving strictly along the electric field $E = E_0 \sin \omega t$, obeying the equation of motion:

$$mc \frac{d\beta}{dt} = qE_0 \sin \omega t + \frac{2q^2}{3c^2} \frac{d^2\beta}{dt^2} . \tag{4.19}$$

It is easy to prove that the solution driven by the field E has the form:

$$\beta(t) \approx -\frac{qE_0}{mc\omega} \left(\cos \omega t - \frac{4\pi r_c}{3\lambda} \sin \omega t \right) \quad \text{for} \quad r_c \ll \lambda . \tag{4.20}$$

The second term, proportional to a small ratio of the emitter classical radius r_c to the wavelength $\lambda = 2\pi c/\omega$, is due to the radiation friction and provides a small phase shift between the electric field and particle acceleration. So, the time-average force proportional to the particle velocity and directed along the incident wave vector is nonzero only because of the radiation friction:

$$\langle F \rangle \approx \frac{E_0^2 r_c^2}{3} = \frac{P_0 \sigma_0}{c} . \tag{4.21}$$

Here P_0 is the density of the incident wave energy flow; σ_0 is the Thomson scattering cross section [1].

At the same time, this result is obtainable directly from the law of conservation of momentum. It is altogether predictable because the radiation deceleration force has been introduced for ensuring the fulfillment of this law.

The radiation friction influence on relativistic particles is especially noticeable. Increasing as the square of the particle energy, the radiation friction can become the major force acting upon the electron.[5] In this case, making use of (4.10), one may present the energy variation of the emitter along its trajectory as a result of the work performed by the radiation friction force only [1]:

$$\frac{d\gamma}{ds} = -\gamma^2 \frac{2q^2 \mathbf{F}_\perp^2}{3m^2 c^3}, \qquad (4.22)$$

where $\mathbf{F}_\perp$ is an external force transverse to the particle trajectory. Choosing the initial condition $\gamma = \gamma_0$ when $s \to -\infty$ one gets

$$\gamma^{-1} = \gamma_0^{-1} + \frac{2q^2}{3m^2 c^3} \int_{-\infty}^{s} F_\perp^2 \, ds .$$

The limit of $\gamma_0 \to \infty$ indicates that the emitter energy at the point s remains finite (Pomeranchuk's theorem). In other words, when a charged particle has passed through an external field, its energy cannot exceed the value:

$$\gamma_{\max} = \frac{3m^2 c^3}{2q^2} \left(\int_{-\infty}^{+\infty} F_\perp^2 \, ds \right)^{-1} . \qquad (4.23)$$

However, to avoid misunderstandings, one should keep in mind the following. The electric field work upon the particle all over its passage through the external field is supposed to be much less than the total radiation losses. As it is easy to see, this phenomenon is conditioned by the strong dependence of the radiation friction force versus the emitter energy.

The fundamental changes in the emitter motion characteristics are somewhat less evident if the particle radiation losses are compensated by an external electric field. In this case, the particle energy remains constant or even, on an average, increases according to a prescribed law. Such conditions are typical of installations for accelerating or storing relativistic electrons or positrons. One of their most important characteristics is the beam brightness defined as the particle density in the phase space. In particular, it is the brightness that determines the possibilities of the beam focusing. In their turn, these characteristics influence the effectiveness of the beam application in the high–energy experiments. Great significance of the beam brightness consideration follows from the fact, purely of the theoretical interest at first sight: the behavior of individual noninteracting particles (or the ones interacting via the self–consistent field) can be described by Hamilton canonical equations:

---

[5] This does not contradict the requirement that the radiation friction force is small, which has to be satisfied only in the reference frame where the particle is in rest.

4.3 Radiation Friction and Charged Particle Dynamics Radiation Cooling

$$\frac{\mathrm{d}\mathbf{r}}{\mathrm{d}t} = \nabla_{\mathbf{P}}\mathcal{H}\,; \qquad \frac{\mathrm{d}\mathbf{P}}{\mathrm{d}t} = -\nabla\mathcal{H}\,. \qquad (4.24)$$

The Hamiltonian $\mathcal{H}(t,\mathbf{r},\mathbf{P})$ depends on time, coordinates, and components of the generalized momentum $\mathbf{P}$, conjugated with the coordinates.

As it will be demonstrated below, this is enough for the brightness to become an exact integral of motion under acting upon the beam by an arbitrary external field. This statement being a fundamental one is well–known as the Liouville theorem [31]. There is its another formulation: the shape of the six–dimensional phase space region where the particle image points are located can be changed almost in an arbitrary manner, but its total volume (the beam emittance) all the same remains constant. The point to be made here is that the formulations used by us relate to the continuum presentation of the beam. [6] This implies that an average distance between the image points is much smaller than the intervals $\mathrm{d}\mathbf{r}$ and $\mathrm{d}\mathbf{P}$, which are already physically small. A somewhat imperative character of the statement concerning the impossibility of changing the brightness is conditioned by the following fact: as external fields are of macroscopic nature, with their help, one cannot introduce a new particle into an already–occupied phase space cell. To do this, one has to move apart the particles, located there (to be more precise, their image points). Consequently, the beam particle stacking in magnetic systems must be accompanied by extending the phase space occupied by the particles (i.e., the beam emittance). It is also easily predictable that pair collisions between the beam particles, being of microscopic nature, violate the Liouville theorem. Such collisions cause the emittance increase and enlarge the system total entropy. [7]

The above–mentioned noncanonical nature of the radiation friction force implies violation of the Liouville theorem as well. In fact, this force is "personalized" i.e., the radiation reaction is considered to be acting directly and solely upon the emitter. However, it is also implied that the total power of radiation emitted by a system of particles is equal, at least on an average, to the sum of powers emitted by each particle (i.e., radiation is regarded as incoherent). This condition is to be examined below. At present, we will limit ourselves to studying the beams rarefied enough to guarantee the absence of the emitter coherence. Generally speaking, compatibility of this limitation with the notion of the continuous distribution of particles implies the radiation wavelength smallness in comparison with the characteristic distance between the emitters.

---

[6] This limitation is not necessary for the general Liouville theorem, formulated for the $6N$-dimensional phase space, N being a number of particles.

[7] However, there remains the possibility of diminishing the phase space of one macroscopic component of the beam system by the corresponding equivalent heightening of the emittance of the rest of the system. A refined method of the heavy–particle beam "electron cooling," bearing no relation to the radiative effects, works on this principle [32].

To quantitatively describe variations in the beam brightness, let us introduce the distribution function of the beam particles in the phase space $\Psi(\mathbf{r}, \mathbf{P}, t)$. This function describes the number of particles within the phase space cell $d\mathbf{r}d\mathbf{P}$ and satisfies the continuity equation of the form

$$\frac{\partial \Psi}{\partial t} + \nabla \cdot \left(\Psi \frac{d\mathbf{r}}{dt}\right) + \nabla_{\mathbf{P}} \cdot \left(\Psi \frac{d\mathbf{P}}{dt}\right) = 0. \tag{4.25}$$

The image point velocities $d\mathbf{r}/dt$ and $d\mathbf{P}/dt$ in the phase space must be presented as functions of $\mathbf{r}$, $\mathbf{P}$, and t.

Introduction of the radiation friction force, non–Hamiltonian by nature, means that the generalized force

$$-\frac{P_0}{c^2}\mathbf{v} = -\frac{P_0}{\gamma m c^2}\left(\mathbf{P} - \frac{q}{c}\mathbf{A}\right) \tag{4.26}$$

must be included into the second equation in this system (4.24) (as before, $P_0(\gamma, t)$ is the power of the radiation losses). Thus, (4.25) may be presented in the form:

$$\frac{\partial \Psi}{\partial t} + \frac{d\mathbf{r}}{dt}\nabla\Psi + \frac{d\mathbf{P}}{dt}\nabla_{\mathbf{P}}\Psi = \Psi \nabla_{\mathbf{P}} \frac{P_0}{\gamma m c^2}\left(\mathbf{P} - \frac{q}{c}\mathbf{A}\right), \tag{4.27}$$

because the terms containing cross derivatives of $\mathcal{H}$ are mutually reducible.

The left–hand side of (4.27) represents by itself the total derivative $d\Psi/dt$ along the particle phase trajectory determined by (4.24). In the absence of the radiation friction, this derivative is equal to zero, which depicts the essence of the Liouville theorem. As regards the right–hand side of this formula, making use of the equality

$$\gamma^2 = 1 + \frac{1}{m^2 c^4}\left(\mathbf{P} - \frac{q}{c}\mathbf{A}\right)^2$$

one may equate

$$\nabla_{\mathbf{P}} \frac{P_0}{\gamma}\left(\mathbf{P} - \frac{q}{c}\mathbf{A}\right) \approx 3\frac{P_0}{\gamma} + \gamma \frac{\partial}{\partial \gamma}\frac{P_0}{\gamma} = \frac{1}{\gamma^2}\frac{\partial P_0 \gamma^2}{\partial \gamma}. \tag{4.28}$$

According to (4.10), $P_0 \propto \gamma^2$. Finally, the equation for Ψ takes the form:

$$\frac{d\Psi}{dt} = \frac{4P_0}{\gamma m c^2}\Psi \tag{4.29}$$

and

$$\Psi \propto \exp\int^t \frac{4P_0}{\mathcal{E}}\,dt, \tag{4.30}$$

where $\mathcal{E}$ is the particle energy.

Thus, it has been demonstrated that the beam brightness exponentially increases in time. In the presence of the radiation friction force and in the

4.3 Radiation Friction and Charged Particle Dynamics Radiation Cooling

absence of counteraction (e.g., the above–mentioned pair collisions), the six–dimensional phase space occupied by the beam particles would tend to zero whatever variations in the particle energy be [33]. The characteristic time of this contraction (or the "radiation cooling") $\tau_c \approx \mathcal{E}/P_0$ by the order of magnitude coincides with the time during which the charged particle would spend its total energy on the radiation emission, provided any compensation were absent.

Importance of the phenomenon of the radiation cooling to physics of accelerators is conditioned by the two factors. First, the radiation cooling substantially limits the influence of various disturbances that in an uncontrollable manner give irreversible rise to the beam emittance. Second, which is more important, the radiation cooling is prospective for accumulating light particles in the circulating beam without any increase of its emittance (surely, the accumulation time has to exceed τ_c).

However, the difficulties in realizing the radiation cooling should be mentioned as well. Surely, for the beam long-time existence in a storage ring or in an accelerator, the total six–dimensional emittance of the beam must not increase. At the same time, there also must not take place any increase of the three partial two–dimensional emittances, corresponding to the particle proper oscillations, i.e., small deviations of the particle from the equilibrium coordinates, independent of one another. In the absence of the radiation emission, the betatron oscillations, transverse to the equilibrium orbit and not related to the particle energy variations, play the role of such proper oscillations. Another independent type of oscillations consists in variations in the particle energy (synchrotron oscillations). Providing correlation between the betatron and synchrotron oscillations, the process of the radiation emission also causes redistribution of the partial emittances between these types of oscillations (the total damping decrement still remains equal to $4P_0/\mathcal{E}$). For instance, in the simplest magnetic system, consisting of a sequence of focusing and defocusing sectors, the partial emittances behave in the following way:

$$\varepsilon_z \propto \exp\left(-\int \frac{P_0}{\mathcal{E}}\, dt\right) \qquad \text{vertical oscillations} \qquad (4.31a)$$

$$\varepsilon_x \propto \exp\left((1-\alpha)\int \frac{P_0}{\mathcal{E}}\, dt\right) \qquad \text{radial oscillations} \qquad (4.31b)$$

$$\varepsilon_s \propto \exp\left((\alpha-4)\int \frac{P_0}{\mathcal{E}}\, dt\right) \qquad \text{synchrotron oscillations}. \qquad (4.31c)$$

Here α is the momentum compaction factor that characterizes the dependence of the equilibrium orbit perimeter versus the particle energy. In particular, in strong–focusing systems ($\alpha \ll 1$), the radial betatron oscillations are excited rather quickly at the expense of damping of the synchrotron oscillations. More detailed information about this effect and the methods of its suppression can be found in [33].

To conclude this section, it should be noted that the considerations above are relevant for a single particle isolated from the others. On the one hand, that seems justified for rare beams where the mean distance between particles exceeds essentially the radiation wavelength. On the other hand, the radiation field decreases with distance rather slowly especially in waveguide systems. So, each particle motion could be influenced by a great many others. This influence being systematic, the proper radiation reaction can be shadowed by the total (collective) radiation field. Then qualitatively new many-particle phenomena may develops, which are the subject of the following sections.

Part II

Radiation by Particles Ensembles

The above–examined mechanisms of radiation emission relate to a single point like charged particle under prescribed conditions of its motion. The latter supposition implies that the radiation field backward influence on the emitter dynamics is negligible. In particular, decrease in the emitter energy which irreversibly goes out of the system in the form of the free field energy was supposed to be small. Actually, the statement that the radiation field backward influence on the single emitter dynamics could be negligible is justified in the overwhelming majority of cases. Only radiation emitted by an ultra-relativistic particle, the radiation losses of which are proportional to the relativistic factor squared, can turn out to be comparable with the emitter total energy during a sufficiently long-time interval, as it takes place in the case of synchrotron radiation. This is the reason why we have not estimated the total radiation power, limiting ourselves to the description of electromagnetic fields. Even being multiplied by a large number of particles simultaneously located in the region of interaction, it turns out to be far below the values that could be of interest for microwave technologies.[8]

The two examples below are sufficient to emphasize the principal difference between the radiation emission by a single particle and by ensembles.

According to the single particle theory, the energy loss in the output cavity of a 10 cm–range klystron–amplifier makes several microelectronvolts. At the same time, the 450 A–current electron beam loses the pulse power of 50 MWt in the same cavity. To generate this power, each electron of the beam transfers about 100 keV of its energy to the microwave field. This is 10^{10} times (!) higher than in the case of an individual electron.

The analogous situation takes place in the free electron maser driven by the accelerator ATA (LLNL). In this experiment, the individual electron loss for undulator radiation should be about 5 microelectronvolts. However, the radiation loss of the electron beam of the pulse current of order of 0.5 kA in the same undulator makes approximately 0.1 GWt of the microwave power. Under such conditions the energy loss of each of the beam electrons is of the order of 0.1 MeV. That is, deceleration by the collective radiation field is nine orders of magnitude larger than that of an individual particle.

Coherent summation of the radiation fields (not powers!) emitted by individual particles is essential for interpreting the effects. As regards this summation, certain mechanisms are necessary for providing the conditions for the emitter ensemble coherence and its automatic self–maintenance. These mechanisms are discussed in the part presented below. With the help of simple physical models we trace the links between coherence of radiation emitted by ensembles of charged particles and properties of emission by individual charged particles, described in Part I.

---

[8] Apropos, a relative smallness of the individual emitter radiation losses evidently contradicts the demand for a high efficiency, declared at the beginning of Part I.

5
Coherence of Individual Emitters

Notion of individual emitters coherence is a keystone for physics of the processes of the cooperative interaction between flows of charged particles and electromagnetic waves in structures used in microwave electronics and charged particle accelerators. As it has been mentioned above, it follows even from the direct comparison of the magnitudes of the energy losses of individual charged particles with those typical for intense flows of the same emitters.

5.1 Spatial Coherence

The term *temporal coherence* applied to oscillations (waves) refers to the existence of certain relations between their phases during a time interval long enough. The latter reservation implies that for coherence of two (or more) waves their frequencies and wave numbers must be sufficiently close.

In an ensemble of particles moving under identical conditions, their individual radiation fields of the same frequency differ in phases depending on particles' positions. The summation of these fields with proper phasers may give striking coherent effects (*spatial coherence*).

Let us consider from this point of view summation of the radiation field with an already-existing external wave $\{\mathbf{E}_0, \mathbf{B}_0\} \exp[i(kz - \omega t)]$ of the same frequency and phase velocity. The latter means that the external wave satisfies the condition of synchronism and, hence, can be absorbed or amplified by the particles. In the first case the radiation field is in antiphase with the external one, decreasing the latter; in the second case being in phase it increases the total field.

In a general case, because of linearity of Maxwell equations, the total wave field must be equal to the sum of the fields:

$$\mathbf{E} = [\mathbf{E}_0 + \mathbf{E}_{\text{rad}} \exp(i\varphi)] \exp[i(kz - \omega t)];$$
$$\mathbf{B} = [\mathbf{B}_0 + \mathbf{B}_{\text{rad}} \exp(i\varphi)] \exp[i(kz - \omega t)].$$

Here $\mathbf{E}_{rad}$ and $\mathbf{B}_{rad}$ are real amplitudes of the emitter radiation fields and φ is the field phase determined by the emitter location on the z-axis. Clearly, the total Poynting vector of the resulting wave is

$$\mathbf{P} = \operatorname{Re} \frac{c}{4\pi} \overline{[\mathbf{E} \times \mathbf{B}]}$$
$$= \mathbf{P}_0 + \mathbf{P}_{rad} + \frac{c}{8\pi} \{[\mathbf{E}_0 \times \mathbf{B}_{rad}] + [\mathbf{E}_{rad} \times \mathbf{B}_0]\} \cos\varphi, \quad (5.1)$$

where $\mathbf{P}_{rad}$ and $\mathbf{P}_0$ are Poynting vectors of the emitter's radiation and that of the external source, respectively.

Thus, the total Poynting vector $\mathbf{P}$ differs from $\mathbf{P}_0$ not only by the small vector $\mathbf{P}_{rad}$, usually negligible (as it is proportional to the radiation field amplitude squared). In addition, there also arises an interference cross term, proportional to the first powers of E_0, E_{rad} and depending on the phase φ. It is this addendum that describes the effect of the external source field coherent amplification (or extinction) by the emitter microwave radiation.

Surely, the result is rather evident. It just indicates that a surplus (positive or negative) energy flow is being spent on the resonance acceleration/deceleration of the charged particle moving synchronously with the wave.[1] In its essence, the consideration above consists in confirming that the energy conservation low is fulfilled. All the same, it is often applied for describing slow variations of the wave field in space and time. For instance, as regards electron linacs, the third (mixed) addendum in (5.1) describes the beam loading (under the supposition that the external field's amplitude essentially exceeds the radiation field one). This term also indicates that there exists a symmetry between the following physical processes:

- charged particle resonant acceleration (the external generator's energy is transferred to the beam bunched with respect to the accelerating field phase) and
- electromagnetic field amplification as a result of the reverse process. The bunch energy is transferred to the wave under amplification.

This illustrative example is more important from the another viewpoint. If the emitters' distribution over phases is uniform, the field energy, on an average, is not being transferred to the particles and vice versa (neglecting the small term P_{rad}). One may say that a half of the particles, characterized by the corresponding (decelerating) phases, emits radiation, whereas another half of them absorbs it. So, this is the phase distribution that determines the resulting effect (the field energy absorption/emission). If $E_0 \gg E_{rad}$, the effect of the wave–beam energy interchange can become extremely substantial (see the above–given examples).

---

[1] To prove this statement, one should demonstrate that the radiation *amplitude* is proportional to the field *work* upon the particle. This dependence follows from the rather general expressions (3.26b) and (3.27) of Part I.

5.1 Spatial Coherence

A very similar picture should be expected during the summation of the radiation fields of a very large number of the emitters in absence of any external field. In this case Poynting vector of a fixed frequency mode of the radiation field may be presented in the form:

$$\mathbf{P} = NC(\omega)\mathbf{P}_{\text{rad}}, \tag{5.2}$$

where

$$\mathbf{P}_{\text{rad}} = \frac{c}{8\pi}[\mathbf{E}_{\text{rad}} \times \mathbf{B}_{\text{rad}}]$$

is Poynting vector of single particle radiation and

$$C(\omega) = \frac{1}{N} \sum_{n,m=1}^{N} \exp\left[\mathrm{i}\left(\varphi_n - \varphi_m\right)\right] \tag{5.3}$$

may be called a coherence factor.

Depending on the phase distribution (the emitters' locations with respect to the common wave), the coherence factor magnitude can vary within $0 \leq C(\omega) < N$. If the particle distribution is uniformly random, only the terms with $n = m$ contribute to the averaged value of the sum in (5.3). In this particular case, $\mathbf{P} = N\mathbf{P}_{\text{rad}}$ and $C = 1$. If the absolute phase fortuity is kept for all wavelengths of interest, the character of the spectral–angular distribution of radiation emitted by an ensemble of particles remains the same as in the case of a single particle. That is, the corresponding spectral density of the radiation power just becomes N times larger. Radiation of this type is defined as totally incoherent or spontaneous (the physical meaning of the latter term is to be discussed below).

If phases of individual emitters were correlated, the pattern could be much more complicated. In particular, were all the emitters completely co-phased (so that the phase difference makes an integral number of 2π), the maximal coherence factor $C = N$ would be realized. Consequently, if $N \gg 1$, the radiation power spectral density would essentially be larger. However, the radiation emitted at other frequencies remains incoherent. Surely, the supposition that individual emitters can become completely cophased for any frequency of interest is justifiable only for the wavelengths substantially exceeding the particle bunch sizes.

Conception of the spatial coherence is also appropriate for a system of large sizes but consisting of a periodic sequence of bunches, small in comparison with the wavelength of the radiation emitted. Of course, for very short wavelengths both the ideas are of a purely academic interest. It is practically impossible to prescribe particle locations with the absolute accuracy. In fact, the phases should be considered as random values with a certain probability distribution.

For example, let neighboring emitters have an average phase shift μ with uncertainty obeying the normal distribution of dispersion δ. To find the coherence factor value, the double sum in (5.3) must be weighted with the distribution:

$$\prod_{s=1}^{N} \frac{1}{\delta\sqrt{\pi}} \exp\left(-\frac{(\varphi_s - s\mu)^2}{\delta^2}\right).$$

The terms with $n = m$ must be calculated separately because any emitter is always correlated with itself. After some algebra, one gets

$$C(\omega) = 1 + \left[\frac{\sin^2 \mu N/2}{N \sin^2 \mu/2} - 1\right] \exp\left(-\delta^2/2\right). \qquad (5.4)$$

If the phase uncertainty δ is small, radiation emitted by individual particles heavily interferes and the above–mentioned maxima and minima in the coherent radiation spectrum appear. They correspond to the frequencies for which the neighboring emitters are strictly cophased (or, respectively, antiphased). The coherence factor in the maxima is close to N, whereas in the minima it goes to zero.

In the limiting case of very short waves, the phase uncertainty δ is always large. Under such conditions, smearing of the maxima occurs, and the radiation emission is incoherent: $C \approx 1$.

Here we do not examine more complicated cases of phase correlation between emitters. In fact, this correlation essentially depends on peculiarities of a particular problem. It is appropriate just to mention that the radiation spectral density is to be described in statistical terms. Really, any particular fixed realization of emitters mutual phasing is, in a sense, coherent. At the same time, it is evident that the physical sense is inherent, as a rule, only in a sequence of many cases observed experimentally and distributed according to a specific statistical law.

5.2 Interference in Regular Lattices

The problem of coherence in beam systems of sizes essentially larger than the radiation wavelength is especially important for various applications in microwave electronics and accelerator physics. The above–submitted considerations indicate that a considerable degree of coherence can be achieved only for regular distributions and under condition of a small uncertainty of particle positions in the wavelength scale. From this viewpoint, absolute coherence can be provided only by a 'point-like' bunch. That is, in the rest reference frame the bunch geometrical sizes must be much smaller than the radiation wavelength. As regards the bunch of finite sizes, it is quite predictable that manifestation of the coherence effects depends on the bunch geometry. In particular, there should be an essential difference between a 3-D bunch, a 2-D one where one of its dimensions is smaller than the wavelength, and an one-dimensional train of particles.

5.2.1 One-Dimensional Distribution

The third case is the simplest for the investigation. It is also a good approximation in many practical situations.

So, let us consider Cherenkov radiation emitted by a periodic train of N particles separated by the distance D from one another. As far as physical conditions of all the emitters are the same, the field phases of the neighboring particles are shifted by kD. The corresponding coherence factor of this ensemble is

$$C = \frac{1}{N}\left|\sum_{j=1}^{N} \exp(ikDj)\right|^2 = \frac{\sin^2(kDN/2)}{N\sin^2(kD/2)}. \quad (5.5)$$

As it has already been mentioned, a nonlinear relation of the wave frequency to its longitudinal wave number $\omega(k)$ exists in systems with dispersion. There is a point of this curve, corresponding to the condition of synchronism. The radiation field wavelength is $2\pi/k$. Thus, the coherence factor reaches its maximum $C = N$, if the distance between the emitters is equal to an integer number of the wavelengths and goes to zero if the total number of wavelengths along the train is an integer of $(1 - N^{-1})$.

The latter statement is apparently paradoxical. Really, it implies that, under a certain location of the particles along the chain, they do not emit any radiation, notwithstanding the fact that the front emitters in this ensemble have no information about motion of backward ones (and even about their existence). In fact, the same paradox arises under the conditions of the total coherence of emitters, that is, when the radiation losses per particle increase N times. The matter is that the radiation emitted is identified with the far–zone fields. Their interference obeys exactly this law. In particular, the far–zone fields experience complete mutual extinction if $C = 0$. At the same time, the decelerating field in the near–field zone of each particle depends on the particle position in the ensemble. If there is an integer number of wavelengths between particles, the field increases linearly from the head to the tail of the train. Then the emitters in the train tail become the principal sources of energy being decelerated in a resonance manner by the total field of the front particles. If $C = 0$, there just takes place an energy interchange between the emitters. Some of them are decelerated (the front ones included), whereas the other particles are accelerated. No amount of the ensemble total kinetic energy is transferred to the radiation field at large distances .

5.2.2 Multidimensional Lattices

As regards the emitter spatial distribution (the two–dimensional and especially the three–dimensional ones), the pattern becomes much more complicated. In general, coherent effects are expressed weaker even in the case of simple regular lattices.[2] Actually, the simple conditions of fields summation or subtraction can be fulfilled for particular directions only, if at all. So, coherence reveals itself in fine-structure angular and spectral characteristics, while the main integral effect – nonlinear power dependence on the number of emitters – is not expressed.

To illustrate this, we consider the simplest system of N one-dimensional oscillators consisting of a cubic lattice of period l. Their coordinates are determined by $3N$ integer numbers j_x, j_y, and j_z:

$$\mathbf{r}_j = (j_x \mathbf{e}_x + j_y \mathbf{e}_y + j_z \mathbf{e}_z)\, l\,; \tag{5.6}$$
$$1 \leq j_x \leq N_x\,; \quad 1 \leq j_y \leq N_y\,; \quad 1 \leq j_z \leq N_z\,; \quad N_x N_y N_z = N$$

where $\mathbf{e}_{x,y,z}$ are unit vectors of the lattice and $N_{x,y,z}$ is a number of the crystallographic planes normal to the vectors. If the oscillators phases are not correlated, the radiation is, of course, completely incoherent.

To simplify calculations and to keep the coherence in this example as high as possible, we suppose that all oscillators have the same amplitudes and that the phase differences of each two neighbors in x, y, z directions are μ_x, μ_y, and μ_z, respectively. This model corresponds to two cases of practical importance. First, such oscillations can be driven by an external plane wave with the propagation vector $\mathbf{n}$. In this case, radiation emitted by particles can be interpreted as a coherent scattering of the incident wave of certain frequency ω with the phase shifts $\mu_{x,y,z} = (\omega l/c)\,(\mathbf{n}\mathbf{e}_{x,y,z})$. Second, the model represents proper plane waves in the lattice of interacting particles where the values $\mu_{x,y,z}$ determine the wave eigenfrequency ω. Then one may speak about radiation induced by proper waves of polarization in the lattice in the direction of

$$\mathbf{n} = (\mu_x \mathbf{e}_x + \mu_y \mathbf{e}_y + \mu_z \mathbf{e}_z)/\mu\,; \qquad \mu^2 = \mu_x^2 + \mu_y^2 + \mu_z^2. \tag{5.7}$$

According to (5.3), the coherence factor of radiation can be presented in the form:

$$C = \left| \sum_{j=1}^{N} \exp\left[i\left(\frac{\mu}{l} \mathbf{n} - \frac{\omega}{c} \mathbf{e} \right) \mathbf{r}_j \right] \right|^2 \tag{5.8}$$

$$= \left| \sum_{j_{x,y,z}=1}^{N_{x,y,z}} \exp\left[i\left(\frac{\mu}{l} \mathbf{n} - \frac{\omega}{c} \mathbf{e} \right) (j_x \mathbf{e}_x + j_y \mathbf{e}_y + j_z \mathbf{e}_z) \right] \right|^2.$$

The radiation angular distribution determined by the coherence factor and by the angular dependence of the single particle radiation is rather complicated and would lead us into the maze of optics of crystals. In what follows we

---

[2] We do not consider here phase-controlled lattices where the phase of each emitter is correlated with its position.

5.2 Interference in Regular Lattices

restrict ourselves by a total coherence condition meaning the approximate equality $C \approx N$.

It is easy to see that in a 3-D lattice ($N_{x,y,z} \gg 1$) the total coherence occurs under three conditions:

$$\left(\mu \mathbf{n} - \frac{\omega l}{c}\mathbf{e}\right)\mathbf{e}_{x,y,z} = 2\pi s_{x,y,z}, \qquad (5.9)$$

where $s_{x,y,z}$ are arbitrary integers.[3] Beside of the radiation direction these equalities determine its frequency. It is interesting that absolutely coherent radiation ($\mu = \omega l/c$) turns out to be possible (if at all) only in those directions where the recoil momentum of a radiated quantum is multiple of the lattice inverse period:

$$\frac{\hbar\omega}{c}(\mathbf{n} - \mathbf{e}) = \frac{2\pi\hbar}{l}(s_x\mathbf{e}_x + s_y\mathbf{e}_y + s_z\mathbf{e}_z). \qquad (5.10)$$

In other words, an incident wave of frequency ω can be scattered coherently only for

$$\frac{\omega l}{\pi c} = \frac{s_x^2 + s_y^2 + s_z^2}{(s_x\mathbf{e}_x + s_y\mathbf{e}_y + s_z\mathbf{e}_z)\mathbf{n}}, \qquad (5.11)$$

the angle of scattering $\alpha = \arccos(\mathbf{en})$ being determined by

$$\sin\frac{\alpha}{2} = \pm(s_x\mathbf{e}_x + s_y\mathbf{e}_y + s_z\mathbf{e}_z)\mathbf{n}\sqrt{s_x^2 + s_y^2 + s_z^2}. \qquad (5.12)$$

Two-dimensional and 3-D lattices are less selective requiring only two (one) equalities of the (5.9) type to be satisfied. The remaining one(s) determine then the scattered wave phase. So, for a fixed frequency there is a direction (cone) of the coherent scattering.

These arguments lead to the conclusion that coherence effects influence mainly differential radiation characteristics rather than the integrated ones. To find a coherence factor for the total scattering cross section, one has to multiply (5.8) by $\sin^2\theta$, where θ is an angle between $\mathbf{e}$ and a unit polarization vector $\mathbf{p}$, and then integrate it over the solid angle. The result is

$$C = \frac{3}{2}\sum_{j,j'}\exp\left[i\frac{\mu c}{\omega l}\mathbf{n}y\right]\left(1 + \frac{1}{y}\frac{d}{dy} + \cos^2\nu\, y\frac{d}{dy}\frac{1}{y}\frac{d}{dy}\right)\frac{\sin y}{y}, \qquad (5.13)$$

where $y = \omega(\mathbf{r}_j - \mathbf{r}_{j'})/c$ and ν is an angle between $\mathbf{p}$ and $\mathbf{y}$. Note that this expression does not imply any particular emitters distribution and the parameter l just plays a role of a scaling factor.

The case of the main interest corresponds to wavelengths larger than the particle-to-particle distance but smaller than the lattice size. The first condition implies a certain coherence which, one can expect, depends on the lattice

---

[3] One may consider only the interval $-\pi < \mu < \pi$.

geometry. For the sake of simplicity, we consider only scattering of a transverse electromagnetic wave putting $\mu/l = \omega/c$ and $\mathbf{n} \perp \mathbf{p}$.

Under conditions listed above, summation in (5.13) can be replaced by integration over all directions of the vector $\mathbf{y}$ and over its length. For $N \gg 1$ one can neglect all edge effects and gets for $|\mathbf{y}_{\max}| \gg 1$

$$C \approx 6\pi N \int_0^{y_{\max}/2} \sin^2 y \, dy \to \frac{2}{3}\pi N y_{\max}. \tag{5.14}$$

Hence, C is proportional to the ratio of the lattice size to the wavelength (of course, one should not take seriously the numerical coefficient). The result shows that scattered radiation keeps a certain degree of coherence, because phases of close neighbors remain correlated for large wavelengths.

In systems of two and one dimensions, the coherent effects are weaker. For example, for emitters located in one plane the coherence factor is proportional to the square root of the mentioned ratio and almost independent of the latter for an one–dimensional chain of emitters.

It has been supposed above that the wavelength exceeds a particle-to-particle distance, so that the lattice is, in a sense, continuous. In the opposite case of short waves, the integral coherence disappears while the interference maxima still exist. Certain computations [34] show that this really takes place for particle separation distances comparable with the wavelength.

The point to be made here is that the above–given examples of ensembles of coherent emitters are purely illustrative. In fact, a number of possible examples of such systems are infinitely large. We would like just to emphasize once more that coherent radiation would be emitted in a macroscopic system with its sizes substantially exceeding the wavelength if the field phases of individual emitters were appropriately correlated. Consequently, the coherent radiation can be emitted only at a certain frequency or in narrow ranges of frequencies and wave numbers. As a rule, this process is accompanied by the radiation extinction, also coherent, at some other modes. In this connection, it is appropriate to dwell on one rather reiterated statement: N particles in a point–like bunch always emit radiation at any frequency as a single particle, its charge being a multiple of N; i.e., the total radiation power becomes N^2 times higher. Generally speaking, this is not correct: a system of interacting particles is irreducible to one elementary particle with its charge and mass heightened. Really, there are the system internal degrees of freedom ('plasma waves'). The smaller are the bunch sizes, the higher would be characteristic frequencies of these waves. The spectral–angular distribution of coherent radiation emitted at these frequencies differs from the case of radiation emitted by a single point–like particle.

6

Spontaneous and Stimulated Emission

We have already mentioned the term 'spontaneous radiation' in Chap. 5. It was used as a synonym with the notion of the random–phase wave fields summation. In terms of statistics, radiation emitted at random phases is characterized by a spectral–angular distribution of the average power flow. The total field phase is also random in this context. At the same time, it is clear that any realization of the emitter ensemble could be, in a sense, coherent if capable of preserving the fixed correlation between individual emitters during a time interval sufficiently long. For instance, the process of the regular wave scattering by a fixed lattice of charged particles meets these conditions (see Sect. 5.2.2). Surely, the oscillation phase of each particle, prescribed by the wave under scattering, remains correlated with the particle location even if the latter is random. Therefore, a certain degree of coherence is inherent in the total radiation, emitted by this ensemble. Naturally, if the emitters are characterized by a regular spatial distribution, the effects of the radiation coherence are more expressive.

The above–given reasoning relates to fixed ensembles of emitters. As regards the systems which are substantially nonequilibrium (e.g., flows of charged particles), there arises a question: to what extent can regular spatial distributions of emitters keep the coherence in time? Such steady regular distributions might radically alter the basic characteristics of spontaneous radiation, i.e., its spectral–angular distribution and total radiation power. On the one hand, it seems that regular distributions of a large number of independent emitters cannot survive. The matter is that, influenced by a lot of uncontrollable disturbing factors (thermal spread of emitter velocities, collisions, etc.), radiation would rather quickly lose its coherence. On the other hand, symmetry of the processes of emission and absorption of radiation by individual particles indicates that identical emitters cannot be independent of one another: each of them reacts to the radiation fields emitted by other particles. As a result of this, the amplitude and, what is more, the phase of the emitter proper radiation field is subjected to certain changes. There takes place such a reaction even if individual emitters are quasi neutral (e.g., atoms) and separated from one another by large distances. As regards charged particles in a dense beam, their interaction can also be influenced by comparatively short–range Coulomb forces. The necessity of taking into account these forces essentially complicates the quantitative description of the radiation emission by the corresponding ensembles.

In rarefied ensembles, Coulomb fields are negligible and collective interaction between individual emitters is realized via long–range microwave fields only. The symmetry between the processes of the radiation emission and absorption means that the processes are mutually complementary. Absorption is, of course, a stimulated process meaning that its rate is proportional to the existing electromagnetic power. So is the complementary process that can be called the stimulated emission. The corresponding notions had been for the first time introduced by A. Einstein [2], who used them for the analytical explanation of the black body equilibrium radiation law. In the course of development of quantum theory, the mechanism of the stimulated radiation emission was explained directly.

The quantum genesis of the notion somewhat hampered its application to classical (nonquantum) systems, in particular to intense flows of high–energy electrons. Perhaps, only elaboration of the theory of free electron lasers has revealed the profound correlation between the stimulated emission and the principle of operation of quite traditional devices of microwave electronics (e.g., klystrons or TWT). To emphasize this inner link, we will discuss briefly the quantum and classical interpretations of notions of the spontaneous and stimulated emission. Peculiarity of the mechanisms of realization of these effects in classical ensembles of emitters (flows of charged particles) is also to be discussed below.

6.1 Semiquantum Interpretation

Einstein's semiphenomenological theory is based on the following consideration: in an equilibrium ensemble of emitters the spectral distribution of the radiation energy density is a result of a mutual balance of elementary acts of emitting and absorbing the field energy quanta by individual emitters. For deriving quantitative characteristics of intensity of these processes, the notions of probabilities of the stimulated ('st') and spontaneous ('st') emission of photons as well as the probability of their stimulated absorption were introduced. Transitions between the energy levels E_m and E_n were stimulated by the external microwave radiation of frequency ω_{mn}. The transition probability was supposed to be proportional to the spectral density of the radiation field energy $\varrho(\omega_{mn})$:

$$w_{mn}^{\text{st}} = B_{mn} \varrho(\omega_{mn}). \tag{6.1}$$

6.1 Semiquantum Interpretation

Probabilities of emission ($m > n$) and absorption ($m < n$) processes per unit time were supposed to be equal so that $B_{mn} = B_{nm}$.[1] The latter statement corresponded to the supposition that both the processes were symmetric.

Under the equilibrium condition of the 'emitters + field' system, the total numbers of the emission and absorption transitions per unit time have to be equal:

$$w_{mn}^{tot} N_m = w_{nm}^{tot} N_n. \qquad (6.2)$$

Here N_m and N_n are numbers of the emitters with energies E_m and E_n, respectively. These numbers in equilibrium are correlated via Boltzman's distribution:

$$\frac{N_n}{N_m} = \exp\left(-\frac{\hbar\omega_{mn}}{kT}\right). \qquad (6.3)$$

(For simplicity, the energy levels themselves are considered to be nondegenerated.) As it follows from (6.2) and (6.3), the transition probabilities are in the ratio:

$$\frac{w_{mn}^{tot}}{w_{nm}^{tot}} = \exp\left(\frac{\hbar\omega_{mn}}{kT}\right). \qquad (6.4)$$

Physically it is evident that absorption transitions can be nothing but stimulated ones:

$$w_{nm}^{tot} = w_{mn}^{st} = B_{nm}\varrho\left(\omega_{mn}\right); \qquad m > n. \qquad (6.5)$$

However, as (6.4) indicates, if the value of $\hbar\omega/kT$ is finite, the total probability of the radiation emission w_{mn}^{tot} is larger than the probability of the stimulated radiation emission w_{mn}^{st}. This conclusion is equivalent to the statement that radiation emission, in contrast to absorption, is possible even at the so-called 'zero' energy of the external microwave field. So, the notion of additional ('spontaneous') radiation, independent of the energy density of the stimulating field $\varrho(\omega_{mn})$, has been introduced:

$$w_{mn}^{sp} = w_{mn}^{tot} - w_{mn}^{st}. \qquad (6.6)$$

It was the absence of any dependence on the external field that conditioned calling such transitions the spontaneous ones (i.e., those taking place without any external influence).

As it follows from the above-given expressions, the density of the radiation field energy is determined by the relation:

$$\varrho\left(\omega_{mn}\right) = \frac{w_{mn}^{sp}}{B_{mn}\left[\exp\left(\hbar\omega_{mn}/kT\right) - 1\right]}. \qquad (6.7)$$

The relation (6.7) should look like the classical Rayleigh–Jeans spectrum when $\hbar\omega_{mn} \to 0$:

---

[1] Later on these parameters were called Einstein's coefficients.

$$\varrho\left(\hbar\omega_{mn} \ll kT\right) = \frac{\omega_{mn}^2}{\pi^2 c^3} kT.$$

Making use of this limit, one can relate the coefficients B_{mn} to the probability of the spontaneous emission w_{mn}^{sp}:

$$B_{mn} = \frac{\pi^2 c^3}{\hbar \omega_{mn}^3} w_{mn}^{\text{sp}}. \tag{6.8}$$

Substitution of this expression into the right-hand side of (6.7) finally yields the classical Planck formula:

$$\varrho\left(\omega_{mn}\right) = \frac{\hbar \omega_{mn}^3}{\pi^2 c^3 \left[\exp\left(\hbar\omega_{mn}/kT\right) - 1\right]}. \tag{6.9}$$

Within the quantum theory, correctness of the semiphenomenological considerations used as a basis for derivation of (6.8) and (6.9) has been completely proved as well as the equality between the probabilities of the stimulated absorption and emission. Really, matrix elements of these transitions have turned out to be modulo equal and proportional to the amplitude of the radiation field at the frequency corresponding to this transition. As the probabilities of the emission and absorption transitions are proportional to the square of the matrix element, these characteristics turn out to be also proportional to the spectral density of the radiation field energy. In this sense, the notion of spontaneous transitions, postulated by Einstein, looks like being somewhat contradictory to the causality principle. Really, the atom, not influenced by any external force, ought to stay on the corresponding excited level. At present, this apparent paradox is only of historical interest because it has been obviated because of development of quantum electrodynamics. Briefly speaking, the essence of the matter is explicable in the following way.

The spectral density of the free radiation field energy may be presented as the sum of energies $\hbar\omega$ of the photons characterized by the corresponding frequency:

$$\varrho\left(\omega_{mn}\right) = \hbar\omega_{mn} n\left(\omega_{mn}\right).$$

Here $n(\omega)$ is the number of such photons. Consequently, the relation (6.6) of the total probability to the probability of the spontaneous transition may be written as

$$w_{mn}^{\text{tot}} = \left[1 + n\left(\omega_{mn}\right)\right] w_{mn}^{\text{sp}}. \tag{6.10}$$

The unity in the square brackets corresponds to the spontaneous emission, which takes place even if $n(\omega) = 0$.

On the other hand, as it is known from quantum electrodynamics, the spectral density of the field total energy is determined by the formula:

$$E = \left[\frac{1}{2} + n\left(\omega\right)\right] \hbar\omega.$$

The first term on the right-hand side of this equation corresponds to the so-called zero oscillations, i.e., exactly to the case when $n(\omega) = 0$.

6.1 Semiquantum Interpretation

An interesting interpretation follows from this expression. The spontaneous radiation emission is not spontaneous in the literal sense: the process is, at least half, stimulated by zero oscillations of the electromagnetic field. A detailed analysis (e.g., see [4]) indicates that this explanation does make sense. And what is more, the second half of the total probability of the spontaneous radiation emission is also stimulated. However, in this case, zero oscillations of the emitter itself play the role of the stimulating factor.

In the experiment, the existence of spontaneous emission transitions has to indicate itself in two ways. First, it is some broadening of spectral lines, which is conditioned by finiteness of the atom life time in the excited state: $\tau_{sp} \propto (w_{sp})^{-1}$. Second, positions of the corresponding energy levels are to be shifted with respect to their locations in the absence of zero oscillations. As a matter of fact, both the effects have been observed experimentally. This fact makes the basic proof of the determining role played by the zero oscillations in the spontaneous radiation emission (even if to say nothing about brilliant conformity of Planck formula with the whole totality of the experimental data).

In general, phases of the spontaneous radiation, directions of the propagation of the corresponding waves, and their polarizations are not correlated because the zero oscillations themselves are of the occasional nature. Physically, it means that the spontaneous radiation emitted by an ensemble is incoherent. Therefore, the above–studied radiation emission by the system of particles, the power of which is equal to the sum of the radiation powers of individual emitters, is also called spontaneous.

There are two suppositions implied in this reasoning. Both of them are inherent in the system of emitters that is in dynamic equilibrium with the isotropic radiation. First, photons are characterized only by the frequency ω without fixation of direction of radiation propagation. Second, the emitter energy distribution is supposed to be thermodynamically equilibrium (Boltzmann distribution). It is easy to generalize the first supposition for the number of photons of a given mode $n(\mathbf{k})$ ($\mathbf{k}$ implies the totality of the indexes, describing the mode).[2] Other suppositions, such as the symmetry of the Einstein coefficients and their link with the spontaneous transition probability, can now be considered as proved by the quantum relations:

$$\begin{aligned} B_{mn} &= B_{nm}; \\ w_{mn}^{tot} &= w_{mn}^{sp}\left[1+N\right]; \\ w_{\mathbf{k}\downarrow}^{tot} &= w_{\mathbf{k}\downarrow}^{sp}\left[1+N\right]. \end{aligned} \qquad (6.11)$$

The arrows here indicate the energy variation as a result of the transition.

These relations are not linked with the thermodynamic equilibrium supposition. So, one may write down a kind of a kinetic equation describing a temporal evolution of an average number of photons of the wave vector $\mathbf{k}$

---

[2] As regards free plane waves, $\mathbf{k}$ is the wave vector.

interacting with emitters of energy $\mathcal{E}$. If the emitters energy distribution is $f(\mathcal{E},t)$, then

$$\frac{\partial n_{\mathbf{k}}}{\partial t} = \int \{f(\mathcal{E},t)w_{\mathbf{k}\downarrow} + n_{\mathbf{k}}f(\mathcal{E},t)w_{\mathbf{k}\downarrow} - n_{\mathbf{k}}f(\mathcal{E}-\hbar\omega,t)w_{\mathbf{k}\uparrow}\}\,\mathrm{d}\mathcal{E}. \qquad (6.12)$$

The first term in the curled brackets describes the probability of the spontaneous emission of a photon $\mathbf{k}$ of energy $\hbar\omega$. The third term corresponds to the absorption process rate, which is proportional to the probability $w_{\mathbf{k}\uparrow}$, to the number of existing photons[3] $n_{\mathbf{k}}$, and to the population of the level $\mathcal{E} - \hbar\omega$ the transition starts from.

The second term describes an inverse process of the induced emission proportional to the number of photons as well. As has been discussed above, this addendum is necessary to provide a stationary distribution with $\partial n/\partial t = 0$ if the emitters energy distribution is Boltzmann one. Really, in the steady state with $f(\mathcal{E}) \propto \exp(-\mathcal{E}/\kappa T)$, the expression in the brackets has to vanish so that

$$n(\mathbf{k}) = \frac{1}{\exp(\hbar\omega/\kappa T) - 1}. \qquad (6.13)$$

Multiplying (6.13) by the quantum energy $\hbar\omega$ and by the phase volume for the isotropic radiation $4\pi|\mathbf{k}|^2\,\mathrm{d}\mathbf{k}$, one gets the Planck formula (6.9).

It is worth to note here that if the energy distribution is inverse, i.e. if $f(\mathcal{E}) > f(\mathcal{E} - \hbar\omega)$, the second term could be predominant. Under this condition, exponential growing of number of photons takes place and all of them are exact copies of the first "initiating" photon. One can easily recognize this as lasing which is due to the stimulated emission. In what follows we consider this effect of main importance in the classic limit omitting effects specific for quantum emitter (i.e., for quantum lasers).

6.2 Classical Limit

Transferring the concept of stimulated and spontaneous emission to classical (nonquantum) systems meets certain difficulties. Obviously, this is the spontaneous emission to be identified with numerous examples in Part I, where no external electromagnetic waves influencing the particle motion were considered. However, there is a small discrepancy in this approach: the radiation phase is strictly determined for a classical particle in contrast with the quantum spontaneous radiation. It is not surprising: in the quantum description with a determined number of photons the field phase is not defined because of the uncertainty principle, while a number of photons are meaningless in classical electrodynamics.

---

[3] We keep the same notation $n_{\mathbf{k}}$ for the quantum average of the number of photons.

The physical meaning of the stimulated emission as additional to the spontaneous one is much less obvious in classical electrodynamics. The extra acceleration caused by an external wave does generate some radiation. However, the process is weak and must be regarded as scattering because wave vectors of the corresponding waves do not coincide with that of the incident wave as it should be for stimulated processes.

In spite of that, the concept above can be definitely applied to classical beam systems. Really, the absorption – radiation symmetry for a given wave can be understood as the equality of phase intervals corresponding to the increase and decrease in the particle energy. The temporal variation of the number of quanta is to be interpreted as variations in the field energy spectral density. As regards the quantum discreteness of energy levels, it was used above only for derivation of Planck formula and was not essential for mentioned lasing. Summing, one should expect that the kinetic relation (6.12) is valid even for $\hbar \to 0$ and that it could describe the stimulated effects in classical nonequilibrium systems with inverse populations.

In the classical limit, the photon energy and the distance between energy levels are negligible. So, the energy population can be considered as continuous and can be presented as the expansion:

$$f(\mathcal{E} - \hbar\omega) \approx f(\mathcal{E}) - \hbar\omega \frac{\partial f}{\partial \mathcal{E}} + \cdots . \tag{6.14}$$

Multiplying (6.12) by $\hbar\omega$ and using (6.14), one gets

$$\frac{\partial W_\mathbf{k}}{\partial t} = \int \left\{ f(\mathcal{E}) + \frac{\partial f}{\partial \mathcal{E}} W_\mathbf{k} \right\} p_\mathbf{k}(\mathcal{E}) \mathrm{d}\mathcal{E}, \tag{6.15}$$

where $p_\mathbf{k}$ is the intensity of the spontaneous emission of the wave $\mathbf{k}$ by an emitter of energy $\mathcal{E}$. The value $W_\mathbf{k} = \hbar\omega n_\mathbf{k}$ is, of course, the electromagnetic energy density of the mode. Integrating the second term in right-hand side of (6.15) by parts and putting, for simplicity, $f(\mathcal{E}_{min}) = 0$, we obtain

$$\frac{\partial W_\mathbf{k}}{\partial t} = P_\mathbf{k} - W_\mathbf{k} \int f(\mathcal{E}) \frac{\partial p_\mathbf{k}}{\partial \mathcal{E}} \mathrm{d}\mathcal{E}, \tag{6.16}$$

where

$$P_\mathbf{k} = \int f(\mathcal{E}) p_\mathbf{k}(\mathcal{E}) \mathrm{d}\mathcal{E}$$

is the spontaneous radiation power integrated over all emitters. In particular, for a monoenergetic ensemble[4] of energy $\mathcal{E}_0$

$$\frac{\partial W_\mathbf{k}}{\partial t} = P_\mathbf{k} - W_\mathbf{k} \left(\frac{\partial P_\mathbf{k}}{\partial \mathcal{E}} \right)_0. \tag{6.17}$$

---

[4] We call an ensemble monoenergetic if its energy distribution is narrow enough but still is much wider than $\hbar\omega$ to ensure the expansion (6.14).

This equation is of the same structure as (6.12) but does not contain $\hbar$. It shows that the energy spectral density variations take place because of the spontaneous and stimulated emission, the last being proportional to the density itself. The lasing effect is also presented. However, the condition of lasing looks different than that in a quantum system. It says that the derivative $\partial P_\mathbf{k}/\partial \mathcal{E}$ has to be negative and large enough. In other words, to get lasing the intensity of spontaneous emission of the given mode should depend sharply on the emitter energy. In quantum systems it is ensured automatically because of the levels discreteness and thus the inverse population is of the main importance. On the contrary, a classical beam system is populated obviously only at large energies so that the inverse population exists for granted and means just the free energy availability. In this case, this is the sharp dependence of the spontaneous emission on energy that must be ensured.

Of course, the relation (6.17) obtained as a limiting case of (6.12) does not explain the mechanism of the stimulated emission in classical electrodynamics. Both of them are based on the energy considerations and contain no information about the field phase. It was completely approved in a steady state when the number of photons was fixed and the phase was random. However, for a temporal evolution of the average number of photons the average phase has to depend on time as well. Really, growing predominance of the emission over absorption (lasing) may take place from the classical point of view only under condition of developing phasing of individual emitters. This phasing is nothing but an appearance of a certain coherence. These arguments lead to the conclusion that the stimulated emission of a given field mode is, in a way, equivalent to the development of its coherence.

Coming to beam systems with the expressed direction of motion along the z-axis a narrow-band character of the spontaneous spectrum is associated with the condition of synchronism which links the phase velocity of the wave and the particle longitudinal velocity. The width of the spectrum is conditioned, first of all, by the finite wave–particle interaction distance. The profile of the spectrum can be obtained from the general considerations. Really, if the particle velocity has a component oscillating with a frequency $\Omega(\gamma)$, the amplitude of a quasi-synchronous spectral harmonic is proportional to

$$k \int_0^L \exp\left[i\left((\omega \pm \Omega)z/v - kz\right)\right] dz = ikL \frac{\exp(-i\mu) - 1}{\mu}, \qquad (6.18)$$

where

$$\mu = \left(1 - \frac{\omega \mp \Omega}{kv}\right) kL \qquad (6.19)$$

is the phase slippage of the particle with respect to the wave at the total distance of interaction L. The spontaneous emission power as a function of μ is proportional to the absolute value of (6.18) squared:

$$P_k(\mu) = P_k(0) \frac{\sin^2(\mu/2)}{(\mu/2)^2}, \qquad (6.20)$$

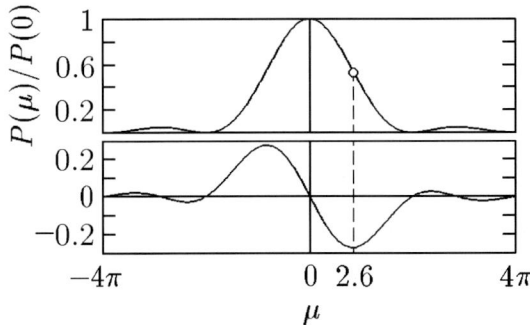

Fig. 6.1. Profile of the spontaneous emission line and its derivative (below)

This universal expression for a profile of the spontaneous radiation spectral line of a single particle at a finite length is shown in Fig. 6.1.

The value of μ depends on the phase velocity and the particle velocity in the fixed combination. So, coming back to (6.17), one can state: for a fixed frequency an equilibrium energy exists ($\mu = 0$) that provides the maximal spontaneous emission. At this energy the induced emission vanishes, but it appears for nonequilibrium energies if $\mu > 0$ where the corresponding derivative in (6.17) is negative. At the opposite side of the resonance ($\mu < 0$), there is the region of the wave absorption. For large μ the particle neither radiates nor absorbs. Treating the particle beam as a medium, one can talk about a band of its optical activity coinciding with the spontaneous emission band. It consists of two symmetric subbands: one of absorbtion and that of stimulated emission (see Fig. 6.1). The latter is possible, of course, only in an active (inversely populated) medium with some intrinsic free energy.

6.3 Stimulated Emission and Beam Phasing

Of course, the arguments above should be considered just as leading ones. First of all, it is unclear to what extent they are valid for nonstationary systems. Second, they do not take into account a degradation of the initial energy distribution, i.e., inevitable saturation effects and beam energy spreading. The last but not the least, the arguments are of a phenomenological character and do not reveal the physics of the correlations developing in the beam of particles. So far as the correlations are related to phasing, we briefly consider below the particles phase dynamics in an external wave.

6.3.1 Phase Dynamics in Quasi-Synchronous Wave

In Chap. 5, we have considered the coherent emission of a single mode by a structure of individual emitters. Now the problem is, in a way, opposite:

this is the development of a spatial structure under action of a monochromatic wave that is of interest. One can call it the second side of the united process of the development of coherence both in the particle motion and in the electromagnetic field.

For the sake of simplicity we consider the electric field of the form

$$\mathbf{E}(z,t) = \mathbf{E}_0 \exp(i\varphi); \qquad \varphi = kz - \omega t; \qquad z \geq 0$$

neglecting its dependence on transverse coordinates and treating the amplitude $\mathbf{E}_0$ as constant. Doing this, we ignore the initial stage of the process when the stimulated monochromatic radiation is just appearing against the spontaneous background exactly as we ignored above the prehistory of the emitters lattice. The constancy of the amplitude implies an input signal large enough to be practically unchangeable by additional radiation. Beside that, we neglect Coulomb interaction of the particles. It is clear, by intuition, that all these simplifications are approved only for low-intensity beams.

The particles motion in the absence of the wave is supposed to be a superposition of the longitudinal velocity βc and of transverse oscillations of frequency ω, small enough not to influence the longitudinal velocity (in a particular case of Cherenkov interaction the oscillation amplitude can be zero). We shall accept these conditions for granted because they are quite obvious in many cases of interest (e.g., for an undulator). The constant particle energy means then the constant rate of the particle phase slipping with respect to the wave

$$\frac{d\varphi}{dz} = k\left(1 - \frac{\omega \mp \Omega}{k\beta c}\right). \qquad (6.21)$$

For a nonzero wave amplitude the energy and the phase slippage undergo variations which can be presented as a superposition of slow (in ω-scale) systematic changes and ripples vanishing in average. Omitting the sign of averaging, one can write down an obvious relation for the systematic part

$$\frac{d\gamma}{dz} = gk\cos\varphi, \qquad (6.22)$$

where g is a maximal possible increase in the particle energy per a wavelength expressed in mc^2 units. This dimensionless amplitude is a small parameter in the overwhelming majority of cases of interest. Of course, it depends on field and trajectory configurations but is always proportional to the wave amplitude.

The radiation reaction, that is, the proper field of a single particle is not included in the equation. This approximation is valid if the width of the spontaneous radiation spectral line is determined mainly by a finite length of the interaction path rather than by particle acceleration.

For a synchronous particle, by definition, the phase slippage is zero because its velocity $\beta_s = (\omega \mp \Omega)/kc$. For small energy deviations from the equilibrium, we can present the phase shift per a unit of length as

$$\frac{d\varphi}{dz} = \alpha k \left(\gamma - \gamma_s\right), \tag{6.23}$$

where the index s denotes synchronous values and

$$\alpha = -\beta_{\rm ph} \left(\frac{\partial}{\partial \gamma} \frac{\omega \mp \Omega}{\omega \beta}\right)_s \tag{6.24}$$

is the phase slippage sensibility to energy variations.

Equations (6.22) and (6.23) describe the phase stability mechanism well known in the theory of accelerators (see, e.g., [33]). For $g = $ const they have an integral of motion (Hamiltonian):

$$\mathcal{H} = (\gamma - \gamma_s)^2 - \frac{2g}{\alpha} \sin \varphi, \tag{6.25}$$

which predicts stable "synchrotron" oscillations[5] around equilibrium values $\varphi_s = \text{sign}(\alpha)\pi/2$ and $\gamma = \gamma_s$. If deviations from the equilibrium values are small, the Hamiltonian (6.25) can be presented as

$$\mathcal{H} = (\gamma - \gamma_s)^2 + \frac{g}{|\alpha|} (\varphi - \varphi_s)^2.$$

This positive quadratic form corresponds to harmonic oscillations of period $2\pi\sqrt{|\alpha|/g}$ in space. Note that this period is expressed in units of the wavelength and usually exceeds the latter.

The synchrotron oscillations are nonlinear, their period increasing with the amplitude. For $\mathcal{H} = 4\pi/|\alpha|$, the period becomes infinitely large. The corresponding phase trajectory

$$(\gamma - \gamma_s)^2 = \frac{4g}{|\alpha|} + \frac{2g}{\alpha} (\sin \varphi - \sin \varphi_s)$$

is called a separatrix, dividing trapped particles oscillating around the equilibrium from nontrapped or librating ones. The latter ones slip in phase with respect to the wave in the positive or negative direction, depending on the sign of $\alpha(\gamma - \gamma_s)$. The separatrix passes through the points $\gamma = \gamma_s$ and $\varphi = \varphi_s \mp \pi$ with a maximum deviation from the φ-axis

$$\gamma_{\max} - \gamma_s \pm \sqrt{4g/|\alpha|} \tag{6.26}$$

taking place at $\varphi = \varphi_s$. The qualitative structure of other phase trajectories (supplied with arrows) is shown in the Fig. 6.2.

6.3.2 Phase Bunching by External Wave (Low-Gain Regime)

Let us consider now evolution of particles initially distributed uniformly over phases and having the same initial energy $\gamma_i > \gamma_s$. In the plane (φ, γ), this

---

[5] Also known in electronics as bounce oscillations.

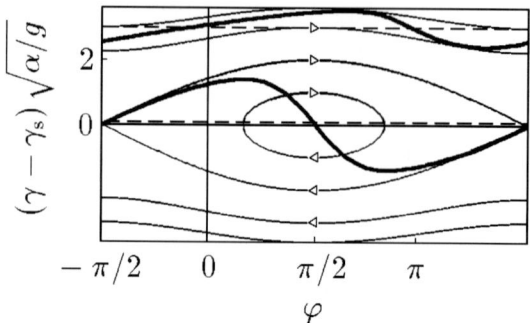

Fig. 6.2. Phase trajectories and beam bunching at $zk < \pi/2\sqrt{\alpha/g}$. Dashed lines show initial distributions

distribution is presented by a straight line $\gamma = \gamma_i$ (see Fig. 6.2). Moving along bent phase trajectories, half of the particles are accelerated and half decelerated (i.e., are absorbing or radiating). At the beginning these processes are quite symmetric what corresponds to the symmetry of Einstein coefficients. But accelerated particles move faster to the right in Fig. 6.2 and overtake the decelerated ones. As a result, the particles begin to bunch in the phase region of deceleration, disturbing the initial symmetry between absorption and radiation in the latter's favor. If the initial energy is lesser than γ_s, the bunching process goes in opposite direction resulting in absorption. One can easily see that this scenario describes the amplification and absorption bands discussed above. If the initial energy is outside the optical activity band, the phase trajectories are almost straight lines, bunching vanishes, and the beam stays transparent.

The exact solution of (6.22) and (6.23) can be obtained in terms of elliptical functions. However, it is rather cumbersome and is not really necessary if we limit ourselves by the initial stage of bunching described above. Because of the smallness of the amplitude g, one can exploit a perturbation theory if the interaction distance is not very large.

In the zeroth approximation ($g = 0$) just kinematic slipping takes place:

$$\varphi = \varphi_i + \zeta; \quad \zeta = \alpha k \delta_i z; \quad \delta = \gamma - \gamma_s.$$

Substituting this in (6.22) gives the induced energy modulation of the first-order

$$\delta - \delta_i = \frac{g}{\alpha k \delta_i} \left[\sin(\varphi_i + \zeta) - \sin \varphi_i \right]$$

and the corresponding dynamical phase slippage:

$$\varphi = \varphi_i + \zeta + \frac{g}{\alpha \delta_i^2} \left[\cos \varphi_i - \cos(\varphi_i + \zeta) - \zeta \sin \varphi_i \right]. \quad (6.27)$$

Note that both values vanish after averaging over initial phases. This means, in particular, that the first-order radiation losses are zero because the numbers of absorbing and radiating particles are equal (one can remind again the symmetry of Einstein coefficients). However, the induced phase shift (6.27) does disturb the symmetry. In the next approximation

$$\frac{d\gamma}{dz} = kg\cos(\varphi_i + \zeta) - \frac{g^2 k}{\alpha \delta_i^2} \sin(\varphi_i + \zeta)\left[\cos\varphi_i - \cos(\varphi_i + \zeta) - \zeta\sin\varphi_i\right]. \quad (6.28)$$

Averaging over initial phases gives now

$$\left\langle \frac{d\gamma}{dz} \right\rangle = -\frac{g^2 k}{2\alpha(\gamma_i - \gamma_s)^2}\left[\sin\zeta - \zeta\cos\zeta\right], \quad (6.29)$$

i.e., an additional induced energy change proportional to the field amplitude squared. The average energy loss for this induced or stimulated radiation can be calculated by integrating (6.29) once more and noting that the kinematic phase shift at the length L is $\mu = \alpha k L \delta_i$:

$$\langle\gamma(L) - \gamma_s\rangle = g^2 \frac{\alpha k^3 L^3}{\mu^3}\left[\cos\mu - 1 + \frac{\mu}{2}\sin\mu\right] = g^2\frac{\alpha k^3 L^3}{4}\frac{d}{d\mu}\left(\frac{\sin\mu/2}{\mu/2}\right)^2. \quad (6.30)$$

To calculate the radiation power emitted by the beam as a whole, one should multiply (6.30) by the number of particles passing the region per unit time, i.e., by I/q where I is the beam current. This power obviously contains the spontaneous radiation spectral line profile (6.20), exactly in the same way as the phenomenological expression (6.16). Beside, the average radiation losses are proportional to g^2, i.e., to the external wave power. So, the stimulated emission in a classical system really can be interpreted as coherence self-organization due to the autophasing mechanism with a consequent increase in the radiation spectral brightness within the optical activity region.

The dependence of the radiation power on the phase slippage parameter deserves a special comment. For a fixed interaction length the power is maximized by $\mu \approx 2.6$, that is, the wave that overtakes the beam almost by a wavelength is the most prosperious.[6] Note that the exactly synchronous wave is not amplified at all while all slow waves are attenuated.

These calculations predict the evolution of the initially monoenergetic beam as well. The method of successive approximations used above works only if the induced phase shift is small enough, or under qualitative conditions:

$$\delta_i \gg \sqrt{g/\alpha}\min\{\mu, 1\}, \quad (6.31)$$

$$kL \ll \frac{1}{\sqrt{\alpha g}}\min\{\mu, 1\}. \quad (6.32)$$

---

[6] This is true for $g = $ const only, i.e., for the low-gain regime.

So, the case $\mu > 1$ corresponds to particles situated mainly or totally outside the separatrix. Note that this is the condition to realize the optimizing value $\mu = 2.6$ (see the footnote on page 109). In the case of $\mu < 1$ almost all particles can be initially trapped.

If the interaction length exceeds the value (6.32), one can predict the faster particles outrunning the slower ones, the distribution "overturning" and filamentation taking place. At the final stage of this mixing, the phase distribution would be symmetric again with a corresponding increase in the energy spread. The larger is the initial μ value the later this filamentation occurs. From the viewpoint of physics the process can be interpreted as a nonlinear saturation of the stimulated emission (or absorbtion) accompanied by beam heating. The corresponding length (by order of magnitude)

$$L_{\text{sat}} = 1/k\sqrt{\alpha g}$$

can be called the distance of saturation.

Although the results of this section are restricted by the fixed field approximation, i.e., do not take into account possible amplitude and phase variations, they can be directly used in some cases of interest. In particular, if the field is "locked" in a cavity of a finite Q-value and the amplification exceeds certain threshold, one may foresee a steady state with a time independent established amplitude. For large Q, this amplitude could be large enough although the transient process takes a long time. This low-gain regime is typical for generators of coherent radiation with low current beams, where the feedback necessary for self-excitation is provided by a cavity.[7] These problems will be considered in detail in Part III.

6.3.3 Spatial Amplification in Particles Flow (High-Gain Regime)

Nevertheless, high gain systems are also important from the general viewpoint as well as for applications. Suppose that there is no feedback and the steady state self-consistent field depends essentially on the longitudinal coordinate being determined by an input signal and by emitted radiation. This regime may be called a high-gain spatial amplification of the input signal by the beam. Naturally, only those input waves could be amplified that are inside the beam optical activity domain.

Of course, the spatial amplification depends on electrodynamic properties of the system as a whole. But, basing on general arguments, one should expect that a quasi-synchronous mode would be mainly amplified if, of course, it is presented in the input signal. This is the mode that, according to the previous considerations, pumps the energy out of the beam most effectively. So, we restrict ourselves below by this one-mode approximation, bearing in mind that modes are independent in a linear system. Other suppositions are the same as above including the near-zone interaction being neglected.

---

[7] Or by a negative group velocity typical for backward wave tubes.

6.3 Stimulated Emission and Beam Phasing

So, we consider now the field amplitude growing with distance but being independent of time. Note, by the way, that the beam represents both active and reactive loads for the wave changing its phase $\psi(z)$ as well as the amplitude $g(z)$. Taking this into account, the equations of phase dynamics (6.22), (6.23) are to be rewritten as

$$\frac{d\delta}{dz} = kg(z)\cos(\phi+\psi), \qquad \frac{d\phi}{dz} = k\alpha\delta. \qquad (6.33)$$

In the same way as in the previous section we get in the first approximation with respect to g:

$$\delta = \delta_i + k\int_0^z g(z')\cos(\varphi_i + k\alpha\delta_i z' + \psi(z'))\,dz';$$

$$\varphi = \varphi_i + k\alpha\delta_i z$$
$$+ k^2\alpha \int_0^z dz' \int_0^{z'} g(z'')\cos(\varphi_i + k\alpha\delta_i z'' + \psi(z''))\,dz''$$

or, with the same precision,

$$\frac{d\varphi_i}{d\varphi} - 1 \qquad (6.34)$$

$$= k^2\alpha \int_0^z dz' \int_0^{z'} g(z'')\sin[\varphi - k\alpha\delta_i(z-z'') + \psi(z'')].$$

Note that this expression describes particles bunching in the φ space.

Now we need the second equation relating field variations to beam bunching. Restricting ourselves by plane motion with the amplitude of the oscillation velocity $\tilde{v}$, we note that a single particle at the phase φ creates a resonant harmonic of the transverse current equal to

$$\frac{q\tilde{v}}{4\pi\beta c}\exp[i(kz-\varphi(z))].$$

Averaging it over all initial phases with the help of (6.34) yields the driving transverse current

$$j_\perp = -j_0 \frac{ik^2\alpha\tilde{v}}{8\pi\beta c}\exp(ikz) \qquad (6.35)$$

$$\times \int_0^z dz' \int_0^{z'} g(z'')\exp[-ik\alpha\delta_i(z-z'') + i\psi(z'')]\,dz''.$$

Looking for a solution of the wave equation for the corresponding transverse component of the electric field

$$\frac{d^2 E}{dz^2} + k^2 E = -i\frac{4\pi k}{c}j_\perp,$$

we present it as $E_0(z)\exp(ikz)$, where the complex amplitude $E_0(z)$ is a slow function of the argument. Neglecting its second derivative, we get

$$\frac{dE_0}{dz} = i\frac{k^2 j_0 \tilde{v}\alpha}{4c^2\beta} \qquad (6.36)$$

$$\times \int_0^z dz' \int_0^{z'} g(z'')\exp\left[i\left(-k\alpha\delta_i(z-z'')+\psi(z'')\right)\right].$$

Note now that, by definition,

$$g = \frac{q\tilde{v}}{2k\beta mc^3}|E_0| \quad \text{and} \quad g\exp(i\psi) = \frac{q\tilde{v}}{2k\beta mc^3}E_0.$$

Substituting that into (6.36) yells the self-consistent equation for the electric field amplitude:

$$\frac{dE_0}{dz} = \frac{i}{L_r^3}\int_0^z dz' \int_0^{z'} E_0(z'')\exp\left[-ik\alpha\delta_i(z-z'')\right]dz'', \qquad (6.37)$$

where

$$L_r = 2\left(\frac{\beta^2 c^2 I_0}{kj_0\alpha\tilde{v}^2}\right)^{1/3} \qquad (6.38)$$

with $I_0 = mc^3/q \approx 17$ kA. For reasons explained below, L_r can be called a radiation length.

Equation (6.35) can be easily reduced to the third-order differential linear equation:

$$\frac{d}{dz}\left(\frac{d}{dz}+ik\alpha\delta_i\right)^2 E_0 = \frac{i}{L_r^3}E_0. \qquad (6.39)$$

The solution of this equation under the initial condition

$$E_0 = 1; \quad dE_0/dz = 0; \quad d^2E_0/dz^2 = 0 \quad \text{for } t = 0 \qquad (6.40)$$

represents a complex amplification coefficient describing amplitude and phase characteristics of the process. [8]

Looking for a solution in the form $\exp(i\nu z)$, one gets the characteristic equation

$$\nu(\nu+k\alpha\delta_i)^2 = -L_r^{-3}. \qquad (6.41)$$

An equation of this type will be investigated in details in Chap. 10. Here we just note that the coefficients in (6.41) are real. So it has either three real roots or one real and two complex conjugated ones. In the first case, all three linearly independent partial solutions are of an oscillatory type. Their linear combination cannot exceed essentially the initial field, meaning that

---

[8] We suppose that the final signal amplitude is small enough to exclude nonlinear processes.

amplification does not occur. In the second case, which takes place under condition

$$\mu_r > -2^{-2/3}3; \qquad \mu_r = k\alpha\delta_i L_r \qquad (6.42)$$

one of the complex conjugated roots has a negative imaginary part and the corresponding partial solution grows exponentially with an e-fold length L_r. Note that μ_r has a meaning of the kinematic phase shift at the length L_r related to the corresponding detuning in initial energy.

If the distance z is essentially smaller than L_r, as it happens for small currents, the amplification coefficient slightly exceeds unity. One can easily see that this leads to the result discussed above: at a fixed length, maximal amplification occurs for the wave with the optimal kinematic phase shift $\mu \approx 2.6$, but the gain itself is small and linearly proportional to the beam current.

Unlike the case of an almost constant amplitude, the high–gain amplification takes place not only at positive μ values but also at small negative ones if the condition (6.42) is fulfilled. The maximal gain is reached for the exactly synchronous wave[9] when the total solution under the initial condition (6.40) is relatively compact:

$$E_0 = \frac{1}{3}\left[\exp\left(iz/L_r\right) + 2\exp\left(-iz/2L_r\right)\cosh\left(\sqrt{3}z/2L_r\right)\right]; \qquad (6.43)$$

$$|E_0| = \frac{1}{3}\sqrt{1 + 4\cosh\left(\sqrt{3}z/2L_r\right)\cos\left(3z/2L_r\right) + 4\cosh^2\left(\sqrt{3}z/2L_r\right)}. \qquad (6.44)$$

It is worth to note that the increment is proportional to $j_0^{1/3}$.

For $\mu = 0$ the characteristic equation (6.41) has the roots

$$\nu_n = L_r^{-1}\exp\left[i\pi n/3\right], \qquad n = 0, 1, 2.$$

The asymptotic behavior of the electric field amplitude is determined by the root with maximal imaginary part, i.e., $\nu_0 = L_r^{-1}\exp\left(i\pi/3\right)$.

The mechanism of the spatial amplification is a basic one for a variety of high–power amplifiers, using high–current electron beams. Besides, it can be used in coherent radiation sources where spontaneous radiation plays the role of the input signal. The selective mechanism of the spatial amplification shares out a narrow spectral line from the spontaneous radiation spectrum. In the theory of free electron lasers, such regime is called SASE (Self-Amplification of Spontaneous Emission) and appears as a direct analog of optical superradiance [35]. The notion of "spontaneous coherent radiation" used sometimes is intrinsically contradictory on our opinion. We will return to these problems in Sect. 10.

Considerations above are related to a flow of harmonic oscillators with amplitude determined by $\tilde{v}$. Of course, the increase in the wave power comes

---
[9] We mean the asymptotic behavior.

from the total energy of particles which must be accompanied by a change of the oscillations amplitude. As it has been proved in Part I, a resonant interaction under conditions of a normal Doppler effect damps the amplitude and the amplification is limited because of oscillation energy exhaustion. Prolongation of interaction would lead to the inverse process of wave absorption. However, under conditions of the anomalous Doppler effect, this saturation does not exist and the oscillations amplitude grows at the account of an additional decrease in the particle longitudinal momentum.

6.4 Dynamic Chaos

All material above was based on the assumption that a particle interacts with a single harmonic wave under synchronous or resonance conditions. Being quite productive for explanation of the induced radiation mechanism as a result of particles self-bunching in the wave, this assumption still needs additional discussion of its applicability, especially in the case of large amplitude fields.

We do not mean here negligible changes of a particle trajectory still governed by external fields while radiation fields can be treated as perturbations. The dimensionless electric field amplitude g introduced by (6.22) remains small in practically all cases of interest. Thereby, all our arguments were based on consideration of resonances (Cherenkov type, Doppler-shifted oscillator resonances, cyclotron ones, etc.,). For small g factors only resonant conditions and a long-term wave–particle interaction can provide a large energy transfer from particles to the wave (amplifiers and oscillators) or vice versa (accelerators). In the case of large g, a particle could get a relativistic velocity during one period of the wave and the resonant conditions would lose their paramount importance. Electrodynamics of so large fields is still in the developmental stage.

Nevertheless, even within the frames of the resonant perturbation theory, one cannot exclude a simultaneous action on the particle of two waves of different frequencies satisfying approximate resonant conditions for two different degrees of freedom, for example, of Cherenkov and cyclotron type. For very small wave amplitudes, when these resonances are reliably separated, provide two well-separated stability regions one can consider them independently. But an increase in field amplitude which is desirable for high-power devices leads to an increase in the resonances width. In a sense they can act together, so that the particle motion in the phase plane becomes unpredictable and close to stochastic one. They call this phenomenon as a dynamic chaos limiting, naturally, the power increase and broadening the spectrum of oscillations. The last plays, of course, an essential role.

Remaining within the frames of the self-consisting theory and neglecting the radiation damping, one can consider fields as external ones and formulate the problem as canonical one. This permits to use the well-developed powerful formalism of Hamiltonian mechanics, especially the resonant perturbation theory. This implies that the Hamiltonian of the dynamical system under consideration (in our case of a particle in the external field) can be presented as a sum of two terms: $\mathcal{H} = \mathcal{H}_0 + \mathcal{H}_1$. Here $\mathcal{H}_0$ is a nonperturbed Hamiltonian and $\mathcal{H}_1$ is a small periodic perturbation. Note that the canonical formulation of the problem implies the applicability of the Liouville theorem which simplifies essentially the description of the energy exchange between the particle and the field.

So, the processes of interest can be investigated using the theory of small perturbations acting over a long period of time. It will be shown below that the particle motion can be reduced to dynamics of either one nonlinear pendulum or that of a system of interacting pendulums. In the first case, the particle motion remains regular, and in the second, it can be chaotic. The conditions of this change will be discussed below, but before we discuss we need the basic notions of the resonant perturbation theory applied to the problems of microwave electronics.

6.4.1 Resonant Perturbation Theory

The main ideas and methods of the resonant perturbation theory will be considered as applied to motion of a particle under action of two waves, denoted below by indices 1 and 2. Let the system be described by a Hamiltonian in action–phase variables

$$\mathcal{H} = \mathcal{H}_0 (J_1, J_2) + \mathcal{H}_1 (J_1, J_2, \theta_1, \theta_2), \tag{6.45}$$

where $\mathcal{H}_0$ is a nonperturbed Hamiltonian and $\mathcal{H}_1$ is a perturbation supposed to be periodic over θ_1, θ_2. So, it can be presented as a Fourier series:

$$\mathcal{H}_1 = \sum_{l,n} H_{l,n} \exp\left[i\left(l\theta_1 + n\theta_2\right)\right], \tag{6.46}$$

where l, n are integers. Suppose a resonance condition is fulfilled, that is, there exists a relation between the proper frequencies of the nonperturbed system ω_1, ω_2:

$$r\omega_1 - s\omega_2 \approx 0, \tag{6.47}$$

where

$$\omega_1 = \frac{\partial \mathcal{H}_0}{\partial J_1}, \quad \omega_2 = \frac{\partial \mathcal{H}_0}{\partial J_2}, \quad \dot{\theta}_1 = \omega_1, \quad \dot{\theta}_2 = \omega_2, \quad r, s \quad \text{integers}.$$

It is necessary to describe slow (in the proper frequencies scale) but systematic variations of the values $J_{1,2}$, which are integrals of motion in the absence of perturbations.

Let us come to the variables denoted by a bar by means of a generating function

$$F_2 = (r\theta_1 - s\theta_2)\bar{J}_1 + \theta_2 \bar{J}_2. \tag{6.48}$$

According to general rules [31], we get

$$J_1 = \frac{\partial F_2}{\partial \theta_1} = r\bar{J}_1, \quad J_2 = \frac{\partial F_2}{\partial \theta_2} = -s\bar{J}_1 + \bar{J}_2, \quad (6.49)$$

$$\bar{\theta}_1 = \frac{\partial F_2}{\partial \bar{J}_1} = r\theta_1 - s\theta_2, \quad \bar{\theta}_2 = \frac{\partial F_2}{\partial \bar{J}_2} = \theta_2.$$

The third equation of the system (6.49) yields

$$\dot{\bar{\theta}}_1 = r\dot{\theta}_1 - s\dot{\theta}_2 = r\omega_1 - s\omega_2 = 0.$$

So, the first new angular variable turns out to be a slow varying one while the second coincides with the original phase ($\bar{\theta}_2 = \theta_2$). Taking into account (6.46) and the resonant conditions (6.47), we obtain the new Hamiltonian:

$$\bar{\mathcal{H}} = \mathcal{H}_0 \left(r\bar{J}_1, -s\bar{J}_1 + \bar{J}_2, r\bar{\theta}_1 - s\bar{\theta}_2, \bar{\theta}_2\right) \quad (6.50)$$
$$+ \sum_{l,n} H_{l,n}(\bar{J}) \exp\left[\frac{i}{r}\left(l\bar{\theta}_1 + (ls + nr)\bar{\theta}_2\right)\right].$$

Note that fast motion in (6.50) is represented by $\bar{\theta}_2$, while $\bar{\theta}_1$ is slow because of the resonant conditions (6.47). So, one can expect that the application of the perturbation theory would not give rise to appearance of new essential resonances. Really, as far as $\bar{\theta}_1$ is a slow variable while $\bar{\theta}_2$ is a fast one, a resonance can occur only at large values r. These are so-called secondary resonances developing under certain peculiar circumstances only. Otherwise resonant terms are not presented in (6.50). Thus, the procedure above reduces the problem to investigation of particle motion in a vicinity of a single chosen resonance.

It follows from (6.50) that there is only one slow varying term $\exp\left[i\bar{\theta}_1 l/r\right]$ in the sum representing the perturbation. In so far as the variable $\bar{\theta}_1$ is slow, one may average (6.50) over the fast variable $\bar{\theta}_2$ (for justification of this method see, for example, [31]):

$$\bar{\mathcal{H}} = \bar{\mathcal{H}}_0(J) + \langle \mathcal{H}_1 \rangle \quad (6.51)$$

where

$$\langle \mathcal{H}_1 \rangle = \frac{1}{2\pi} \int_0^{2\pi} \mathcal{H}_1 d\bar{\theta}_2.$$

Taking into account that the remaining term corresponds to $ls + nr = 0$, the perturbed Hamiltonian takes the form:

$$\langle \mathcal{H}_1 \rangle = \sum_{p=-\infty}^{\infty} H_{-rp,sp}(\bar{J}) \exp\left[-ip\bar{\theta}_1\right] \quad \text{where} \quad p = -\frac{l}{r}, \quad s = \frac{n}{l}r. \quad (6.52)$$

6.4 Dynamic Chaos

The new averaged Hamiltonian is independent of $\bar{\theta}_2$. So, the new canonical action is an integral of motion:

$$\bar{J}_2 = J_2 + s\bar{J}_1 = J_2 + \frac{s}{r}J_1 = \text{const}. \tag{6.53}$$

Hence, the original system (6.50) of two degrees of freedom is reduced to the system (6.51) with one degree of freedom. The new canonical momentum $\bar{J}_2$=const is an integral of motion and can be considered in what follows as a parameter.

Let us consider now the system described by the Hamiltonian (6.51). Traditionally, the first step is determination of stationary points. According to general rules, the equation for the stationary points has the form:

$$\dot{\bar{J}}_1 = -\frac{\partial \bar{\mathcal{H}}}{\partial \bar{\theta}_1} = 0; \quad \dot{\bar{\theta}}_1 = \frac{\partial \bar{\mathcal{H}}}{\partial \bar{J}_1} = 0. \tag{6.54}$$

For the overwhelming majority of systems of interest, members of the series (6.52) decrease with increasing p. Bearing this in mind, one can leave only three terms with numbers $p = 0, \pm 1$. It should be noted also that the perturbation is a real value, that is, $H_{-l,m} = H_{m,-l}$. In result the Hamiltonian (6.51) can be presented in the form:

$$\bar{\mathcal{H}} = \bar{\mathcal{H}}_0(\bar{J}) + H_{0,0}(\bar{J}) + 2H_{r,-s}(\bar{J})\cos\bar{\theta}_1 . \tag{6.55}$$

It follows from (6.54) and (6.55) that the stationary points are disposed at $2H_{r,-s}(\bar{J})\sin\bar{\theta}_1 = 0$, $\bar{\theta}_{10} = 0$, $\bar{\theta}_{11} = \pi$. The value of the new canonical action at these points is determined by the equation:

$$\frac{\partial \bar{\mathcal{H}}_0}{\partial \bar{J}_1} + \frac{\partial H_{0,0}}{\partial \bar{J}_1} + 2\frac{\partial H_{r,-s}}{\partial \bar{J}_1}\cos\bar{\theta}_1 = 0. \tag{6.56}$$

It should be noted also that as far as

$$\frac{\partial \bar{\mathcal{H}}_0}{\partial \bar{J}_1} = \frac{\partial \bar{\mathcal{H}}_0}{\partial \bar{J}_1}\frac{\partial J_1}{\partial \bar{J}_1} + \frac{\partial \bar{\mathcal{H}}_0}{\partial \bar{J}_2}\frac{\partial J_2}{\partial \bar{J}_1} = \omega_1 r - \omega_2 s \approx 0$$

the second term in (6.56) vanishes and the equation for stationary points takes the form:

$$\frac{\partial H_{0,0}}{\partial \bar{J}_1} \pm 2\frac{\partial H_{r,-s}}{\partial \bar{J}_1}\cos\bar{\theta}_1 = 0, \tag{6.57}$$

where the sign plus is to be chosen for $\bar{\theta}_1 = 0$ and the sign minus for $\bar{\theta}_1 = \pi$.

It should be noted that variations in the canonical action $\bar{\theta}_1$ can be large while changes in the variable $\bar{J}_1$ are small (proportional to perturbations). Hence, to describe the system near the resonance, the Hamiltonian (6.55) can be expanded into a Taylor series over powers of a small deviation from the stationary value of the canonical action $\bar{J}_{10}$:

$$\bar{\mathcal{H}} = \bar{\mathcal{H}}_0\left(\bar{J}_{10}\right) + \frac{\partial \mathcal{H}_0}{\partial \bar{J}_1}\left(\bar{J}_1 - \bar{J}_{10}\right)$$
$$+ \frac{1}{2}\frac{\partial^2 \mathcal{H}_0}{\partial \bar{J}_1^2}\left(\bar{J}_1 - \bar{J}_{10}\right)^2 + 2H_{r,-s}\left(\bar{J}_{10}\right)\cos\bar{\theta}_1 + \cdots \quad (6.58)$$

Then
$$\Delta\mathcal{H} \equiv \bar{\mathcal{H}} - \bar{\mathcal{H}}_0(\bar{J}_{10}) = G(\Delta\bar{J})^2 - F\cos\bar{\theta}_1, \quad (6.59)$$

where
$$G = \frac{1}{2}\frac{\partial^2 \mathcal{H}_0}{\partial \bar{J}_1^2} \quad \text{and} \quad F = 2H_{r,-s}\left(\bar{J}_{10}\right).$$

The Hamiltonian (6.59) known as a standard one describes the nonlinear system dynamics in the vicinity of a resonance. The most interesting cases of resonant interaction, including the particle phasing in a single external wave considered above, can be reduced to its analysis.

Note that exactly the same arguments could be applied to the variable θ_1. In other words, the approximation under consideration describes two independent resonances and yields the phase trajectory schematic pattern shown in Fig. 6.3a.

6.4.2 Randomization of Motion

Strictly speaking, the concept of a single isolated resonance is adequate to 1-D conservative systems only. There are two exact integrals of motion in such cases, namely an action and a phase of oscillation, which provide totally determined motion. In the overwhelming majority of cases, such conditions cannot be supported. Particularly, this relates to short-wave radiation where the electrodynamic system permits a coexistence of a large number of proper modes interacting with various degrees of particles freedom. Even for two degrees of freedom, the resonances interact, in a sense. This coupling leads to a principally new phenomena – to development of a dynamic chaos.[10] It should be emphasized that in microwave electronics this phenomenon mainly relates to a nonlinear character of a particle motion in a wave. However, sometimes a nonlinearity of interacting waves plays its role as well.

The possibility of chaotic regimes is of essential interest for both general theory and applications. Really, a chaotic electrodynamic system should generate wide spectrum radiation. This can be used for the development of powerful generators of electromagnetic noise. On the other hand, such noisy regime is definitely deleterious for narrow spectrum highly coherent radiation. In both cases, the region of parameters corresponding to the dynamic chaos is of a principal importance.

---

[10] This relates as well to open one-dimensional systems where an external force is in a certain resonant ratio with proper oscillations. Such systems can be described in a 3-D phase space and are usually used for simple illustration of the chaos development. However, such examples are hardly peculiar for our problems.

The germs of the dynamic chaos are those points of the phase space where two phase trajectories cross each over (homocline points). In our case they are the saddle points which belong to a separatrix. The reason is rather obvious: particles spend long time in the vicinity of these points and are subjects of small but long-acting perturbations. Then a so-called local instability appears when two points originally close together go away rapidly their separation increasing exponentially with time. This is exactly what happens when two points initially close but coming to a saddle point along different sides of a separatrix have quite different destinations.

Under action of a small perturbation provided by other resonances, the outgoing branches of a separatrix split and oscillate with increasing amplitude. As a result, they cross each over creating new homocline points where the process is repeated behaving like an avalanche. Then a kind of a particle trajectories web appears and the motion becomes indistinguishable from the stochastic one. Practically at every point phase trajectories run away from each other.

This scenario looks rather apocalyptically and can open to question the possibility of dynamical description of nonlinear systems in general. Fortunately, the reality is not so bad. There is a remarkable theorem by Kolmogorov–Arnold–Moser (KAM theorem) [36] telling that the dynamic chaos takes place only in a close vicinity of a homocline trajectory if the perturbation is small enough. Only a few phase trajectories leave this domain. So, the concept of almost independent resonances stated above still has a right for existence. But in the case of large enough perturbations, numerous computer simulations really show practically stochastic wandering of representing points over all phase space in the absence of external stochastic forces. This remarkable and comparatively new result of classical mechanics actually opens a way to the understanding of irreversible character of real physical processes in nondissipating systems. Moreover, if the dissipation does exist, a lot of remarkably new effects are predictable including so-called strange attractors, a fractal structure of phase portraits etc. This theory is intensively developing now and can be found in a row of specialized monographies [37]. Here we have to return to our main question – to the criterion of the dynamic chaos regime and to its consequences. Unfortunately, KAM theorem itself does not answer the question.

6.4.3 Criteria of Dynamic Chaos

The main problem in this context is a definition of the parameters domain where the system becomes chaotic under action of several resonances. There are several more or less formal criteria. In what follows we stay only on two simplest ones.

Lyapunov's Criterion

Consider a dynamic system described by the ordinary differential equations:

$$\dot{x}_i = f_i(\mathbf{x}), \qquad i = \{1...N\}, \qquad (6.60)$$

where $\mathbf{x}(t)$ is an arbitrary trajectory of the system (6.60). Let $\mathbf{x}_1(t) = \mathbf{x}(t) + \delta\mathbf{x}(t)$ be another trajectory being in a close vicinity of $\mathbf{x}(t)$. Then for a small deviation $\delta\mathbf{x}(t)$, one gets the following system of ordinary differential equations:

$$\delta\dot{\mathbf{x}} = \mathsf{M}\delta\mathbf{x}, \qquad (6.61)$$

where

$$\mathsf{M} = \{a_{ik}\}; \qquad a_{ik} \equiv \frac{\partial f_i}{\partial x_k}.$$

Generally, the matrix M depends on time, but we restrict ourselves at the moment by the case when it has constant elements. A validity of this approximation will be considered below.

For a constant matrix M, the general solution of (6.61) has the form:

$$\delta\mathbf{x} = \sum C_j \mathbf{e}_j \exp[\lambda_j t], \qquad (6.62)$$

where $\mathbf{e}_j$ are the proper vectors of the matrix M, λ_j are the corresponding eigenvalues, and C_j are constants. It follows from (6.62) that for negative $\mathrm{Re}\,\lambda_j < 0$ the small deviations under consideration damp. They say that the system is locally stable in this case. However, if just one of the eigenvalues has a positive real part, the small deviations rise exponentially as well as deviations of the phase trajectories. In these cases, the system is locally unstable.

Using this criterion, we identify the local instability with the possibility of the dynamic chaos. The eigenvalues λ_j depend on the system (6.61) and those parameters which give $\mathrm{Re}\,\lambda_j > 0$ just for one eigenvalue determine the dynamic chaos domain.

Mathematically, this criterion is immaculate but difficult for applications if the matrix M depends on time. Then every point of the phase trajectory requires a separate investigation of the eigenvalues.

Chirikov's Criterion

Now we stay with another criterion based on simple physical arguments. In spite of its semi-intuitive character, it describes qualitatively a great variety of systems. One should not, of course, expect an exact quantitative result all the more that dynamic chaos appearance is sometimes rather sensitive to governing parameters.

Bearing in mind our particular interests, we shall consider now the motion of a particle in the fields of two waves satisfying close resonance conditions.

Let us start with one isolated resonance (e.g., Cherenkov one) for certain ω and k. As it was stated above, the equations of motion in this case look like the equations of a nonlinear pendulum:

$$\frac{d\gamma}{dz} = g\cos\varphi; \qquad \frac{d\varphi}{dz} = \alpha k(\gamma - \gamma_s), \qquad (6.63)$$

where g is a maximal increase in the particle energy (in units of mc^2) per unit of length which can be obtained from the wave. Note that we use the energy deviation instead of the action in the general theory 6.4.1. In a case of neighboring resonances, it does not matter but is more convenient for physical interpretations.

The analysis of Sect. 6.3.1 shows that the regions of vibration and librations (phase slippage) are separated by a which passes via points $\gamma = \gamma_s$ and $\varphi - \varphi_s \mp \pi$ and has a spread along the energy axis

$$\gamma_{\max} - \gamma_s = \pm\sqrt{4g/k|\alpha|}$$

realized at $\gamma = \gamma_s$ and $\varphi - \varphi_s \mp \pi$. This spread should be identified as the nonlinear resonance width. The phase trajectory pattern was shown in Fig. 6.2.

Suppose now that the particle at the same time is in the vicinity of another resonance that produces the same picture by itself. If their interaction is weak enough, the resulting picture would look like that presented in Fig. 6.3

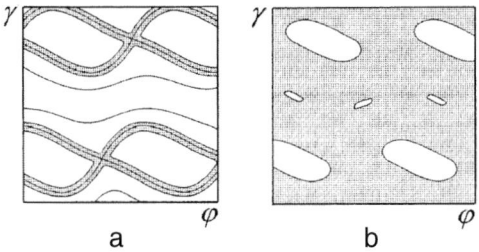

Fig. 6.3. A phase portrait of a nonlinear system under action of two resonances. (**a**) Almost independent resonances. (**b**) Overlapping resonances. The stochastic regions are shadowed

Note that according to the previous section the separatrix of an almost independent resonance has a narrow stochastic layer also shown in the figure.

According to the arguments above Fig. 6.3a describes satisfactory the motion if the waves phase velocities are essentially different. Otherwise, the resonances are close to each other, and the procedure of averaging above is not justified anymore. One can foresee a situation when the stochastic layers of both resonances are partially overlapped and a lot of additional homocline points appear. The corresponding set of parameters can be identified as a criterion of the dynamical chaos development[11] (see Fig. 6.3b). This condition suggested by B. Chirikov [38] is sometimes quoted as the resonances overlapping criterion. Numerous analytic and computer investigations show that it is in a good qualitative (and sometimes quantitative) agreement with reality for a wide variety of systems.

---

[11] The small additional islands of stability appear because of secondary resonances mentioned above.

Using this criterion to find the dynamic chaos condition, one has to find a nonlinear resonance width and a distance from the nearest adjacent resonance. In our particular case the latter is supposed to be a cyclotron one.

The distance between two resonances is determined by a difference between the corresponding equilibrium particle velocities or by a difference of equilibrium energies. If this distance turns out to be lesser than a sum of semi-widths of the resonances (including their stochastic layers), the particle dynamics becomes stochastic. Labeling parameters of the Cherenkov and cyclotron resonances by indices 1 and 2, one has for the equilibrium values

$$\omega_1 - k_1 v_1 = 0; \qquad \omega_2 - k_2 v_2 = \pm \Omega_0/\gamma_2, \qquad (6.64)$$

where $\Omega_0 = qB/mc$ is Larmor frequency in the magnetic field B. Assume for simplicity that the resonances are close to each other. Then the difference of the equilibrium energies is

$$|\Delta\gamma| = |\gamma^3 \beta \Delta\beta| = \left| \frac{\gamma^3 \beta (\beta - \beta_g)}{1 \pm \gamma \beta^2 \Omega_0/\omega} \frac{\Delta k}{k} \right|, \qquad (6.65)$$

where $\beta_g c = \partial \omega / \partial k$ is the group velocity in the vicinity of the resonances. For semiwidths of the resonances, one can take the expression (6.26). If they are equal to each other, the condition of a stochastic regime takes a simple form:

$$\sqrt{g} > \left| \frac{\sqrt{\alpha} \gamma^3 \beta (\beta - \beta_g)}{4 (1 \pm \gamma \beta^2 \Omega_0/\omega)} \frac{\Delta k}{k} \right|. \qquad (6.66)$$

Note that if the group velocity of the wave is close to the phase one, the limiting amplitude can be rather small. For nonequal widths, one should take into account that for a cyclotron resonance the phase slippage coefficient is equal to

$$\alpha = -\beta \frac{\partial}{\partial \gamma} \frac{\omega \mp \Omega_0/\gamma}{\omega \beta} = \frac{1 \mp \Omega_0 \gamma/\omega}{\gamma^3 \beta^2}. \qquad (6.67)$$

In more detail, this effect will be considered in use to cyclotron resonance masers (Sect.9) and to free electron lasers (Sect.10).

7
Proper Waves in Flows of Charged Particles

Up to this point it has been assumed that individual charged particles interact only via their collective radiation field. That is, the short–range interaction via Coulomb fields has been neglected. However, it is physically evident that this approach is justified in the only case of low–density beams. As regards many modern microwave devices, the Coulomb interaction must be taken into account (especially in the high–current electronics). In these cases, Coulomb fields can exert substantial influence on the process of grouping charged particles into bunches emitting coherent radiation. At least, even the trivial Coulomb 'repulsion' of charged particles in an inhomogeneous beam exerts influence on the processes of the beam spatial modulation.

7.1 Proper Waves in Beams of Interacting Particles

Making use of general considerations, one can intuitively guess that this interaction must influence mainly the beam inertial properties. Really, the motion of every individual particle cannot be regarded as independent of the motion of others. What is more, certain 'elasticity' has to be inherent in distributions of the spatial charge and current. Suppose that a high–current beam exists as a relatively equilibrium stable physical object. Then if this equilibrium is locally disturbed, the beam tends to restitute its state. The beam particles are bound so that in the process of this restitution any local disturbance spreads all over other beam particles with some finite velocity. This velocity is determined by the type of the interaction and, of course, by the boundary conditions. Thus, the notion of proper waves of the space charge and current density can be interpreted as spatially-temporal variations of these parameters. After being excited by some disturbance, these variations are freely propagating along the beam at large distances from their origin. Surely, in the beam of charged particles, space charge waves and current waves are inevitably related to corresponding electromagnetic fields. The point to be made is that some of these fields are not the radiation fields in the direct sense of the word existing only within the beam volume or in its vicinity.

Thus, we make transition from the model of discrete particles to their continuous description in terms of the space charge density ϱ and current density **j**. Logically, this transition is validated in the case when the average distance between the particles is smaller than the wavelength of emitted radiation. However, the criterion itself does not indicate that the collective interaction is of any principal importance. This criterion just emphasizes the fact that collective ensembles of charged particles are the real sources of the radiation field emission. The degree of correspondence of dynamics of the collective ensembles with dynamics of individual particles depends on interaction between particles. If it is strong enough, beam dynamics can dramatically differ from dynamics of collective ensembles. For instance, the motion of the sea surface has nothing in common with the motion of individual water molecules – as regards either its quantitative characteristics or direction. However, at the microscopic level, the sea surface motion is also reducible to the molecular motion.

In this connection, there are two points to be made here. First, the continuous description of the beam implies the complete refusal to take into account the spontaneous incoherent radiation emission unless we take into account microscopic fluctuations in the charge and current distributions with correlation distances much shorter than the radiation wavelength. Second, it is implied that the total field acting on a beam particle is determined by a large number of other particles. In a sense, it is an 'external' field; i.e., it is independent of the motion of this very particle. This approach called the self-consistent field approximation enables to solve Maxwell equations and dynamic equations as a common system. Their canonical properties remain preserved (in particular, the Liouville theorem is fulfilled). If the number of charged particles per wavelength is small, the self-consistent field approximation becomes dubious. One should keep in mind that even Lagrangian presentation of dynamics of two interacting particles is possible only in the case of their nonrelativistic relative motion not mentioning the above discussed problems of the self-accelerated motion of a particle and its proper field taken into account.

7.1.1 Dispersion Relations

As regards the proper waves, one can make certain general conclusions not going into details of the interaction between the particles. If the beam equilibrium is stationary, the temporal dependence of a structure proper wave can be nothing but harmonic. So, a wave is characterized by a certain frequency. This already follows from the fact that any physical value participating in the wave process can differ at two moments of time (t_1 and t_2) only in the phase factor, which depends on the difference $t_2 - t_1$ and is independent of choosing the moments themselves:

$$A(t_2) = \exp\left[-i\omega\left(t_2 - t_1\right)\right] A(t_1); \qquad \omega = \text{const}.$$

The only function that meets this condition has the form: $A(t) \propto \exp(-i\omega t)$, which confirms the statement given above. By analogy, one can show that in a system homogeneous with respect to z-axis the dependence of the wave on z is described with the harmonic function $\exp(-ikz)$. Therefore, in a homogeneous stationary system, proper waves always have the form

$$A = A(\mathbf{r}_\perp) \exp[i(kz - \omega t)],$$

with the wave amplitude depending only on transverse coordinates. By the way, all these facts just follow as particular cases of so-called Noether theorem [39]. According to this theorem, any type of a system symmetry corresponds to a certain conservation low. That is, for electromagnetic waves, a temporal homogeneity corresponds to a constant value of the quantum energy $\hbar\omega$; homogeneity in a coordinate corresponds to preservation of the respective component of the momentum $\hbar\mathbf{k}$, the system azimuthal symmetry is responsible for conservation of the angular momentum $\hbar m$ (here m is the azimuthal wave number), etc. An extension of this principle to proper waves of periodic systems was given in Part I.

The parameters $k_0 = \omega/c$ and k are not independent. In each specific system, they are related to one another by the so-called dispersion relation that determines inertial properties of the system (analogously, the kinematic relation of the energy of a free particle to its momentum $\gamma = \sqrt{1+p^2}$ determines the particle inertia). Formally, the roots of the dispersion relation $k_0(k)$ can be complex for real values of k. In this case, the amplitude of a system's proper wave exponentially decreases or increases in time. In the second case, the wave gains energy from the beam motion. From the viewpoint of physics, this fact means the coherent radiation emission. Thus, the spectrum of the structure proper waves determines not only the inertial properties of the ensemble of interacting particles but also processes of self–organization of this ensemble. The phasing of particles and the induced radiation emission, described above in the one–particle approximation, belong to the processes of this type.

Physical interpretation of the beam instability depends on the reference frame. To be more precise, it depends on a formulation of the problem in the laboratory coordinate system. From the viewpoint of physics, existence of complex frequencies at the real values of k means a so-called absolute instability. Physical values involved in this dynamic process (such as field components, those of the beam particle velocity as well as space charge density, etc.) exponentially increase in time remaining spatially harmonic in the intervals of wave numbers where the roots of the dispersion relation $k_0(k)$ have positive imaginary components. Such instability is known as an absolute one.

For applications of instability for wave generation or beam bunching, another viewpoint is more appropriate. Instead of the initial conditions, one can suppose that some harmonic signal of a fixed frequency is supported externally at $z = 0$. If the dispersion equation has complex roots, this perturbation should increase along the z-coordinate remaining harmonic in time. The moment of the start of the excitation can be related to the remote past. This type of a spatial increase of a signal along the beam is called a convective instability.

One just must keep in mind that the signal field strength vanishes at very large distances along z because of the causality principle (any disturbance cannot get to such distances even with the velocity of light). Therefore, it is mathematically legitimate to present the signal as a Fourier integral, i.e., as a set of the spatial harmonics in the form $\exp(ikz)$. More adequate formal approach to this problem should involve the Laplace transform instead of the Fourier used here.

In a stable system, all roots of the dispersion relation are real at real values of k. As it is clear, at any point $z > 0$ the signal may be presented as a superposition of the system proper waves. These proper waves are characterized by the same frequency, equal to the frequency of the driving signal ω. Their discrete real wave numbers correspond to the points of intersection of the line $k_0 = \text{const}$ with the branches of the dispersion curve. The amplitude of each of the travelling harmonics is constant in time. Hence, one can consider the system of the travelling harmonics as that transferring the power of the source of excitation. The velocity of the energy transport is defined by the group velocity $\partial \omega / \partial k$, which must be less than c for all stable systems. If there are no intersections for a given frequency, the wave cannot propagate in the system at all because of a total inner reflection. For instance, this takes place in waveguides below cutoff in the absence of a beam.

So, a general statement can be made: the inequality $\partial \omega / \partial k < c$ is a necessary and sufficient stability condition. Otherwise, the wave is unstable and the group velocity loses the sense of the velocity of energy and signal transportation.[1]

However, the situation radically changes if the system is unstable, that is in the case of the dispersion equation possessing complex roots with $\operatorname{Im} k_0(k) > 0$ in a certain interval of wave numbers. Then the corresponding spatial harmonics are increasing in time gaining energy from the beam motion (such a process is principally impossible in passive electrodynamic structures without a beam). In the case of a sufficiently long interaction, the spatial harmonic, for which the frequency imaginary component is positive and has the maximal absolute value, is dominant. Thus, while propagating, the signal asymptotically obtains a quasi–harmonic structure and exponentially increases in space. The instability of this type is called a convective one.

Under many circumstances the instability appearance is a mixed one. Let us suppose, for example, that at the system input ($z = 0$) the signal has the form of a prescribed pulse. In the case of stability this pulse propagates along the system (generally speaking, in both directions along the z-axis). It undergoes deformation according to the law of dispersion of its frequency components. At the same time, the energy transferred by the signal is preserved (of course, if there is no dissipation in the system). Making use of arguments

---

[1] The latter is valid for stable dissipative systems as well.

above, one can imagine the temporal and spatial evolution of the signal in the amplifying system. At a fixed and sufficiently large distance from the input, the signal initially increases in time. Then it drops while the propagating part gradually takes the form of a quasi–harmonic wave packet growing and spreading in the self-frame.

Even for this rather specified problem, the signal evolution (if not asymptotic) depends on specific boundary conditions. The matter is that the input signal can be presented as a superposition of perturbations of various kinds (those of the electromagnetic field, the beam density, the velocity of the beam particles, etc.). The system responds to various perturbations in a different manner. This response consists in excited waves that propagate in various directions with specific matching to the input signal. Generally speaking, the instability of this type is developing in time as well as in space. Existing a feedback – that is, a signal propagating in the backward direction – can change relations of the spatial increment of the instability to its temporal increment. Thus, the convective instability can be even transformed into the absolute one, when disturbances simultaneously increase in time exponentially within the volume of interaction. Surely, this fact is of great importance if one deals with specific devices (e.g., FEL operating in the master–oscillator or low-gain mode; see Sect. 10) Strictly speaking, this classification is somewhat ambiguous. It is sufficient to mention that the instability character is not invariant with respect to the Lorentz transformation. That is, in a certain moving reference frame, a convective instability can look like an absolute one.

Nevertheless, from the viewpoint of possible applications, the character of the instability in the laboratory frame is of the main importance. General criteria can be elaborated on the theory of complex variables basis (see, for example, [40, 41]).

Of course, any instability in a beam system can be considered as a radiation source given for nothing. But the main attention should be paid to organization of a controlled instability at a desirable frequency to provide the temporal coherency. Possible ways of governing roots of the dispersion equation are discussed below.

Surely, determining the characteristics of the system proper waves and their dispersion in the general form is not easier than to solve the equations for a large number of interacting particles. However, reasonable division of the system into weakly interacting, relatively simple subsystems could be very useful. Effective exchange of energy among them can take place under the condition of synchronism. Then analysis of a system is reduced to the qualitative and quantitative study of the interaction of a few partial waves, that is, of the proper waves of the simple subsystems. Of course, the principle of dividing the totality of proper waves and their presentation as a result of interaction between partial waves depend on the physical specificity of a particular problem.

We deal with the effects that, one way or another, are of the perturbation nature because they are reducible to slow[2] temporal or spatial changes in the electromagnetic wave amplitude and phase due to the presence of the beam. Therefore, it would be natural to choose free waves of a 'cold' electrodynamic structure and beam waves of the space charge and current in an unlimited free flow as the partial subsystems. The beam density plays then the role of the natural coupling parameter of the free waves with the beam waves while the beam motion serves as the energy source for a possible instability.

Within the framework of this approach, the processes of radiation emission and absorption can be interpreted as a weak systematic energy interchange between the electromagnetic field and the beam waves – i.e., between the system proper partial waves. This implies the following. First, there must exist a certain mechanism for waves coupling. Second, the frequencies of the interacting waves as well as their wave numbers must be approximately equal to one another. If values of these parameters differ, the proper waves of the system as a whole (i.e., the normal waves) decompose into two independent groups. These wave groups will be called partial beam and electromagnetic waves, each of these has its own dispersion characteristics.

Under condition of synchronism, that is, in vicinity to the points of intersection of the corresponding dispersion curves, normal waves differ from partial waves still preserving the characteristic features of both the subsystems. In particular, they can escape from the beam being readily transformed into purely electromagnetic waves. That is, the radiation emission is possible. Besides, the normal mode frequencies slightly differ from the frequencies of the partial waves. They can also have an imaginary frequency component at a fixed wavenumber. This corresponds to the emission of radiation that is coherent in time and/or in space.

Existence of approximately defined regions of two or three partial waves interaction simplifies essentially solution and interpretation of the dispersion equation. Near the point of intersection of the partial waves dispersion curves, each of them can be approximated by a straight line passing through the point with a slope equal to the group velocity. Such simplified dispersion equations will be called local ones and can be reduced to quadratic or cubic algebraic equations.

Considering small deviations from the intersection point denoted by a star

$$\kappa_0 = k_0 - k_0^* \quad \text{and} \quad \kappa = k - k^*,$$

one gets the local dispersion equation for the two-wave interaction in the general form

$$(\kappa_0 - \beta_1 \kappa)(\kappa_0 - \beta_2 \kappa) = -C. \tag{7.1}$$

The corresponding dispersion curves are two-branch hyperbolas reclining the asymptotes determined by the partial group velocities β_1 β_2 of the

---

[2] In the scale of the wave period and the wavelength.

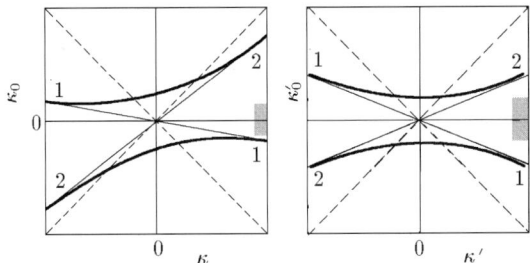

Fig. 7.1. Splitting of two partial waves for $C < 0$. Right: the same in the moving frame. Shading shows a stop–band

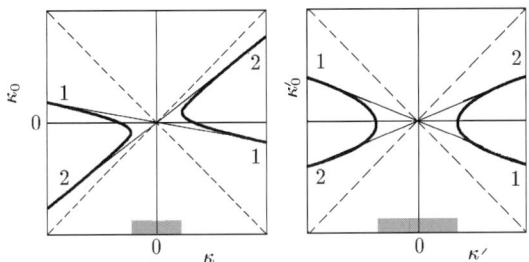

Fig. 7.2. The same as in Fig. 8.1 for $C > 0$. Shading shows an instability band

interacting waves (see Figs. 7.1 and 7.2). These solutions have qualitatively different physical sense, depending on the coupling constant's C sign.

Really, the direct solution of the quadratic equation is

$$\kappa_0 = \frac{\beta_1 + \beta_2}{2} \pm \sqrt{\frac{(\beta_1 - \beta_2)^2 \kappa^2}{4} - C} \,. \tag{7.2}$$

The case $C < 0$ corresponds to two real frequencies for any real value of κ, that is, to the stable proper waves. At the same time, a fixed real frequency does not guarantee real κ. Nevertheless, the system is stable because all the group velocities remain below c. One may use the reference frame moving with the velocity $(\beta_1 + \beta_2)/2$. In this frame $\beta_1' = -\beta_2'$ (see the right side of Fig. 7.1). The band between the branches is just a stop-band. Really, the group velocities of both normal waves is zero at the crossover ($\kappa' = 0$). That corresponds to a total inner reflection because the energy cannot be transported along the system as it should be in a case of instability.

The case of $C > 0$ and its appearance in the moving frame is illustrated in Fig. 7.2. The last presentation is not really necessary because the main result – the complex κ existence at arbitrary real κ_0 – is obviously invariant with respect to the Lorentz transformation. As it has been mentioned above, the instability can develop as absolute, convective, or mixed one, depending on initial/boundary conditions. It takes place within the band

$$|\kappa| < \frac{2C^{1/2}}{|\beta_1 - \beta_2|}$$

and has the maximal temporal increment

$$\operatorname{Im} \kappa_0 |_{\max} = -C^{1/2}$$

at the middle of the band.

The three-wave interaction takes place usually if one of the interacting waves is dually degenerated or is very close to degeneration. Then the dispersion equation has the form

$$(\kappa_0 - \beta_1 \kappa)(\kappa_0 - \beta_2 \kappa)^2 = -C \ .$$

We will not investigate it here because an equation of that type will be met in Part III. Just note that the maximal increment is proportional to $C^{1/3}$ instead of $C^{1/2}$ as it was for a two-wave interaction. Bearing in mind that the coupling constant is usually proportional to the beam current, that essentially influences quantitative characteristics of the process.

7.1.2 Partial Beam Waves

The subsystem of partial electromagnetic waves has been defined above as a totality of free waves in then electrodynamic structure in the beam absence. In fact, expansion of the current over such waves has been already applied in Part I being used as a method of determining the radiation field. One should just keep in mind that, formally speaking, this system is incomplete because it contains only solenoidal eigenfunctions of the wave equation. At the same time, the current density generally does contain a potential component. The corresponding potential component of the field of a moving charged particle has not been taken into account up to now. It does not play role in the process of the radiation emission as far as the particle motion is supposed to be prescribed (not self–consistent).

This fact becomes clear if, by trivial procedures, one reduces Maxwell equations to the inhomogeneous wave equation for the electric field:

$$\triangle_\perp \mathbf{E} + \left(\varepsilon k_0^2 - k^2\right) \mathbf{E} = \frac{4\pi}{\varepsilon} \left(\nabla_\perp \varrho + \mathrm{i} k \mathbf{e} \varrho\right) - \mathrm{i}\frac{4\pi k_0}{c}\mathbf{j} \ . \qquad (7.3)$$

Here $\triangle_\perp$ and $\nabla_\perp$ are the transverse parts of the Laplace and gradient operators, respectively; ε is the medium permittivity, $\mathbf{e}$ is the unit vector along z-axis. As it is easy to see, the right–hand side of (7.3) contains the gradient terms.

Naturally, dealing with the self–consistent approach, one needs a second equation providing relation of microwave fields with currents. That is, an equation of motion in the continuous presentation has to be used. As it is easy to see, a beam response (in the sense of the currents induced) is to be

7.1 Proper Waves in Beams of Interacting Particles

also of the wave type, at least for sufficiently small amplitudes of the wave–like fields. Its amplitude and phase depend on k_0 and k. At certain frequencies, the currents can be finite even for an infinitesimal driving field. Such free waves with a specific dependence $k_0(k)$ are the partial beam waves. Note that, generally speaking, equations for partial beam wave dynamics cannot be reduced to the standard form of a wave equation. Hence, the notions of completeness or orthogonality of the corresponding subsystem is inapplicable. In fact, the self–consistent approach implies that only the complete system of the structure proper waves preserves these properties.

Just now we take it for granted that variations in the local velocity $\mathbf{v}(\mathbf{r}, t)$ and in the corresponding local momentum per unit rest mass $\mathbf{p}(\mathbf{r}, t)$ can describe the flow dynamics. The mutual dependence of these values is supposed to be the same as for an individual relativistic particle:

$$\frac{\mathbf{v}}{c} = \frac{\mathbf{p}/c}{\sqrt{1 + p^2/c^2}} . \tag{7.4}$$

Thus, we identify the flow velocity at a given point and at a given moment of time (the hydrodynamic velocity) with the velocity of a particle that is located at the same point at the same time. As it is intuitively clear, in this way we have completely neglected the possibility of the local spread of velocities of the flow particles and intersections of their trajectories; i.e., the flow is regarded as laminar. As a rule, this approximation is called the cold hydrodynamic approximation. Limitations on its applicability will be discussed below.

As it is known (see, e.g. [2]), dynamics of a flow is described – within the hydrodynamic approximation – with the continuity equation:

$$\frac{\partial}{\partial t}\varrho(\mathbf{r}, t) + \operatorname{div}\left[\mathbf{v}(\mathbf{r}, t)\,\varrho(\mathbf{r}, t)\right] = 0 , \tag{7.5}$$

together with the Euler dynamic equation:

$$\frac{\partial}{\partial t}\mathbf{p} + (\mathbf{v} \cdot \nabla)\mathbf{p} = \frac{q}{m}\mathbf{E} + \frac{q}{mc}\left[\mathbf{v} \times \mathbf{B}\right] , \tag{7.6}$$

Here the ratio q/m must be taken the same as for an individual particle. As it has been stipulated, the system proper waves are regarded as small linearly independent perturbations in the equilibrium flow. Therefore, (7.5) and (7.6) can be linearized with respect to small deviations from the equilibrium state. Preserving the notation $\mathbf{v}$, $\mathbf{p}$ for the wave vector amplitudes, we introduce the equilibrium velocity $\beta = j_0/c\varrho_0$ where zeroes denote the steady state values. So, the continuity equation may be rewritten as

$$\varrho = \frac{k\varrho_0(\mathbf{e} \cdot \mathbf{v})}{(k_0 - \beta k)c} \quad \text{or} \quad \mathbf{j} = \varrho_0 \mathbf{v} + \frac{k\beta\varrho_0}{k_0 - \beta k}\mathbf{e}(\mathbf{e} \cdot \mathbf{v}) . \tag{7.7}$$

Respectively, the Euler equation takes the form:

$$\mathbf{p} = \frac{iq}{mc(k_0 - \beta k)} \left\{ \mathbf{E} + \beta\left[\mathbf{e} \times \mathbf{B}\right] - \frac{B_0}{c}\left[\mathbf{e} \times \mathbf{v}\right] \right\}. \qquad (7.8)$$

We now find the mutual dependence of the small variations $\mathbf{v}$ and $\mathbf{p}$. For this purpose, in (7.4), we substitute $\gamma\beta c\mathbf{e} + \mathbf{p}$ for $\mathbf{p}$ and $\beta c\mathbf{e} + \mathbf{v}$ for $\mathbf{v}$ (here $\gamma = 1/\sqrt{1-\beta^2}$). Thus, we get

$$\mathbf{v} = \frac{1}{\gamma}\left(\mathbf{p} - \beta^2 \mathbf{e}\left(\mathbf{e} \cdot \mathbf{p}\right)\right). \qquad (7.9)$$

This relation depicts the well–known relativistic effect: the change in the particle longitudinal velocity is much less sensitive (γ^2 times) to the increase in the longitudinal momentum than the corresponding transverse components. Exclusion of $\mathbf{p}$ from (7.7) and (7.8) yields

$$-i(k_0 - \beta k)\mathbf{j} + k_c\left[\mathbf{e} \times \mathbf{j}\right]$$
$$= \frac{k_p^2}{4\pi}\left\{\mathbf{E} + \mathbf{e}\beta\frac{k - \beta k_0}{k_0 - \beta k}(\mathbf{E} \cdot \mathbf{e}) + \beta\left[\mathbf{e} \times \mathbf{B}\right]\right\}. \qquad (7.10)$$

Here, $k_c = qB_0/mc^2\gamma$ is the cyclotron frequency and $k_p = \sqrt{4\pi q\varrho_0/(mc^2\gamma)}$ is the Langmuir frequency (both in units of c).

It is easy to express the derived algebraic relation via the longitudinal and transverse components of the microwave current:

$$j_\| = \frac{ik_p^2}{4\pi\gamma^2(k_0 - \beta k)^2}E_\|; \qquad (7.11a)$$

$$\mathbf{j}_\perp = \frac{k_p^2 c}{4\pi}\frac{k_c\left\{\left[\mathbf{e} \times \mathbf{E}_\perp\right] - \beta\mathbf{B}\right\} + i(k_0 - \beta k)\left\{\mathbf{E}_\perp + \beta\left[\mathbf{e} \times \mathbf{B}\right]\right\}}{(k_0 - \beta k)^2 - k_c^2} \qquad (7.11b)$$

yielding

$$\varrho = \frac{k}{k_0}j_\| - \frac{i}{k_0}\nabla_\perp \mathbf{j}_\perp. \qquad (7.12)$$

Judging by the form of the denominators in (7.11a) and (7.11b), one can see that the above–defined beam waves can be classified into two subclasses: the waves of the longitudinal current and of the charge density. Their phase velocity for low densities is approximately equal to the beam velocity ($k_0 \approx \beta k$). These waves are called the space charge waves (SCW). The waves of the transverse current, described with the partial dispersion equation $k_0 \approx \beta k \pm k_c$. These waves are called the cyclotron waves (the fast and slow ones). They do not cause disturbances in the space charge density within the transverse uniformity of the flow. There arise such disturbances only at the beam boundary.

7.1.3 Proper Waves in Flow

We now investigate the proper waves in a boundless homogeneous flow of a finite particle density. The flow is propagating along a uniform magnetic field

7.1 Proper Waves in Beams of Interacting Particles

B_0 in a medium with the dielectric permittivity ε. Surely, practical applicability of this model is rather limited. However, it helps understand better more complicated cases (they are to be examined below). Certainly, the notion of the flow boundless in the transverse direction implies that the flow transverse sizes are much larger than the wavelength (to be more exact, much larger than $\lambda\gamma$). At the same time, longitudinal extent of the flow is unlimited. Furthermore the results that are to be obtained will be applied to the case of the infinitely long beam with the limited transverse sizes. Therefore, we consider here the waves propagating along z-axis only.

As the system is homogeneous along z, all the proper waves belong either to TE–types or to TM–types. Those of the first kind have no longitudinal electric field. Consequently, they do not induce any longitudinal current. There are also no disturbances in the particle density (excluding the lateral bounds, infinitely removed). Besides, in the transversely homogeneous wave $B_\parallel = 0$. Hence, the proper TE–waves for sure belong to the TEM–type.

The transverse components of fields obey the equations:

$$k\,[\mathbf{e} \times \mathbf{B}_\perp] + \mathrm{i}\frac{4\pi}{c}\mathbf{j}_\perp + k_0\varepsilon\mathbf{E}_\perp = 0 \, ; \tag{7.13a}$$

$$k\,[\mathbf{e} \times \mathbf{E}_\perp] - k_0\mathbf{B}_\perp = 0 \, ; \tag{7.13b}$$

$$-\mathrm{i}\,(k_0 - \beta k)\,\mathbf{j}_\perp + k_\mathrm{c}\,[\mathbf{e} \times \mathbf{j}_\perp] - \frac{k_\mathrm{p}^2 c}{4\pi}\,(\mathbf{E}_\perp + \beta\,[\mathbf{e} \times \mathbf{B}_\perp]) = 0 \, . \tag{7.13c}$$

They make a homogeneous algebraic linear system for the three vector variables. We now exclude

$$\mathbf{B}_\perp = \frac{k}{k_0}\,[\mathbf{e} \times \mathbf{E}_\perp] \, , \qquad \frac{4\pi}{c}\mathbf{j}_\perp = \mathrm{i}\frac{k_0^2\varepsilon - k^2}{k_0}\mathbf{E}_\perp \, . \tag{7.14}$$

Thus, it is easy to reduce this system to one homogeneous equation

$$(k_0 - \beta k)\,(k_0^2\varepsilon - k^2 - k_\mathrm{p}^2)\,\mathbf{E}_\perp + \mathrm{i}k_\mathrm{c}\,(k_0^2\varepsilon - k^2)\,[\mathbf{e} \times \mathbf{E}_\perp] = 0 \tag{7.15}$$

with the condition of solvability[3]:

$$D\,(k_0,k) \equiv (k_0 - \beta k)^2\,(k_0^2\varepsilon - k^2 - k_\mathrm{p}^2)^2 - k_\mathrm{c}\,(k_0^2\varepsilon - k^2)^2 = 0 \, . \tag{7.16}$$

This expression splits into two independent cubic equations. They correspond to the waves with the clockwise and counterclockwise circular polarizations (in a gyrotropic medium the linear polarization is impossible, and the beam in the magnetic field represents a medium of this type):

$$(k_0^2\varepsilon - k^2)\,(k_0 - \beta k \pm k_\mathrm{c}) = k_\mathrm{p}^2\,(k_0 - \beta k) \, . \tag{7.17}$$

If $k_\mathrm{p}^2 \to 0$, (7.17) describes two cyclotron waves polarized in the direction of rotation of electrons in the magnetic field and interacting with electromagnetic ones of different circular polarization. The fast wave interacts with the

---

[3] It is convenient to consider $\mathbf{E}_\perp$ and $[\mathbf{e} \times \mathbf{E}_\perp]$ as two independent variables and get one equation with the vector multiplication of (7.15) by $\mathbf{e}$.

electromagnetic wave of the same polarization, while the slow one is synchronous with the wave of opposite polarization.

If the beam density is finite, there takes place recombination of different branches of these curves. The results are different in the pre–Cherenkov ($\beta\sqrt{\varepsilon} < 1$) and Cherenkov ($\beta\sqrt{\varepsilon} > 1$) regions. In the first case, all roots of the dispersion relation are real and the system remains stable. The wave amplification takes place in the Cherenkov region. Perhaps, it would be more adequate to talk about an unstable hybrid of the slow cyclotron wave and the forward slow electromagnetic wave with reverse polarization. The latter should not embarrass as the particles that outstripping the wave are under action of the effective electric field rotating in the proper direction providing the wave–particle synchronism.

As regards the proper TM–waves, the situation is the following. Because of splitting out of the longitudinal component of the Euler equation, these waves in the unlimited flow belong to TEM–waves, or their dispersion relation describes purely longitudinal SCW:

$$(k_0 - \beta k)^2 = \frac{k_p^2}{\gamma^2 \varepsilon} \ . \tag{7.18}$$

Thus, stable plane SCW of two types can propagate in the flow. One of them is faster than the beam, another is slower. Their splitting in frequency is equal to the doubled 'weighted' plasma frequency (Langmuir frequency divided by γ). The reader should keep in mind that we talk about longitudinal oscillations of particles. Respectively, the external magnetic field does not affect the characteristics of such waves.

The existence of fast and slow SCW becomes perfectly evident in the reference frame moving with the velocity of the flow. In this system, both SCW are just two plasma waves propagating symmetrically in opposite directions. Their frequencies are equal (it is the Langmuir frequency in the moving reference frame).

In the examined model of the unlimited hydrodynamic flow, SCW do not interact with TEM–waves. The reason is that SCW have longitudinal component of the electric field. At the same time, this very component is absent in the TEM–wave. Therefore, the questions of possible instability of SCW, as well as clarification of physical sense of the slow cyclotron wave instability, will be postponed until investigating a more realistic model of the transversely limited beam.

7.1.4 Proper Waves in Transversely Limited Beam

As it has been already mentioned above, in a transversely limited system, there arises the transverse electric field component in SCW. This is conditioned by finiteness of the beam transverse sizes. At the same time, the cyclotron waves must contain the longitudinal component of the electric field because charges

7.1 Proper Waves in Beams of Interacting Particles

appear at the beam lateral surface. By introducing the medium dielectric permittivity ε into the model, one can vary the phase velocity of electromagnetic waves. That is, it becomes possible to investigate their interaction with beam waves both in the Cherenkov region and in the pre–Cherenkov region. In addition, the model of the limited flow is of importance in itself. Really, there exists the case when the beam transverse size is comparable with the system proper wavelength (and the more so with $\lambda\gamma$). This situation is not at all exceptional in practice.

Within the framework of the hydrodynamic approach, the equations of motion are in no way influenced by the presence of the lateral walls or by finiteness of the beam transverse size. As regards Maxwell equations, in the system homogeneous with respect to z-axis, they still preserve the important property, i.e., separation of the system proper waves into the subsystems of TE– and TM–waves. [4]

Wave Equations

The scheme of determining the proper wave spectrum in the coasting beam is obvious but somewhat bulky. For TE–waves, all variable parameters must be expressed via the magnetic field longitudinal component by making use of the equation of motion and Maxwell equations, as the boundary conditions for this component (e.g., zero value of the derivative normal to a conductive surface) determines the spectrum of eigenvalues of the wave equation. For TM–waves, the same procedure is necessary, but $E_\parallel$ must be substituted for $B_\parallel$. Correspondingly, the boundary condition must be changed on a metallic surface).

According to this scheme, for TE–waves the longitudinal electric field and the longitudinal current can be right away equated to zero. All the same, (7.13b) and (7.13c) remain unchanged. Instead of (7.13a), one gets

$$k\left[\mathbf{e}\times\mathbf{B}_\perp\right] + k_0\varepsilon\mathbf{E}_\perp + \mathrm{i}\frac{4\pi}{c}\mathbf{j}_\perp = -\mathrm{i}\left[\mathbf{e}\times\nabla_\perp B_\parallel\right], \qquad (7.19)$$

and

$$\mathbf{e}\left[\nabla_\perp\times\mathbf{E}_\perp\right] \equiv -\nabla\left[\mathbf{e}\times\mathbf{E}_\perp\right] = \mathrm{i}k_0 B_\parallel . \qquad (7.20)$$

Respectively, (7.15) also becomes inhomogeneous:

$$(k_0-\beta k)\left(k_0^2\varepsilon - k^2 - k_\mathrm{p}^2\right)\mathbf{E}_\perp + \mathrm{i}k_\mathrm{c}\left(k_0^2\varepsilon - k^2\right)\left[\mathbf{e}\times\mathbf{E}_\perp\right]$$
$$= -\mathrm{i}(k_0-\beta k)k_0\left[\mathbf{e}\times\nabla B_\parallel\right] - k_0 k_\mathrm{c}\nabla B_\parallel . \qquad (7.21)$$

It possesses the solution

$$\left[\mathbf{e}\times\mathbf{E}_\perp\right] = D^{-1}\left\{\mathrm{i}k_0 D_\mathrm{TE}\nabla B_\parallel + k_0 k_\mathrm{c}k_\mathrm{p}^2(k_0-\beta k)\left[\mathbf{e}\times\nabla B_\parallel\right]\right\}, \qquad (7.22)$$

---

[4] In transmission lines, there can also exist TEM–waves. It is possible to correspond them to any of these subclasses.

where
$$D_{\text{TE}} = (k_0 - \beta k)^2 \left(k_0^2 \varepsilon - k^2 - k_p^2\right) - k_c^2 \left(k_0^2 \varepsilon - k^2\right) . \quad (7.23)$$

To avoid cumbersome expressions, further we restrict ourselves to the case when the beam particle density is uniform over the beam cross section. We now make the scalar multiplication of (7.22) by ∇. The point to be made is that the second term in the curly brackets goes to zero. Thus, the wave equation takes the form:

$$D_{\text{TE}} \triangle_\perp B_\| + D B_\| = 0 . \quad (7.24)$$

For TM–waves, a similar procedure differs from the described one just in details of the algebraic calculations. It yields the following expressions:

$$i k \left[\mathbf{e} \times \mathbf{B}_\perp\right] + i k_0 \varepsilon \mathbf{E}_\perp = \frac{4\pi}{c} \mathbf{j}_\perp ; \quad (7.25\text{a})$$

$$k \left[\mathbf{e} \times \mathbf{E}_\perp\right] - k_0 \mathbf{B}_\perp = -i \left[\mathbf{e} \times \mathbf{E}_\|\right] ; \quad (7.25\text{b})$$

$$\nabla \left[\mathbf{e} \times \mathbf{B}_\perp\right] = i k_0 \varepsilon E_\| - \frac{4\pi}{c} j_\|$$

$$= i k_0 \varepsilon \left(1 - \frac{k_p^2}{\varepsilon \gamma^2 (k_0 - \beta k)^2}\right) E_\| . \quad (7.25\text{c})$$

Thus, instead of (7.22), we get

$$\left[\mathbf{e} \times \mathbf{B}_\perp\right] = D^{-1} \left\{-i k_0 D_{\text{TM}} \nabla E_\| + k k_c k_p^2 \left(k_0 \varepsilon \beta - k\right) \left[\mathbf{e} \times \nabla E_\|\right]\right\} . \quad (7.26)$$

The wave equation for $k_p = \text{const}$ takes the form:

$$D_{\text{TM}} \triangle E_\| + D \left(1 - \frac{k_p^2}{\varepsilon \gamma^2 (k_0 - \beta k)^2}\right) E_\| = 0 , \quad (7.27)$$

where

$$D_{\text{TM}} = \frac{1}{k_0 \varepsilon} \left[k_0 \varepsilon (k_0 - \beta k) - k_p^2\right] (k_0 - \beta k) \left(k_0^2 \varepsilon - k^2 - k_p^2\right)$$
$$- k_c^2 \left(k_0^2 \varepsilon - k^2\right) . \quad (7.28)$$

To derive the dispersion relation in the case the transversely uniform beam density, one needs just to substitute the eigenvalue $-k_{\text{c.o.}}^2$ for the Laplace operator. The physical sense of this parameter is the cutoff frequency of the corresponding electromagnetic wave in units of c. Thus,

$$D = D_{\text{TE}} k_{\text{c.o.}}^2 \qquad \text{for TE–waves,} \quad (7.29\text{a})$$

$$D \left(1 - \frac{k_p^2}{\varepsilon \gamma^2 (k_0 - \beta k)^2}\right) = D_{\text{TM}} k_{\text{c.o.}}^2 \qquad \text{for TM–waves.} \quad (7.29\text{b})$$

Surely, the parameters $k_{\text{c.o.}}$ are different in these cases.

Local Dispersion Relations

Dispersion relations (7.29a and b) have been derived without concretizing the magnitude of k_p. That is, the magnitude of the beam current is not prescribed. However, one needs to investigate the roots of the algebraic equation of the 6-th or 8-th order in the presence of five free parameters (k_p^2, k_c^2, β, $k_\text{c.o.}^2$, and ε). The prospects to perform such calculations qualitatively (and the more so quantitatively) look, at least, not very attractive. Therefore, furthermore we regard k_p^2 as a small parameter. This does not contradict the principal subject to which the monograph is dedicated. That is, we proceed from the assumption that the wave–particle interaction is weak; that is, changes in the field amplitude and phase within the distances of the order of wavelength are small. As a rule, this condition is provided by postulating smallness of the beam current in comparison with the magnitude of the Alfven current mc^3/q.

In the first nonvanishing order in k_p^2, the dispersion relation for TM–waves takes the form:

$$\left[(k_0 - \beta k)^2 - k_\text{c}^2\right] \left(k_0^2 \varepsilon - k^2\right) \left(k_0^2 \varepsilon - k^2 - k_\text{c.o.}^2\right)$$
$$= k_\text{p}^2 (k_0 - \beta k)^2 \left[\left(k_0^2 \varepsilon - k^2\right) + \left(k_0^2 \varepsilon - k^2 - k_\text{c.o.}^2\right)\right] . \qquad (7.30)$$

If $k_\text{p}^2 = 0$, (7.30) splits out. It describes two direct and two backward electromagnetic branches. Fast and slow cyclotron branches are also presented. We are interested in investigating the area of intersection of one of the electromagnetic waves with one of the cyclotron waves, i.e., the vicinity of the points k_0^* and k^*. These points are the solutions of the equations:

$$k_0^* - \beta k^* = k_\text{c} , \quad \text{and} \quad k_0^* \varepsilon = k^{*2} ,$$

or

$$k_0^* = \frac{k_\text{c}}{1 - \beta \sqrt{\varepsilon}} , \quad k^* = \frac{k_\text{c} \sqrt{\varepsilon}}{1 - \beta \sqrt{\varepsilon}} . \qquad (7.31)$$

Positive/negative values of k_c correspond to the fast/slow cyclotron wave (intersection with the slow wave is possible only in the Cherenkov area $\beta^2 \varepsilon > 1$).

So far as k_p is small, the values corresponding to the intersection points can be substituted into the right–hand side of (7.30). One should note that for the electromagnetic branches ratio, $\beta_\text{g} = k^*/k_0^* \varepsilon$ represents the reduced group velocity. Thus, one gets the local dispersion relation for the small deviations $\mu = k_0 - k_0^*$ and $\kappa = k - k^*$:

$$(\mu - \beta \kappa)(\mu - \beta_\text{g} \kappa) = \frac{k_\text{p}^2 k_\text{c} \beta_\text{g}}{4 k^*} . \qquad (7.32)$$

It should be also noted that, for both electromagnetic branches, (7.32) has the same form. Surely, it goes without saying that the parameters k^* and β_g are different. We now write the solution to the derived quadratic equation:

$$\kappa_0 = \kappa \frac{\beta + \beta_g}{2} \pm \sqrt{\frac{(\beta - \beta_g)^2 \kappa^2}{4} + \frac{k_p^2 k_c \beta_g}{4k^*}} . \qquad (7.33)$$

It indicates that all roots remain real in the vicinity of intersection of the system proper waves with the fast cyclotron wave. One can also see that, in the vicinity of intersection of the system proper waves with the slow cyclotron waves, there exists a band of instability:

$$(k - k^*)^2 < \frac{k_p^2 |k_c| \beta_g}{k^* (\beta - \beta_g)^2} . \qquad (7.34)$$

The increment of this instability is maximum at the intersection of partial waves. There it is equal to

$$\operatorname{Im} \kappa_0 c = \mathrm{i} \frac{k_p c}{2} \sqrt{\frac{|k_c|}{k^*} \beta_g} . \qquad (7.35)$$

It is worth mentioning that this parameter is proportional to the square root of the beam current.

As regards TM–waves, the corresponding dispersion relation is by two orders higher. This is conditioned by the fact that it also comprises SCW. In the first nonvanishing approximation, this dispersion relation has a form:

$$(k_0 - \beta k)^2 \left[(k_0 - \beta k)^2 - k_c^2 \right] \left(k_0^2 \varepsilon - k^2 \right) \left[k_0^2 \varepsilon - k^2 - k_{\mathrm{c.o.}}^2 \right]$$
$$= \frac{k_p^2}{\varepsilon \gamma^2} \left[(k_0 - \beta k)^2 - k_c^2 \right] \left(k_0^2 \varepsilon - k^2 \right)^2$$
$$+ k_p^2 (k_0 - \beta k)^4 \left[\left(k_0^2 \varepsilon - k^2 - k_{\mathrm{c.o.}}^2 \right) + \left(k_0^2 \varepsilon - k^2 \right) \right] . \qquad (7.36)$$

In the case of interaction between electromagnetic and cyclotron waves, the first term on the right-hand side of (7.36) can be neglected. Thus, the result coincides with the case of TE–waves (surely, values of the corresponding parameters are different). As it has been already mentioned, interaction of the system proper waves with SCW is realizable only in the Cherenkov area. Under these conditions, the second term on the right-hand side of (7.36) needs to be neglected. After some simple calculations, one gets

$$(\kappa_0 - \beta_g \kappa)(\kappa_0 - \beta \kappa)^2 = \frac{k_p^2 k_0^{*2} \beta_g}{2\gamma^2 k^*} \left(1 - \frac{\beta_g}{\beta} \right) . \qquad (7.37)$$

Both the fast and slow SCW participate in the interaction with the proper waves. Therefore, at least for small k_p, the local dispersion relation is cubic. The standard research on the subject of existence of the complex roots of this equation demonstrates the slow wave instability. Its increment is maximal at the point of intersection with the electromagnetic wave:

$$\operatorname{Im} k_0 c = i\frac{\sqrt{3}}{2}\left[\frac{k_p^2 k_0^{*2}\beta_g}{2\gamma^2 k^*}\left(1-\frac{\beta_g}{\beta}\right)\right]^{1/3}. \qquad (7.38)$$

In contrast to the case of cyclotron waves, the increment is proportional to the cube root of the beam current.

Splitting of partial waves is topologically unequivocal and independent of concrete methods of retarding of the electromagnetic wave. So, we kept "local" parameters k_0^*, $k*$ and β_g in (7.38) instead of $k_{c.o.}$ and ε.

7.2 Negative Energy Waves

From the first sight, the conclusion about stability of fast cyclotron waves contradicts to the effect of the proper wave spatial amplification in the flow of moving oscillators considered in Sect. 6.3.3 (to be more precise, in the given case they are rotators). In spite of all these considerations, all the roots of the dispersion relation (7.36) are real in vicinity of the crossing point. That is, there does not exist any amplification. This seeming contradiction is conditioned by the following fact. Under conditions of the normal Doppler effect (we are talking about the fast cyclotron wave), the oscillator energy certainly has to decrease. At the same time, this very oscillator energy completely does not exist in the approximation of the cold hydrodynamic equilibrium. In other words, the stimulated radiation can be emitted only when there exists microscopic rotational motion of the beam particles. This fact is very important to practical applications because fast waves can be excited in the simplest regular waveguide systems without applying any complicated periodic waveguides or other slowing down structures. However, more general kinetic approach to the beam description is required for the complete investigation of this effect. This is to be postponed till Sect. 7.3.3.

By analogy, one should expect a radically different evolution of the slow waves – whatever the wave type be (either SCW or cyclotron waves). For realizing this change, their dispersion curve must intersect the branch of a slow free electromagnetic wave in the dispersion diagram. Really, elementary emitters under the condition of the Doppler effect ($k_0 = k\beta_0 - \Omega/c$) could gain energy from the longitudinal motion increasing their oscillating energy as well as the radiation emission. In other words, we suppose that slow waves, the velocities of which are lower than the beam velocity, can propagate in the system without a beam. In this case, the system self–excitation should be expected in vicinity to the point of the intersection of a slow wave with the dispersion curve of the partial SCW or that of the cyclotron wave. That is, there arises the stimulated radiation emission even in the case of a 'cold' hydrodynamic beam.

Surely, the above–given examples are rather particular: each case is characterized by its own dependence of the system increment versus the system parameters. Notwithstanding this fact, the general conclusion is rather convincing.

That is, if a partial electromagnetic wave in the electrodynamic structure is synchronous with one of the slow partial beam waves, the system is unstable. This means that there take place both the stimulated radiation emission and spatial amplification in a certain frequency band. Because of this fact, beam partial slow waves are called in a somewhat unusual manner: the negative energy waves. The physical sense of this term is the following. Under the conditions of the wave excitation process, the energy in the system is lower than that in the equilibrium state. Therefore, the wave gains in energy due to any loss of the system electromagnetic energy (e.g., for radiation emitted away from the system). In the case of cyclotron waves, this is almost evident even judging by the one–particle approximation. In fact, as it has been already mentioned above, the energy of the emitter transverse motion is increasing when the particle emits a slow quantum under the conditions of the Doppler anomalous effect. The statement seems, in a way, unexpected because of a habitual motion of waves in a thermodynamically equilibrium medium. In this case the wave energy really can be shown to be positive. However, even in the simplest case of a moving medium of a scalar permittivity ε situation is much less clear.

Really, let us consider an oscillating dipole moving in a medium under conditions of the anomalous Doppler effect. As a natural result of radiation, the oscillator increases its amplitude decreasing the longitudinal velocity. In the proper coordinate frame moving with the dipole, the energy balance looks differently. The fact of radiation and of an increase in amplitude, of course, take place as well. However, the longitudinal motion as an energy source is absent. Under these conditions, one has to consider the outgoing wave as a negative energy one requiring for its excitation taking the energy from the moving medium rather than supplying it.

The conditions for the existence of negative–energy waves are explicable with the help of a rather simple argument. Let us consider a wave propagating in a medium. The wave energy can be presented as $\mathcal{E} = N\hbar\omega$, where $\hbar\omega$ is the energy of one quantum; N is their number. Suppose that the medium is moving with the velocity βc. In the medium rest frame, the wave energy takes the form: $\mathcal{E} = N\hbar(\omega - k\beta c)$. That is, if the wave phase velocity $v_{\rm ph} = \omega/k$ is lower than the velocity of the medium motion, the wave energy becomes negative.

Let us consider now a criterion of the negative energy wave existence in a medium in rest which is characterized by a tensor permeability $\hat{\varepsilon}(\omega)$. As is known from electrodynamics, the value

$$\frac{1}{4\pi}\left(\mathbf{E}\frac{\partial \mathbf{D}}{\partial t} + \mathbf{B}\frac{\partial \mathbf{B}}{\partial t}\right) = \frac{\partial U}{\partial t}$$

has a sense of a time derivative of the electromagnetic energy density. The latter is equal to $\left(\varepsilon|\mathbf{E}|^2 + |\mathbf{B}|^2\right)/8\pi$ in absence of the medium dispersion, i.e. if $\partial \mathbf{D}/\partial t = \varepsilon \mathbf{E}/\partial t$. To get an analogous relation for a dispersive medium

7.2 Negative Energy Waves

and a smoothly varying wave of an almost definite frequency ω_0, the following arguments are used.

Let the electric field strength and induction be dependent on time as

$$\mathbf{E} = \mathbf{E}_0(t)\exp(-i\omega_0 t); \qquad \mathbf{D} = \mathbf{D}_0(t)\exp(-i\omega_0 t),$$

where $\mathbf{E}_0(t)$ and $\mathbf{D}_0(t)$ are slow functions. These are the real parts of the expressions that are to be considered as physical values. So, following the general rules, one gets after averaging over the period $2\pi/\omega_0$

$$\frac{\partial U}{\partial t} = \frac{1}{8\pi}\operatorname{Re}\left(\mathbf{E}_0^* \frac{\partial \mathbf{D}_0}{\partial t}\right). \qquad (7.39)$$

For a logical notion of an electric permeability, the Fourier transforms

$$\mathbf{E}(t) = \int_{-\infty}^{+\infty} \mathbf{E}(\omega)\exp(-i\omega t)\,d\omega;$$

$$\mathbf{D}(t) = \int_{-\infty}^{+\infty} \hat{\varepsilon}(\omega)\mathbf{E}(\omega)\exp(-i\omega t)\,d\omega$$

should be used. Hence,

$$\frac{\partial \mathbf{D}(t)}{\partial t} = -i\int_{-\infty}^{+\infty} \hat{\varepsilon}(\omega)\omega\mathbf{E}(\omega)\exp(-i\omega t)\,d\omega. \qquad (7.40)$$

As far as the electric field is quasi monochromatic, its Fourier transform has a sharp maximum at $\omega \approx \omega_0$. The function $\omega\hat{\varepsilon}(\omega)$ can be presented as a Taylor expansion in the vicinity of this point:

$$\omega\hat{\varepsilon}(\omega) \approx \omega_0\hat{\varepsilon}(\omega_0) + \left.\frac{\partial \omega\hat{\varepsilon}(\omega)}{\partial \omega}\right|_{\omega_0}(\omega - \omega_0) \qquad (7.41)$$

so that

$$\frac{\partial \mathbf{D}}{\partial t} = -i\omega_0\hat{\varepsilon}(\omega_0)\mathbf{E}(t) - i\left.\frac{\partial \omega\hat{\varepsilon}(\omega)}{\partial \omega}\right|_{\omega_0}\int_{-\infty}^{+\infty}(\omega-\omega_0)\mathbf{E}(\omega)\exp(-i\omega t)\,d\omega$$

$$\qquad (7.42)$$

$$= -i\omega_0\hat{\varepsilon}(\omega_0)\mathbf{E}(t) + \left.\frac{\partial \omega\hat{\varepsilon}(\omega)}{\partial \omega}\right|_{\omega_0}\exp(-i\omega_0 t)\frac{\partial}{\partial t}\mathbf{E}_0(t).$$

Substituting (7.42) in (7.39), we suppose that the tensor $\hat{\varepsilon}(\omega)$ is Hermitian; that is, the medium is not absorbing.[5] Then the first term vanishes because $\mathbf{E}_0^*\hat{\varepsilon}(\omega_0)\mathbf{E}_0$ is a real value. The second term yields

---

[5] This supposition is contradictious, strictly speaking. A medium dispersion is always linked to absorption at certain frequencies. To be more precise, one should consider a transparency band where the imaginary components of $\hat{\varepsilon}$ are very small.

7 Proper Waves in Flows of Charged Particles

$$\frac{\partial U}{\partial t} = \frac{1}{8\pi} \text{Re} \frac{\partial}{\partial t} \mathbf{E}_0^* \left. \frac{\partial \omega \hat{\varepsilon}(\omega)}{\partial \omega} \right|_{\omega_0} \mathbf{E}_0 . \quad (7.43)$$

Calculating the magnetic energy $|\mathbf{B}_0|^2/8\pi$ does not meet such complications as far as the magnetic permeability equals unity. So, the wave energy in a dispersive medium is

$$U = \frac{1}{8\pi} \text{Re} \left[\mathbf{E}_0^* \left. \frac{\partial \omega \hat{\varepsilon}(\omega)}{\partial \omega} \right|_{\omega_0} \mathbf{E}_0 + |\mathbf{B}_0|^2 \right] . \quad (7.44)$$

Components of a plane wave satisfy the relations

$$k_0 \mathbf{B}_0 = k \mathbf{E}_{0\perp} ; \quad \left(\omega_0^2 \hat{\varepsilon} - k^2 c^2 \right) \mathbf{E}_{0\perp} = 0 ,$$

the latter being just the wave equation. Substituting them in (7.44) yields

$$U = \frac{1}{8\pi} \mathbf{E}_{0\perp}^* \left[\frac{\partial}{\partial \omega} \omega \hat{\varepsilon} + \frac{k^2 c^2}{\omega^2} \right]_{\omega_0} \mathbf{E}_{0\perp} \quad (7.45)$$

$$= \frac{1}{8\pi} \mathbf{E}_{0\perp}^* \left[\frac{1}{\omega} \frac{\partial}{\partial \omega} \omega^2 \hat{\varepsilon} \right]_{\omega_0} \mathbf{E}_{0\perp} .$$

In particular, for an isotropic medium characterized by a scalar $\varepsilon(\omega)$ the condition of a negative energy wave existence is

$$\left. \frac{1}{\omega} \frac{\partial}{\partial \omega} \varepsilon \omega^2 \right|_{\omega_0} < 0 . \quad (7.46)$$

For a longitudinal wave

$$|\mathbf{B}_0| = 0 \quad \text{and} \quad \varepsilon_\| = \mathbf{e} \hat{\varepsilon} \mathbf{e} = 0 ,$$

where $\mathbf{e}$ is a unit vector in the propagation direction. Therefore,

$$U = \frac{1}{8\pi} |\mathbf{E}_{0\|}|^2 \left[\frac{\partial}{\partial \omega} \omega \varepsilon_\| \right]_{\omega_0} , \quad (7.47)$$

where ω_0 is the frequency at which $\varepsilon_\| = 0$.

To apply the general theory to an electron beam in an external magnetic field, one can start with the general definition of the electric induction

$$ik_0 \mathbf{D} = ik_0 \mathbf{E} + \frac{4\pi}{c} \mathbf{j} , \quad (7.48)$$

where the current density should be expressed in terms $\mathbf{E}$, using equations of motion. Actually, it has been done in the hydrodynamic description. Using (7.11a) and (7.11b), we change in the latter $k_0 \mathbf{B}_\perp$ for $k [\mathbf{e} \times \mathbf{E}_\perp]$. Taking into account that the vector product with $\mathbf{e}$ is equivalent to the multiplication from the left by the tensor

7.2 Negative Energy Waves

$$\begin{pmatrix} 0 & -1 \\ 1 & 0 \end{pmatrix},$$

yields

$$\hat{\varepsilon}_\perp = 1 + \frac{k_{\rm p}^2 \left(k_0 - \beta k\right)}{k_0^2 \left[k_{\rm c}^2 - \left(k_0 - \beta k\right)^2\right]} \begin{pmatrix} k_0 - \beta k & -ik_{\rm c} \\ ik_{\rm c} & k_0 - \beta k \end{pmatrix}; \quad (7.49)$$

$$\varepsilon_\parallel = 1 - \frac{k_{\rm p}^2}{\gamma^2 \left(k_0 - \beta k\right)^2}. \quad (7.50)$$

For longitudinal waves $k_0 = \beta k \pm k_{\rm p}^*$ the criterion of a negative energy has a simple form:

$$k_{\rm p}^* > \frac{(k_0 - \beta k)^3}{k_0 + \beta k}. \quad (7.51)$$

For small $k_{\rm p}^*$, this inequality can be fulfilled for slow space charge waves only. For the cyclotron waves differentiation with respect to ω the dispersion equation $\left(\omega_0^2 \hat{\varepsilon} - k^2 c^2\right) \mathbf{E}_{0\perp} = 0$ taken into account gives

$$U = \quad (7.52)$$

$$\frac{|E_{0\perp}|^2}{8\pi} \left\{ 1 + \frac{k_{\rm p}^2}{2k_0} \left[\frac{k_0 - \beta k}{k_{\rm c}^2 - (k_0 - \beta k)^2} + k_{\rm c}^2 \frac{k_0 - \beta k + k_{\rm c} \sin \phi}{\left[k_{\rm c}^2 - (k_0 - \beta k)^2\right]^2} \right] \right\}.$$

Here ϕ is the phase difference between the two transverse field components equal to $\phi = \pm \pi/2$, depending on a circular polarization direction. For small $k_{\rm p}$, the energy sign turns out to be negative within a narrow band in vicinity of the cyclotron resonance. Note that it happens for a circularly polarized wave which effective electric field rotates in the same direction as the beam particles.

It is worth to say several words about the mechanism of the energy transfer from the directed motion to the radiation wave. It is almost evident for cyclotron waves from the one-particle theory because the energy of rotation increases if emitting of a slow quantum takes place. For longitudinal space charge waves, it is less clear. According to (7.7), density oscillations in a slow wave ($k_0 < \beta k$) are in antiphase with oscillations of the longitudinal velocity. So, there are a lesser number of particles at the beam regions where the mechanical energy is large. The mechanical energy density related to the wave is proportional to

$$\delta W = \frac{\gamma + \delta \gamma}{2} \left(\varrho \exp\left[i\left(kz - \omega t\right)\right] + \text{c.c.}\right) + \delta \gamma \varrho_0,$$

where the variation $\delta \gamma$ can be expressed via the velocity variation:

$$\delta \gamma = \frac{1}{c} \frac{d\gamma}{d\beta} \left(v_z \exp\left[i\left(kz - \omega t\right)\right] + \text{c.c.}\right) \quad (7.53)$$

$$+ \frac{1}{2c^2} \frac{d\gamma}{d\beta} \left(v_z \exp\left[i\left(kz - \omega t\right)\right] + \text{c.c.}\right)^2 + \cdots$$

Averaging over time abolishes all terms containing exponents. With a precision of the first nonvanishing terms, one get for small amplitudes:

$$\langle \delta W \rangle = \frac{\varrho_0}{c^2} \frac{d^2\gamma}{d\beta^2} |v_z|^2 + \frac{1}{2c} \frac{d\gamma}{d\beta} (\varrho v_z^* + \varrho^* v_z) .$$

Substituting

$$\frac{d\gamma}{d\beta} = \gamma^3 \beta ; \qquad \frac{d^2\gamma}{d\beta^2} = 3\gamma^5 - 2\gamma^3$$

and expressing ϱ via v_z from (7.7), yields

$$\langle \delta W \rangle = \frac{\gamma^3 \varrho_0}{2} \frac{|v_z|^2}{c^2} \left[\gamma^2 (1 + 2\beta^2) + \frac{2k\beta}{k_0 - \beta k} \right] .$$

The first addendum in the square brackets is always positive, but it is the second one that dominates if the density is small. It is positive for a fast wave ($k_0 = k\beta + k_p^*$) and really is negative for a slow one ($k_0 = k\beta - k_p^*$).

Introduction of the notion of the negative–energy waves does not consist just in introducing a new original notation. In fact, it has an important heuristic value. Really, if the negative–energy wave excitation releases a certain energy, any possibility of giving this energy away (e.g., due to the coupling with the partial electromagnetic wave of the purely positive energy) has to cause an avalanche-like process. Both the waves are self–excited at the expense of the energy of the particle longitudinal motion. This has been demonstrated above by particular examples. What is more, the following conclusion logically suggests itself. Let us consider the condition of the electromagnetic energy going away from the system (e.g., this can be conditioned by dissipation of energy in metallic surfaces of a finite conductivity). Hence, there has to take place self–excitation of the slow SCW. A primitive scheme of such an oscillator consists just of a dissipative waveguide and an electron beam, freely propagating along the waveguide axis. In principle, this scheme can operate. Actually speaking, it has been even checked experimentally once upon a time. Unfortunately, the device characteristics (the gain, noise, efficiency, etc.) cannot compete with other schemes. Still the point to be made that, as a rule, the wave self–excitation is an injurious effect in accelerating devices. In this branch, elimination of not very strong but broadband resistive instabilities is rather an actual problem.

7.3 Kinetic Effects

Let us recall now the above–mentioned contradictions between the single–particle approach (anticipating spatial amplification of the electromagnetic wave synchronous with the beam particles under the condition of normal Doppler effect) and stability of the proper hybrid wave at the crossing point of the electromagnetic wave branch and the fast cyclotron wave branch. Explanation of this effect consisted in absence of the transverse free energy that must decrease if the stimulated radiation takes place. We consider below the more adequate model that takes the effect of the microscopic motion into account.

7.3.1 Kinetic Equation

In its essence, the microscopic motion is the particle Larmor gyration around field lines of the external magnetic field. In the steady state, this gyration does not induce charges or currents. To take into account this motion, we must describe the beam by the kinetic equation in the six–dimensional space of coordinates and momenta.[6] The distribution function $\Psi(\mathbf{r}, \mathbf{p}, t)$ in this space obeys the kinetic equation

$$\frac{\partial \Psi}{\partial t} + \mathbf{v} \nabla_r \Psi + \mathbf{F} \nabla_p \Psi = 0 \ . \tag{7.54}$$

Here $\mathbf{v} = \mathbf{p}/m\gamma$ and the Lorentz force is

$$\mathbf{F} = q\mathbf{E} + q\left[\mathbf{v} \times (\mathbf{B} + B_0 \mathbf{e}_z)\right] \ .$$

It should be noted that in contrast to the hydrodynamic approach, now $\mathbf{p}$ is an independent variable (as well as t and $\mathbf{r}$). We normalize the distribution function to the charge density

$$\int \Psi \, d\mathbf{p} = \varrho(\mathbf{r}, t) \ .$$

So, the macroscopic current density may be written as

$$\mathbf{j} = \int \mathbf{v} \Psi \, d\mathbf{p} \ .$$

According to the accepted model, we suppose that in the steady state there are no fields except the homogeneous magnetic field $B_0 \mathbf{e}_z$. An arbitrary function of one–particle integrals of motion, having physical sense, can play the part of the equilibrium distribution function. For the present, we consider the longitudinal momentum $p_\parallel$ and the module of the transverse momentum $p_\perp$, measured in the units of mc, to be these integrals of motion. The point to be made is that the components of the transverse momentum $\mathbf{p}_\perp$ are not integrals in the presence of the external magnetic field.

If there exist weak perturbing electromagnetic fields

---

[6] Strictly speaking, this has to be done in the space of coordinates and canonically conjugated generalized momenta, i.e., in the phase space. However, in the given case of the homogeneous flow, this difference is of no importance.

$$\mathbf{E}, \mathbf{B} \propto \exp\left[i\left(kz - k_0 ct\right)\right]$$

the distribution function is also subjected to deviations from the steady state of the form of a travelling wave:

$$\Psi(\mathbf{r}, \mathbf{p}, t) = \Psi_0(p_\perp, p_\parallel) + \psi(\mathbf{p})\exp\left[i\left(kz - k_0 ct\right)\right].$$

The complex amplitude $\psi(\mathbf{p})$ satisfies the equation

$$-i(k_0 - \beta k)\psi + \frac{k_c}{c}[\mathbf{v} \times \mathbf{e}] \cdot \nabla_{\mathbf{p}} \psi = -\frac{q}{mc^2}\left\{\mathbf{E} + \frac{1}{c}[\mathbf{v} \times \mathbf{B}]\right\} \cdot \nabla_{\mathbf{p}} \Psi_0. \quad (7.55)$$

It is convenient to proceed with further calculations in the cylindrical coordinate system in the momentum space with the polar angle φ in the plane of transverse momenta that the function Ψ_0 is independent of. In these variables:

$$\nabla_{\mathbf{p}} = \frac{\mathbf{p}_\perp}{p_\perp}\left(\frac{\partial}{\partial p_\perp}\right)_{p_\parallel,\varphi} + \frac{[\mathbf{e} \times \mathbf{p}_\perp]}{p_\perp^2}\left(\frac{\partial}{\partial \varphi}\right)_{p_\perp, p_\parallel} + \mathbf{e}\left(\frac{\partial}{\partial p_\parallel}\right)_{p_\perp,\varphi}; \quad (7.56)$$

$$[\mathbf{v} \times \mathbf{e}] \cdot \nabla_{\mathbf{p}} = -\frac{c}{\gamma}\left(\frac{\partial}{\partial \varphi}\right)_{p_\perp, p_\parallel}. \quad (7.57)$$

Then the equation for deviation of the distribution function from the equilibrium state takes the simple form:

$$i(k_0 - \beta k)\psi + k_c \frac{\partial \psi}{\partial \varphi} = \frac{q}{mc^2}\left\{\mathbf{E}\nabla_{\mathbf{p}}\Psi_0 - \frac{\mathbf{v}_\perp}{c}[\mathbf{e} \times \mathbf{B}]\left(\frac{\partial \Psi_0}{\partial p_\parallel}\right)_\gamma\right\}, \quad (7.58)$$

where

$$\left(\frac{\partial \Psi_0}{\partial p_\parallel}\right)_\gamma = \left(\frac{\partial \Psi_0}{\partial p_\parallel}\right)_{p_\perp} - \frac{\beta c}{v_\perp}\left(\frac{\partial \Psi_0}{\partial p_\perp}\right)_{p_\parallel}.$$

To get partial space charge and space current densities $\varrho'(p_\parallel, p_\perp)$ and $\mathbf{j}'(p_\parallel, p_\perp)$ induced by particles with particular $p_\parallel$ and $p_\perp$, we multiply (7.58) by the velocity $\mathbf{v}$ in the corresponding power (the zeroth or the first) an average it over φ. Doing this, one has to take into account that

$$\frac{1}{2\pi}\int \mathbf{v}\frac{\partial \psi}{\partial \varphi}\,d\varphi = -\frac{1}{2\pi}\int \psi[\mathbf{e} \times \mathbf{v}]\,d\varphi;$$

$$\frac{1}{2\pi}\int \mathbf{v}_\perp(\mathbf{v}_\perp \mathbf{A}_\perp)\,d\varphi = \frac{v_\perp^2}{2}\mathbf{A}_\perp,$$

where $\mathbf{A}_\perp$ is an arbitrary transverse constant vector. The terms in (7.58) linear with respect to $\mathbf{v}_\perp$ vanish after averaging. Then

$$\varrho' = -\frac{iqE_\parallel}{mc^2(k_0 - \beta k)}\left(\frac{\partial \Psi_0}{\partial p_\parallel}\right)_{p_\perp} \quad (7.59)$$

and

$$i(k_0 - \beta k)\mathbf{j}' - k_c [\mathbf{e} \times \mathbf{j}'] \qquad (7.60)$$
$$= \frac{q}{mc}\left\{\beta E_\parallel \mathbf{e}\left(\frac{\partial \Psi_0}{\partial p_\parallel}\right)_{p_\perp} + \frac{v_\perp}{2c}\mathbf{E}_\perp\left(\frac{\partial \Psi_0}{\partial p_\perp}\right)_{p_\parallel} - \frac{v_\perp^2}{2c^2}[\mathbf{e}\times\mathbf{B}]\left(\frac{\partial \Psi_0}{\partial p_\parallel}\right)_\gamma\right\}.$$

These equations are of the same structure as hydrodynamic ones but their coefficients depend now on the kinetic variables $p_\perp$ and $p_\parallel$. To find the total current coming in Maxwell equation, the partial current $\mathbf{j}'(p_\parallel, p_\perp)$ should be expressed via electromagnetic fields and integrated over all momenta. This procedure is rather bulky, so we illustrate the importance of kinetic effects in the case of a transversely homogeneous flow. As has been mentioned above, the eigenfunctions in this case belong either to TM or to TE class.

7.3.2 Dispersion Relations

TM (Space Charge) Waves

Let us consider, first, the waves with a longitudinal electric field component, that is, the space charge waves. The wave equation for this component is the most convenient form in this case. For fixed charges and currents

$$\left(k_0^2 \varepsilon - k^2\right) E_\parallel = \frac{4\pi i k}{\varepsilon}\rho - \frac{4\pi i k_0}{c}j_\parallel. \qquad (7.61)$$

The expressions for the space charge and current

$$\varrho = -\frac{iq}{mc^2}E_\parallel \int \left(\frac{\partial \Psi_0}{\partial p_\parallel}\right)_{p_\perp}\frac{d\mathbf{p}}{k_0 - \beta k}; \qquad (7.62)$$

$$j_\parallel = -\frac{iq}{mc}E_\parallel \int \left(\frac{\partial \Psi_0}{\partial p_\parallel}\right)_{p_\perp}\frac{\beta d\mathbf{p}}{k_0 - \beta k} \qquad (7.63)$$

follow directly from (7.59) and (7.60). So, the dispersion relation is

$$k_0^2 \varepsilon - k^2 = -\frac{4\pi q}{mc^2 \varepsilon}\int \left(\frac{\partial \Psi_0}{\partial p_\parallel}\right)_{p_\perp}\frac{k_0 \beta \varepsilon - k}{k_0 - \beta k}d\mathbf{p}. \qquad (7.64)$$

After integration by parts, the kinematic relation

$$\left(\frac{\partial \beta}{\partial p_\parallel}\right)_{p_\perp} = \frac{1-\beta^2}{\gamma}$$

taken into account, the equation can be written as

$$1 - \frac{4\pi q}{mc^2 \varepsilon}\int \Psi_0 \frac{(1-\beta^2)\,d\mathbf{p}}{\gamma(k_0 - \beta k)^2} = 0. \qquad (7.65)$$

Transverse (Cyclotron) Waves

As regards the transverse current waves, it is more convenient to express them via transverse electric field strength. Then (7.60) gives

$$\mathrm{i}\left(k_0 - \beta k\right)\mathbf{j}'_\perp - k_c\left[\mathbf{e} \times \mathbf{j}'_\perp\right] \tag{7.66}$$

$$= \frac{qv_\perp}{2mc^3 k_0}\mathbf{E}_\perp \left\{k_0 c\left(\frac{\partial\Psi_0}{\partial p_\perp}\right)_{p_\parallel} + kv_\perp\left(\frac{\partial\Psi_0}{\partial p_\parallel}\right)_\gamma\right\}$$

with a solution:

$$\mathbf{j}'_\perp = \frac{q}{2mc^3 k_0}\int \frac{\mathrm{i}\left(k_0-\beta k\right)\mathbf{E}_\perp + k_c\left[\mathbf{e}\times\mathbf{E}_\perp\right]}{k_c^2 - \left(k_0-\beta k\right)^2} \tag{7.67}$$

$$\times \left\{k_0 c\left(\frac{\partial\Psi_0}{\partial p_\perp}\right)_{p_\parallel} + kv_\perp\left(\frac{\partial\Psi_0}{\partial p_\parallel}\right)_\gamma\right\}\mathrm{d}\mathbf{p}\,.$$

It is easy to show that the condition of its compatibility with the wave equation for the transverse field

$$\left(k_0^2\varepsilon - k^2\right)\mathbf{E}_\perp = \mathrm{i}\frac{4\pi}{c}k_0\mathbf{j}_\perp$$

is the dispersion equation:

$$\left(k_0^2\varepsilon - k^2\right) = \tag{7.68}$$

$$\frac{2\pi q}{mc^2}\int\left\{k_0 c\left(\frac{\partial\Psi_0}{\partial p_\perp}\right)_{p_\parallel} + kv_\perp\left(\frac{\partial\Psi_0}{\partial p_\parallel}\right)_\gamma\right\}\frac{v_\perp \mathrm{d}\mathbf{p}}{\left(k_0-\beta k\right)\mp k_c}\,. \tag{7.69}$$

The upper (lower) sign here corresponds to the same (opposite) rotation of particles and of the electric field of the wave (for positive k_c).

In a limiting case of a cold beam $\Psi_0 = \varrho_0\delta\left(p_\parallel - \gamma_0\beta\right)\delta\left(p_\perp\right)/p_\perp$, Eqs. (7.65) and (7.68) coincide, correspondingly, with (7.18) and (7.17). Generally, they differ in a special averaging of the resonant denominators that are characteristic for the beam modes. From a viewpoint of physics, this operation means a summation of incomes of quasi-synchronous particles of different detuning.

7.3.3 Kinetic Effects and Landau Damping

Surely, the calculations performed are somewhat cumbersome. This is atoned for the simplicity and the general character of results that predict three main groups of the kinetic effects. We consider them separately using simplified models. Of course, as a rule, all of them can be met together in a real situation.

Cyclotron Resonance

First of all, the results confirm the necessity of the internal energy associated with gyration to be a source of a fast cyclotron wave instability. Actually, let us suppose for simplicity that all particles have the same energy γ_0 and the same longitudinal momentum $\gamma_0 \beta_< \sqrt{\gamma_0^2 - 1}$, and, hence, the same Larmor radii. Although the particles are gyrating, the azimuthal current is zero in the steady state if the axes of rotation are distributed uniformly across the beam. Such a beam cannot be called a hydrodynamical one because at any point all directions of the transverse velocity are presented by the particles multitude.

Taking into account that $p_\perp dp_\perp dp_\parallel = \gamma d\gamma dp_\parallel$, this flow is descried with a distribution function

$$\Psi_0 = \frac{\rho_0}{\gamma_0 p_{\perp 0}} \delta\left(p_\parallel - \gamma_0 \beta\right) \delta\left(\gamma - \gamma_0\right).$$

Integrating (7.68) by parts yields

$$\left(k_0^2 - k^2\right)\left[(k_0 - \beta k) - k_c\right] \qquad (7.70)$$
$$= k_p^2 \left[(k_0 - \beta k) - \frac{v_{\perp 0}^2 \left(k_0^2 - k^2\right)}{2c^2 (k_0 - \beta k - k_c)}\right].$$

The second term in the curly brackets changes essentially the topology of the dispersion branches intersection, however small it is. The algebraic dispersion equation is now cubic rather than quadratic as it is in hydrodynamics:

$$(\mu - \beta\kappa)^2 (\mu - \kappa) = \frac{k_p^2 (1-\beta)}{2} \left[\mu - \beta\kappa - \frac{\beta_{\perp 0}^2}{1-\beta}(\mu - \kappa)\right]. \qquad (7.71)$$

It happens because the partial cyclotron wave is double degenerated what is invisible in the absence of the particles rotation. Actually the branch is double, and two complex conjugated roots appear in the vicinity of the cyclotron resonance (see Fig.7.3).

The imaginary part of the roots is proportional to $I^{1/2}$ being equal to at the middle of the resonance band (i.e., for $\kappa = 0$).

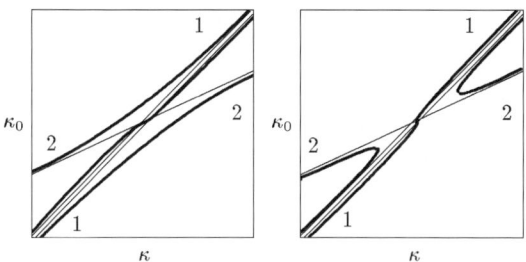

Fig. 7.3. Kinetic interaction with fast cyclotron wave. 1 – electromagnetic wave; 2 – fast cyclotron wave

150   7 Proper Waves in Flows of Charged Particles

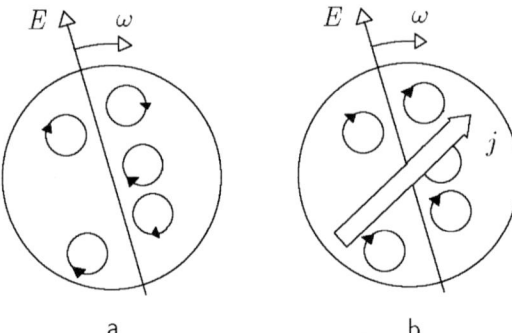

Fig. 7.4. Excitation of a cyclotron wave. Left: noncoherent gyration. Right: cophased gyration. Wide arrow–macroscopic current.

Neglecting the dependence of the longitudinal velocity on the small transverse one, the formal explanation above can be visually supported with Fig. 7.4, where projections of particles trajectories are shown in the beam cross section.

For random Larmor gyration, microscopic currents are compensated at every point of the flow (but at remote boundaries) and the macroscopic current is absent. Under action of a low-intensity plane electromagnetic wave of the proper polarization, one half of the particles turned out to be in phases of acceleration gain energy and decrease their rotational frequency. The other half rotates faster. So for the wave frequency slightly exceeding the initial rotation frequency, the particles begin to group in a decelerating phase; that is, the simulated emission and the spatial amplification take place. This effect perfectly matches the one-particle concept of Part I with the only difference: now the waves of a frequency above the resonance value are amplified.[7]

Two Beams Instability

The relative freedom of choice of the steady state distribution function reveals possibilities of other, sometimes rather exotic, kinetic instabilities. As an example, we consider briefly the interaction of two monoenergetic flows with close longitudinal velocities. For the sake of simplicity, we will neglect the transverse motion.

According to a qualitative dispersion diagram drawn in Fig. 7.5 with the sign of energy of partial space charge waves taken into account, one can foresee a new instability appearance. It takes place near the crossing point of the slow wave of the faster beam and a fast wave of the slower one. Denoting the beams

---

[7] This difference of phase motion in cases of linear and circular motions is well known in the accelerators theory. It is stipulated for increasing of the linear velocity with energy, while the angular velocity in the uniform magnetic field is inversely proportional to the energy.

7.3 Kinetic Effects

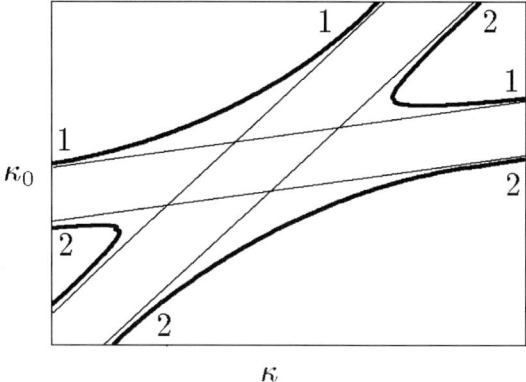

Fig. 7.5. Interaction of partial waves of two parallel flows. 1 – fast space charge waves; 2 – slow space charge waves

velocities as β_1 and β_2 and their plasma frequencies as k_{p1}^* k_{p2}^*, we get at the crossing point

$$k_0 = \frac{k_{p1}^* \beta_2 + k_{p2}^* \beta_1}{\beta_1 - \beta_2}; \qquad k = \frac{k_{p1}^* + k_{p2}^*}{\beta_1 - \beta_2}.$$

One can see that for the velocities difference small enough – a typical case for relativistic beams – the hybrid wave frequency can exceed essentially the plasma frequencies and has a noticeable imaginary part. This two–beam instability plays an important role in particle accelerators and can be of interest for electronics because it does not require a special retarding electromagnetic system.

To calculate its increment, we limit ourselves by beams with equal parameters $k_{p1}^{*2} = k_{p2}^{*2} = k_p^{*2}/2$. Substituting the two–beam distribution function

$$\Psi_0 = \frac{\varrho_0 \delta(p_\perp)}{2p_\perp} \left[\delta\left(p_\| - p_1\right) + \delta\left(p_\| - p_2\right)\right]$$

into the general dispersion relation (7.65) yields

$$1 = \frac{k_p^{*2}}{2} \left[\frac{1}{(k_0 - \beta_1 k)^2} + \frac{1}{(k_0 - \beta_2 k)^2}\right].$$

This equation can be readily reduced to a biquadratic one with a solution

$$\left(k_0 - \frac{\beta_1 + \beta_2}{2} k\right)^2 = \frac{(\beta_1 - \beta_2)^2}{4} k^2 + \frac{k_p^{*2}}{2} \pm \frac{k_p^*}{2} \sqrt{k_p^{*2} + 2k^2 (\beta_1 - \beta_2)^2}.$$

Complex roots appear when $k < 2k_p^*/(\beta_1 - \beta_2)$ and the right-hand side becomes negative. The maximal value of the increment

$$(\operatorname{Im} k_0)_{\max} = k_p^*/2^{3/2}$$

is achieved at $k = \sqrt{3}k_p^*/\sqrt{2}(\beta_1 - \beta_2)$.

A specific feature of this instability is a wide range of frequencies and wavenumbers. That poorly matches the task of generation of a "clean" monochromatic signal with low-level noises. Instead, the instability readily creates obstacles to transportation of many-component beams. By the way, its pedigree starts with the well-known beam–plasma instability [43], if one treats the plasma as the second beam of zero velocity.

Proper Waves in a Kinetic Beam

We consider now the influence of "smeared" resonant denominators in the dispersion relations on proper waves characteristics. First of all, one can readily foresee that the hidden intrinsic energy of microscopic motion has to influence the electron beam inertial properties. At least, the plasma frequency entering almost all formulas above is to suffer these alterations. Supposing that the longitudinal velocity spread around the hydrodynamical value β_0 is small, the integral in the longitudinal dispersion relation (7.64) can be presented as

$$\int \left(\frac{\partial \Psi_0}{\partial p_{\|}}\right)_{p_\perp} \frac{d\mathbf{p}}{k_0 - \beta k} \tag{7.72}$$

$$\approx \int \left(\frac{\partial \Psi_0}{\partial p_{\|}}\right)_{p_\perp} \left[\frac{1}{k_0 - \beta_0 k} + \frac{k(\beta - \beta_0)}{(k_0 - \beta_0 k)^2} + \cdots\right] d\mathbf{p}.$$

The first addendum vanishes after integration over $p_{\|}$ while the second one yields

$$\int \left(\frac{\partial \Psi_0}{\partial p_{\|}}\right)_{p_\perp} \frac{d\mathbf{p}}{k_0 - \beta k} \approx -\frac{k \varrho_0}{(k_0 - \beta_0 k)^2} \left\langle \left(\frac{\partial \beta}{\partial p_{\|}}\right)_{p_\perp}\right\rangle \tag{7.73}$$

$$= -\frac{k \varrho_0}{(k_0 - \beta_0 k)^2} \left\langle \frac{1 - \beta^2}{\gamma}\right\rangle. \tag{7.74}$$

So, the dispersion relation for the space charge waves keeps the former structure

$$k_0 = \beta_0 k \pm k_p^*, \tag{7.75}$$

but the parameter k_p^{*2} depends on the transverse momentum distribution. It can vary within rather wide limits: from $4\pi q \varrho_0 \langle \gamma^{-3} \rangle/m$ for a cold beam up to $4\pi q \varrho_0 \langle \gamma^{-1} \rangle/m$ for a beam of a very low longitudinal velocity rotating in a magnetic field. The higher terms of the expansion over powers of $\beta - \beta_0$ also provide a certain dependence $k_p^*(k)$, well known for a warm plasma [15].

The criterion of the spread smallness is worth of additional remarks. It cannot be reduced to the inequality

$$\left\langle (\beta - \beta_0)^2 \right\rangle k^2 \ll k_{\mathrm{p}}^{*2}$$

following from (7.72). The rather dubious operation of the resonant denominator expansion over powers of $\beta - \beta_0$ has a direct physical meaning if the denominator itself is nonzero in all points of the beam phase space. In other words, the beam does not contain particles which could be in resonance with plasma waves; otherwise, the integrand has a pole on the real axis. Hence, the results above are valid only for "chopped off" distributions. Moreover, the criterion of smallness can be established only when the dispersion equation is solved. The correct method of taking the quasi–resonant particles into account is in the basis of the so-called Landau damping, which is an important kinetic effect essentially influencing the proper waves characteristics.

Landau Damping

There is a mathematical slipshod in the calculations above. The Fourier transform of the distribution function $\psi(k_0, k)$ is defined in the upper half-plane of the complex variable k_0. In fact, if $\mathrm{Im}\, k_0 < 0$ the inverse Fourier transformation

$$\psi(t,k) \propto \int \psi(k_0,k) \exp(-ik_0 t)\, dk_0$$

does not determine the perturbation for $t \to -\infty$, which must be negligible for the causality reasons. In other words, the integrals in the expressions for macroscopic currents should be calculated under condition of $\mathrm{Im}\, k_0 > 0$ and then they must be analytically extended to the lower half-plane. The best way to accomplish the program is putting $\mathrm{Im}\, k_0 = +0$ and integrating over the longitudinal momentum in the complex plane below the real axis. This method actually implies an infinitesimal dissipation in the system and takes into account the causality principle. The latter, in our case, coincides with the radiation principle, telling that the radiation field is leaving a source and no incident waves exist. As a matter of fact we have already used the method in Part I when calculating the temporal structure of a radiation pulse from a single particle. Note that in the hydrodynamic approximation these details were not necessary because the beam conductivity is defined everywhere in the complex plane of k_0.

Following this way, the dispersion integral for longitudinal waves can be presented everywhere as

$$\int \left(\frac{\partial \Psi_0}{\partial p_\parallel}\right)_{p_\perp} \frac{d\mathbf{p}}{k_0 - \beta k} \qquad (7.76)$$

$$= \mathrm{P.V.} \int \left(\frac{\partial \Psi_0}{\partial p_\parallel}\right)_{p_\perp} \frac{d\mathbf{p}}{k_0 - \beta k} - i\frac{\pi}{k} \int \left(\frac{\partial \Psi_0}{\partial \beta}\right)_{\substack{p_\perp \\ \beta = k_0/k}} d\mathbf{p}_\perp,$$

where P.V. stands for a principal value.

154   7 Proper Waves in Flows of Charged Particles

The appearance of the imaginary unit in the dispersion equation indicates, first, that its roots are complex and, second that they are not complex conjugated. It is worth to say here that the effect is directly linked with a derivative of the distribution function at the point of synchronism. That is, this is a difference in a number of particles slightly slower and slightly faster than the space charge wave that plays the important role. Supposing that the corresponding kinetic correction is small and using again the expansion (7.72) to calculate the integral principal value we get

$$k_0 = \beta_0 k \pm k_\mathrm{p}^* \left[1 + \mathrm{i} \frac{2\pi^2 q}{mc^2 k^2} \int \left(\frac{\partial \Psi_0}{\partial \beta} \right)_{\substack{p_\perp \\ \beta = k_0/k}} \mathrm{d}\mathbf{p}_\perp \right]. \qquad (7.77)$$

Note that for plain distributions with a single maximum at $\beta = \beta_0$ the derivative $(\partial \Psi_0 / \partial \beta)_{\beta > \beta_0}$ is negative so that fast quasi-resonant waves turn out damping ones: $(\mathrm{Im}\, k_0 < 0)$. This result is in accordance with the concept of particles interaction with faster waves developed above. Really, in the case of a fast SCW the main part of particles is comparatively slow, so they absorb the wave. The case of a slow wave seems slightly paradoxical from this viewpoint. The change of the sign of the square brackets changes the sign of the mentioned derivative, so the slow wave should damp as well according to (7.77). However, there is no contradiction: the main part of resonant particles does supply energy to the wave but the latter does damp being the wave of a negative energy type. Our arguments above related to positive energy waves so that an increase, in energy and in amplitude were just synonyms.

The proper waves damping under consideration takes place in the absence of energy dissipation in the system. It has been predicted by L. Landau [44] regarding waves in a thermally equilibrium plasma. One should keep in mind, however, that real beams are rather far from the equilibrium and the particles velocity distribution is determined mainly by initial conditions. Even keeping the term temperature for a corresponding distribution moment the latter, as a rule, turns out sharply anisotropic, and the distribution itself is very far from Maxwellian. Therefore, the choice of the proper distribution function for Landau damping evaluation is a rather ambiguous problem.

It is worth to note once more that Landau damping is determined by the slope of the distribution function at the point where the particles velocity equals the wave phase velocity. So, its decrement depends on the wavenumber and vanishes at flat domains of the distribution function independently of its width.

Let us consider, as an example, a beam of particles that are uniformly distributed over longitudinal momentums within the limits $p_0 \pm \Delta/2$ and have no transverse velocities. The direct calculation of the integral over momentums yields the dispersion relation:

$$k_0 = k \frac{\beta_+ + \beta_-}{2} \pm \sqrt{\frac{(\beta_+ - \beta_-)^2}{4} k^2 + k_\mathrm{p}^{*2}}, \qquad (7.78)$$

where $\beta_\pm$ correspond to $p_0 \pm \Delta/2$ and $\beta_+ - \beta_- \approx \Delta/\gamma_0^3$. It is easy to see that (7.78) does not make a hint about Landau damping. However, for $k_0 = k\left(\beta_0 \pm \Delta/2\gamma^3\beta_0\right)$ the integral diverges. This means that Landau damping is strong but influences at very narrow frequency band where the phase velocities are approximately equal to $\beta_\pm$.

So far as the particles distribution over velocities is poorly known (unlike the equilibrium plasma case) further theoretical modelling would be of a very limited interest. For estimations one may put

$$\frac{\partial \Psi_0}{\partial \beta} \approx \beta_0 \gamma_0^3 \frac{p_0}{\Delta}$$

to get for the decrement:

$$\mathrm{Im}\, k_0 \approx \frac{k_\mathrm{p}^3 \beta_0}{2k^2 \Delta}.$$

Of course, this is valid only for

$$\frac{\Delta}{p_0} \ll \frac{k_\mathrm{p}^2}{k^2}.$$

As it has already been noted, Landau damping is related to some redistribution of energy via the wave field rather than to energy dissipation. Its influence on the instabilities considered above is easily foreseen. Roughly speaking, they can develop if the hydrodynamic increment exceeds the Landau decrement. So, Landau damping is of a principal importance, determining threshold currents even in ideal systems.

Landau damping influence on fast cyclotron waves is slightly more complicated. If the distribution over transverse momentums decreases monotonically (as thermally equilibrium distribution does), Landau damping is positive. However, if a group of particles has an externally or initially excited transverse momentum, the energy distribution turns out to be inverse. Then the waves with phase velocities belonging to the domain of $\partial \Psi_0/\partial p_\perp$ experience negative Landau damping, that is, are kinetically unstable. One can easily see that this effect is just a generalization of the cyclotron instability considered above. As regards negative energy slow cyclotron waves, Landau damping can cause a threshold of their instability.

The detailed discussion of the Landau damping physics and of its possible manifestations in various systems can be found in the excellent survey [45].

Part III

Certain Modern Applications

8
Cherenkov Radiation in Beam–Plasma Systems

In this chapter, we will consider those processes of waves amplification that are conditioned by the elementary mechanism of the Cherenkov radiation emission. To distinguish this effect from others, we assume that the external magnetic field strength is very high. In this case, the cyclotron frequency of the electron rotation around the magnetic field lines essentially exceeds all characteristic frequencies emitted by the system in question. So, motion of the electrons can be treated as one-dimensional and directed along the magnetic field lines.

As it has been mentioned, the induced emission of Cherenkov radiation made a real basis for the first devices with distributed interaction of TWT or BWT type, in spite of their actual invention and development being originated from single–particle considerations. The literature on the subject is enormous and is evidently out of the scope of this book. However, the development of microwave electronics tends at present toward applications of relativistic electron beams (REB) and requires a more general approach.

Here we briefly dwell on several advantages of applying REB in Cherenkov oscillators and amplifiers. First, during the beam energy transfer to the wave, the energy of the beam particles decreases essentially. At the same time, the velocity of relativistic particles is subjected just to insignificant changes. This permits preserving the Cherenkov synchronism between the wave and the beam particles for a much longer time interval. Therefore, the effectiveness of the beam energy transfer into the wave energy becomes essentially higher. Second, for obtaining synchronism between the REB particles and the wave, the wave phase velocity must be close to the velocity of light. As it is known, it is considerably easier to transform such waves into free–space waves, that is, to radiation.

To increase generated power, larger beam currents are, of course, required. The limitations imposed by space charge effects could be removed by filling the electrodynamic structure with a plasma. As a rule, the plasma density should be chosen so that it would not considerably change the spatial structure of the field interacting with the beam. At the same time, the plasma has to be dense enough to neutralize the charge and current of the beam.

The cases when the plasma itself plays the part of a retarding electrodynamic structure are of a special interest. Under such conditions, the electron beam interacts with the plasma proper waves. Their excitation can be treated as the plasma–beam instability that for the first time was predicted in [40, 43]. At the same time it can be considered as the induced emission of plasmons [13]. Publication of these papers attracted interest to the plasma–beam systems and the corresponding new branch of plasma electronics is developing nowadays.

Taking this into account, we consider in this chapter only the process of beam–plasma interaction. On the one hand, this permits to avoid cumbersome mathematics involved in the periodic retarding systems theory (see Part I) and not related directly to the questions under consideration. On the other hand, the literature on plasma electronics is rather limited.

Cherenkov radiation of a single particle in cold plasma has been considered in Part I. We determine below increments of the corresponding collective radiation instability and their dependence on plasma and beam parameters.

8.1 Dispersion Equation

We will describe dynamics of the beam and plasma electrons with the help of their distribution function $f(\mathbf{r}, \mathbf{p}, t)$. This function satisfies Vlasov kinetic equation:

$$\frac{\partial f}{\partial t} + \mathbf{v}\frac{\partial f}{\partial \mathbf{r}} + q\left\{\mathbf{E} + \frac{1}{c}[\mathbf{v} \times \mathbf{B}]\right\}\frac{\partial f}{\partial \mathbf{p}} = 0. \tag{8.1}$$

The fields are described by Maxwell's equations:

$$\text{rot}\mathbf{E} = -\frac{1}{c}\frac{\partial \mathbf{B}}{\partial t}; \quad \text{rot}\mathbf{B} = \frac{1}{c}\frac{\partial \mathbf{E}}{\partial t} + \frac{4\pi}{c}\mathbf{j}; \tag{8.2}$$

$$\text{div}\mathbf{E} = 4\pi\varrho; \quad \text{div}\mathbf{B} = 0. \tag{8.3}$$

Here ϱ is the total charge density of the beam and plasma electrons; $\mathbf{j}$ is their current density:

$$\varrho = \int f \, d\mathbf{p}; \quad \mathbf{j} = \int f\mathbf{v} \, d\mathbf{p}.$$

The system (8.1)–(8.3) describes both the linear and nonlinear stages in the beam instability development. Below in this section we will give analysis only to the linear stage.

In the linear approximation, making use of the system (8.1)–(8.3), one can derive the general dispersion relation for the Fourier components of the perturbations. This expression determines a relation of frequencies of the proper waves to their wave vectors. However, it can hardly be analyzed in a general form. Below we will investigate the simplest particular cases only. When

8.1 Dispersion Equation

choosing them, we aim at simplifying mathematical formalism as much as possible. At the same time, the essence of the basic physical processes must not be misrepresented. All basic physical pecularities studied with the help of these simplest models also manifest themselves in more general cases.

Thus, we proceed from the following assumptions. First of all, we will investigate dynamics of the Cherenkov instability of a relativistic electron beam in uniform cold plasma limiting ourselves by plane waves of a small amplitude characterized by a longitudinal wave number k. The undisturbed beam of density n_b is supposed to be compensated with respect to the charge and current.

Under these conditions, harmonic field perturbations can be described by the wave equation for the longitudinal electric field component E following from (8.2) and (8.3). The continuity equation taken into account can be written as

$$E\left(k_0^2 - k^2 - \kappa_\perp^2\right) = 4\pi i \varrho \frac{k^2 - k_0^2}{k}, \tag{8.4}$$

where ϱ is a space charge density determined by the perturbation of the distribution function $\tilde{f}(k_0, k, p)$:

$$\varrho = \int \tilde{f} \mathrm{d}p.$$

The "transverse" wavenumber $\kappa_\perp$ determines a direction of the wave propagation in the case of transversely uniform plasma or a proper mode if boundary conditions exist. In the last case the values of $\kappa_\perp$ are discreet and all perturbations are to be treated as amplitudes of the corresponding membrane functions.

The distribution function perturbation obeys the kinetic equation and is related to the longitudinal field only:

$$(\omega - kv)\tilde{f} = iqE\frac{\partial f_0}{\partial p}. \tag{8.5}$$

It gives the following dispersion relation:

$$k_0^2 - k^2 - \kappa_\perp^2 = -\frac{4\pi q}{k}\left(k_0^2 - k^2\right)\int \frac{\partial f_0/\partial p}{\omega - kv}\mathrm{d}p. \tag{8.6}$$

The distribution of the electrons over momenta has two narrow maxima. The first one is situated at $p = 0$ and corresponds to electrons of the cold plasma of density n_p. The corresponding integral in (8.6) is equal to $qn_b/\omega^2 m$. The second maximum (related to the beam) is in the vicinity of $m\gamma\beta c$. Neglecting for a while the width of these maxima (cold beam and plasma) and taking into account that $\mathrm{d}v/\mathrm{d}p = (m\gamma^3)^{-1}$, we get the dispersion relation in the form:

$$\kappa_\perp^2 + (k_0^2 - k^2)\left[-1 + \frac{k_p^2}{k_0^2} + \frac{k_b^{*2}}{(k_0 - \beta k)^2}\right] = 0. \tag{8.7}$$

Here

$$k_{\rm p}^2 = \frac{4\pi n_{\rm p} q^2}{m} \quad \text{and} \quad k_{\rm b}^{*2} = \frac{4\pi \varrho_0 q}{m\gamma^3}$$

are the squares of Langmuir frequencies of the plasma and of the beam, respectively.

8.2 Cold Beam Instability

Solving concrete problems, one has to add initial and boundary conditions. So, for a particular problem the functions $k_0(k)$ and/or $k(k_0)$ can be of importance. In the first case we shall look for time dependence of perturbations initially distributed in the interaction space (absolute instability). The second case corresponds to spatial amplification of a fixed frequency signal entering the system. The latter is more adapted to the microwave amplification problem. Nevertheless, we start below with the first case typical for problems of temporal stability and self-excitation of oscillations.

8.2.1 Absolute Instability

In accordance with the scheme of the paragraph under Sect. 7.1.4 the dispersion equation is to be written in the form:

$$\left(k_0^2 - k_+^2\right)\left(k_0^2 - k_-^2\right)\left(k_0 - k\beta\right)^2 = \left(k_0^2 - k^2\right) k_{\rm b}^2 k_0^2, \tag{8.8}$$

where the right-hand side is proportional to the beam density. In what following it will be considered as a small parameter:

$$k_{\rm b}^{*2}/k^2 \ll 1 \, .$$

This is justified for the majority of practical problems. In the nonrelativistic case, the inequality above is equivalent to the smallness of the so-called Pierce parameter:

$$\left(\frac{4\pi q \varrho_0}{mk^2\beta^2}\right)^{1/2} \ll 1.$$

However, for relativistic beams this parameter can reach a large value of order of γ^2 still preserving smallness of the coupling coefficient.

The squares of the partial frequencies in (8.8) are equal to

$$k_{\pm}^2(k) = 1/2 \left[\kappa_\perp^2 + k_{\rm p}^2 + k^2 \pm \sqrt{\left(\kappa_\perp^2 + k_{\rm p}^2 + k^2\right)^2 - 4k^2 k_{\rm p}^2} \right]. \tag{8.9}$$

They represent two branches of partial plasma waves with frequencies that are larger than $\sqrt{\kappa_\perp^2 + k_{\rm p}^2}$ and smaller than $k_{\rm p}$ correspondingly. This is the

8.2 Cold Beam Instability

second (slow) wave, which provides induced Cherenkov radiation meeting the space charge wave $k_0 = k\beta$ twice degenerated for $k_b^* = 0$.

Now we have to find the Cherenkov resonance point (k_{0s}, k_s) defined by the equality of the wave and the particles velocities. Solving

$$k_-^2(k_s) = k_s^2 \beta^2$$

for k_s yields

$$k_s^2 = \frac{k_p^2}{\beta^2} - \kappa_\perp^2 \gamma^2; \quad k_{0s}^2 = k_-^2(k_s) = k_p^2 - \kappa_\perp^2 \gamma^2 \beta^2; \quad k_+^2(k_s) = \frac{k_p^2}{\beta^2}. \quad (8.10)$$

So, the local dispersion equation taking into account the three waves interaction can be written in the form:

$$\kappa_0^3 + \kappa^2 \delta + \kappa_0 2\gamma^2 (2\beta^2 - 1) \frac{K^3}{k_p} \sqrt{\frac{1 - \beta \beta_g}{\beta - \beta_g}} = K^3. \quad (8.11)$$

Here

$$\kappa_0 = k_0 - k\beta; \quad K^3 = \frac{k_p k_b^{*2}}{2} \frac{(1 - \beta_g/\beta)^{3/2}}{(1 - \beta \beta_g)^{1/2}}$$

and the partial wave parameters are expressed in terms of its group velocity at the crossing point:

$$\beta_g = \frac{dk_-}{dk} = \frac{\kappa_\perp^2 \gamma^4 \beta^3}{k_p^2 + \kappa_\perp^2 \gamma^4 \beta^4}.$$

Note that in spite of the declared smallness of κ_0/k_s and $(k - k_s)/k_s$ one has to keep the third term at the l.h.s of (8.11) because of the potentially large factor γ^2. The value

$$\delta = (\beta - \beta_g)(k - k_s) + \frac{k_b^{*2} \gamma^2 (6\beta^2 - 1)}{2k_p} \sqrt{(1 - \beta_g/\beta)(1 - \beta \beta_g)} \quad (8.12)$$

related to the deviation of k from the resonance will be referred as detuning. By the way, the expressions above show immediately that the instability cannot develop for $\kappa_\perp^2 > k_p^2/\gamma^2 \beta^2$ when the value $k_-^2(k_s)$ is negative. This inequality corresponds to plasma waves propagating at large angles $> \gamma^{-1}$ or to high transverse modes of a plasma-filled waveguide. For a fixed plasma frequency, their phase velocity along z exceeds that of light making the Cherenkov resonance impossible.

The cubic algebraic equation (8.11) can be solved immediately, but the solution containing several independent parameters still is rather nondescriptive. Instead, we shall consider two characteristics of the main interest – a threshold of the instability and optimizing detuning which corresponds to the maximal increment, that is, to the maximal value of $\mathrm{Im}\,\kappa_0$. To do this, we introduce normalized variables

$$I = \operatorname{Im} \kappa_0/K \ ; \qquad R = \operatorname{Re} \kappa_0/K$$

and normalized detuning $\Delta = \delta/K$. Supposing that $I \neq 0$ and separating the real and imaginary parts of (8.11), one gets the system

$$R^3 - 3RI^2 + \Delta\left(R^2 - I^2\right) + GR - 1 = 0, \qquad (8.13)$$
$$3R^2 - I^2 + 2\Delta R + G = 0 \qquad (8.14)$$

with

$$G = 2\gamma^2 \left(2\beta^2 - 1\right)\sqrt{\frac{1-\beta\beta_g}{\beta-\beta_g}}\frac{K}{k_p} = \frac{\gamma^2\left(2\beta^2-1\right)}{\beta^{1/2}\left(1-\beta\beta_g\right)^{1/3}}\left(\frac{2k_b^*}{k_p}\right)^{2/3}.$$

To find the maximal value of detuning $\Delta_{\max}$ corresponding to instability, one has to put here $I \to 0$ yielding the parametric dependence $\Delta_{\max}(G)$:

$$\Delta_{\max} = -2R - \frac{1}{R^2}, \qquad G = R^2 + \frac{2}{R}. \qquad (8.15)$$

The real roots of the dispersion equation (8.11) for moderate values of G are displayed in Fig. 8.1.

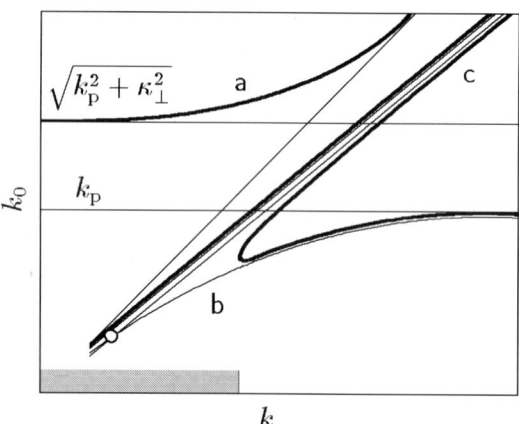

Fig. 8.1. Real roots of the dispersion equation. Shading shows the region of absolute instability. (**a**) Fast plasma wave; (**b**) slow plasma wave, and (**c**) space charge waves. A circle indicates the crossing point

For $\Delta < \Delta_{\max}$ there are two complex conjugated roots.[1] Note that for the particular case of uniform plasma, the instability takes place for all long waves, but its increment depends essentially on detuning. To find the maximizing

---

[1] For $G > 3$, an additional band of stability appears at $\Delta < -3$.

value of the latter, one has to differentiate (8.13) and (8.14) with respect to Δ, to put $dI/d\Delta = 0$, and to exclude $dR/d\Delta$. Then the third equation

$$\Delta = -\frac{RG}{I^2 + R^2} \tag{8.16}$$

is to be added to (8.13) and (8.14) to define the optimizing value Δ_{opt}, the maximal increment I_{max}, and the corresponding value of R, if necessary. Below we shall investigate only the limiting cases of small and large G.

Low-Intensity Regime

For moderately relativistic particles and low beam intensity, the parameter G is small and the boundary value can be presented as

$$\Delta_{\text{max}} \approx \frac{3}{2^{2/3}}. \tag{8.17}$$

The maximal increment is reached for $\Delta \to 0$:

$$I_{\text{max}}^2 \approx \frac{3}{4} \quad \text{or} \quad \text{Im}\,\kappa_0 = -\frac{\sqrt{3}}{2}K + \cdots. \tag{8.18}$$

Note that it is proportional to the cubic root of the beam current what is typical for low intensity traveling wave tubes. By the way, in the theory of free electron lasers (see 10.), an analogous approximation for some reasons is called a Compton regime, although the name does not correspond to the case under consideration. The notion of a "single particle instability" used in [13] also can hardly be applied to a description of the collective process.

The dependence of the increment on detuning can be easily found using Cardan formula. For $G = 0$:

$$I = \frac{\sqrt{3}}{2}\left[\left(\frac{1}{2} - \sqrt{\frac{1}{4} - \left(\frac{\Delta}{3}\right)^3}\right)^{2/3} - \left(\frac{1}{2} + \sqrt{\frac{1}{4} - \left(\frac{\Delta}{3}\right)^3}\right)^{2/3}\right]. \tag{8.19}$$

This dependence for fixed K is presented in Fig. 8.2. It shows a peak of induced Cherenkov radiation on the background of a long tail of low frequency waves. The latter is due to the negative electric permeability for $k_0 < k_{\text{p}}$, which locks the excited field inside plasma. The beam modulation comes from the mutual electrostatic attraction of charges of the same sign in such a medium, but the process can hardly be called radiation. Anyway, the corresponding increment is small and does not play an essential role.

High-Intensity Regime

The opposite case of $G \gg 1$ can be met with relativistic beams in spite of declared smallness of their relative density. In this limit

166   8 Cherenkov Radiation in Beam–Plasma Systems

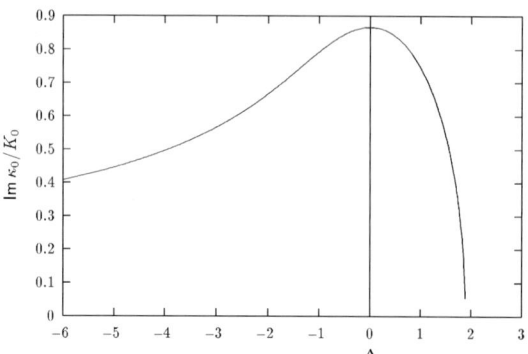

Fig. 8.2. Increment vs. detuning for small G

$$\Delta_{\max} \approx 2G^{1/2}; \quad \Delta_{\text{opt}} \approx G{-}1; \quad I_{\max} \approx -G^{1/2} \qquad (8.20)$$

yielding

$$\operatorname{Im}\kappa_0 = G^{1/2}K \propto k_{\mathrm{b}}^{*}.$$

Note that now the maximal increment is proportional to the square root of the beam current and is achieved at zero detuning. The dependencies of the maximal and optimal detunings and of the maximal increment on the parameter G are shown in Fig. 8.3 covering both cases above.

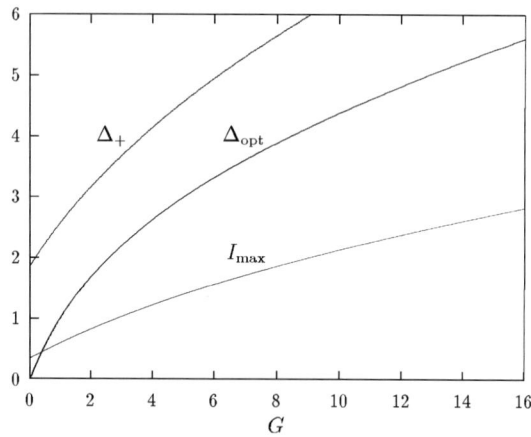

Fig. 8.3. Maximal and optimal detuning and maximal increment vs. parameter G

By the way, the case under consideration is usually identified as Raman regime. Really, the e-fold time of the low-current instability decreases as $k_{\mathrm{b}}^{*-2/3}$ while the period of a beam plasmon goes as k_{b}^{*-2}. So, for intensities large enough, the e-fold time becomes sufficient for excitation of beam proper oscillations, exactly as it happens in the case of Raman scattering. Naturally, this influences the process of plasma waves radiation.

8.2.2 Convective Type Instability

Now we consider the inverse dependence $k(k_0)$, bearing in mind applications to amplifiers when the input frequency is fixed. The imaginary part of this dependence determines the spatial growth of the input signal along the beam.

The crossing point, of course, remains the same as in (8.10):

$$k_{0s}^2 = k_p^2 - \kappa_\perp^2 \gamma^2 \beta^2 = k_p^2 \frac{1 - \beta_g/\beta}{1 - \beta\beta_g} \; ; \quad k_s = k_{0s}/\beta \tag{8.21}$$

but now one has to expand (8.7) up to the third power of $\kappa = k - k_0/\beta$, keeping the first nonvanishing terms in the expansion coefficients for small $\kappa_0 = k_0 - k_{0s}$. This procedure leads to the three-wave dispersion equation in the form:

$$\left(\frac{\kappa}{K_0}\right)^3 - \left(\frac{\kappa}{K_0}\right)^2 \Delta_0 + G_0 \frac{\kappa}{K_0} = -1 \tag{8.22}$$

with

$$K_0^3 = \frac{k_b^{*2} k_p (1 - \beta/\beta_g)^{3/2}}{2(1 - \beta\beta_g)^{1/2} \beta_g} \; ; \quad G_0 = \frac{\gamma^2}{\beta}\left(\frac{4k_b^{*2}(1 - \beta\beta_g)}{k_p^2 \beta_g}\right)^{1/3}$$

and

$$\Delta_0 = \beta\left(1 - \frac{\beta_g}{\beta}\right)\frac{\kappa_0}{K_0} - \gamma^2\beta\left(\frac{k_b^{*2}(1 - \beta\beta_g)}{2k_p^2 \beta_g}\right)^{2/3}.$$

The corresponding equations for real variables $R = \operatorname{Re} \kappa/K_0$ and $I = \operatorname{Im} \kappa/K_0$

$$R^3 - 3RI^2 - \Delta_0\left(R^2 - I^2\right) + G_0 R + 1 = 0, \tag{8.23}$$

$$3R^2 - I^2 - 2\Delta R + G_0 = 0 \tag{8.24}$$

are to be completed with the third equation

$$\Delta_0 = \frac{RG_0}{I^2 + R^2} \tag{8.25}$$

for calculation of $\Delta_{0\mathrm{opt}}$ and $I_{\max}$. The real roots of (8.22) are displayed in Fig. 8.4.

In spite of the different signs of the coefficients, these equations in the limits of $G_0 \ll 1$ and of $G_0 \gg 1$ give the same functional dependencies as for the previous case:

$$\Delta_{0\max} = \frac{3}{2^{2/3}}, \quad I_{\max} = \frac{3}{4}, \quad \Delta_{0\mathrm{opt}} \to 0 \tag{8.26}$$

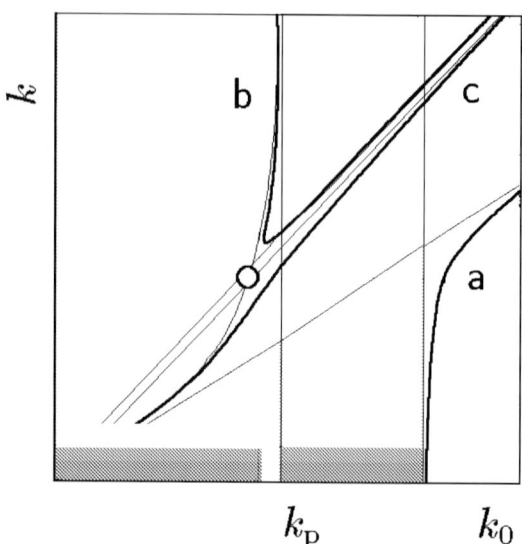

Fig. 8.4. Real roots of the dispersion equation. Shading shows the regions of convective instability. (**a**) Fast plasma wave; (**b**) slow plasma wave, and (**c**) space charge waves. A circle indicates the crossing point

for $G_0 \ll 1$ and

$$\Delta_{0\mathrm{max}} = 2G_0^{1/2}, \quad I_{\mathrm{max}} = -G_0, \quad \Delta_{0\mathrm{opt}} = G_0^{-1} \qquad (8.27)$$

for $G_0 \gg 1$. So, Fig. 8.3 qualitatively illustrates the convective instability as well.

An additional remark should be made concerning two bands of convective instability shown in Fig. 8.4. The left one is, of course, originated by induced Cherenkov radiation combined with electrostatic attraction mentioned above. The second one corresponds to the frequency stop-band for $k_\mathrm{p} < k_0 < \sqrt{k_\mathrm{p}^2 + \kappa_\perp^2}$. Really, if the beam were absent, the input signal at that frequency would be locked near the point of excitation and would not penetrate plasma. The exponentially growing solution vanishes then because of the boundary conditions. The modulated beam transports the signal into plasma, and amplification does take place. However, the excited plasma oscillations cannot propagate and remain in the vicinity of the beam. So, real radiation may exist only in warm plasma with a nonzero group velocity. As noted in [13], the corresponding increment is low and the instability is strongly limited by nonlinear effects.

8.3 Warm Beam Instability

A well-known weak point of microwave devices based on Cherenkov interaction is a comparatively high sensitivity to deviations of particle velocities from the designed value. Really, particles of different longitudinal velocities interact with different waves spreading the radiation spectrum and, hence, decreasing the gain. For evaluation of this effect, we consider below the instabilities of a "warm" beam supposing, of course, that the thermal velocity spread is much smaller than the velocity itself.

Suppose that the beam particles velocity distribution is a Maxwellian one with a small dispersion[2] $c^2 \beta_T^2$:

$$f_0 = \frac{\varrho_0}{\sqrt{2\pi} c \beta_T m \gamma^3} \exp\left(-\frac{(v - \beta c)^2}{2 c^2 \beta_T^2}\right), \qquad (8.28)$$

where the factor $m\gamma^3$ comes from the kinematic relation $dp = m\gamma^3 dv$. If $\beta_T \to 0$, (8.28) takes the form:

$$f_0 = \frac{\varrho_0}{m\gamma^3} \delta(v - \beta c),$$

which corresponds to the cold beam approximation.

After substituting the undisturbed distribution (8.28) function into (8.6), the dispersion relation again can be presented as (8.7) with the only change of $(k_0 - k\beta)^{-2}$ for

$$J = \frac{1}{\sqrt{2\pi} c \beta_T} \int_C \exp\left[-\frac{(v - \beta c)^2}{2 c^2 \beta_T^2}\right] \frac{dv}{(k_0 - kv/c)^2}. \qquad (8.29)$$

According to the general rule, the integral is to be taken in the v-plane along a contour C passing from $-\infty$ to $+\infty$ below the pole $v = ck_0/k$ on the real axis. It can be expressed [43] in terms of a probability integral (Kramp's function) of an imaginary argument. Unfortunately, such representation is not very descriptive and can be analytically traced just in the limiting cases discussed below.

Changing the variable $v = \beta c + \sqrt{2} c \beta_T \delta_T x$ transforms (8.29) to

$$J = \frac{1}{2\sqrt{\pi} \delta_T \beta_T^2 k^2} \int_C \frac{\exp\left(-\delta_T^2 x^2\right) dx}{(1 - x)^2} \qquad (8.30)$$

with a dimensionless detuning

$$\delta_T = \frac{k_0 - \beta k}{\sqrt{2} \beta_T k}.$$

---

[2] Particle velocity cannot exceed c, so the velocity spread is to be much smaller than $c\beta\gamma^{-2}$.

For $\delta_T \gg 1$ the integral can be estimated using the saddle point method. The standard procedure [46] yields the following asymptotic series:

$$J \asymp (k_0 - k\beta)^{-2} \times \sum_{n=0}^{\infty} \frac{(2n+1)!}{2^{2n} n! \delta_T^n}. \tag{8.31}$$

The first term of the expansion represents the cold beam approximation discussed in the previous paragraph. It is worth to be mentioned that an exponentially small imaginary term that reflects Landau damping is omitted in (8.31).

There exists another limiting case of small values of the parameter $\delta_T \ll 1$. Here large values of x provide the main income to the integral (8.30) and one can expand the denominator:

$$(1-x)^{-2} = \sum_{s=0}^{\infty} \frac{s+1}{x^{s+2}}. \tag{8.32}$$

Keeping in mind that

$$\int_C \frac{\exp\left(-\delta_T^2 x^2\right)}{x^{n+2}} dx = \delta_T^{n+1} \times \begin{cases} \frac{(-1)^{n/2+1}\sqrt{\pi}}{(n/2+1)!} & \text{for even } n \\ \frac{i\pi(-1)^{(n+1)/2}}{((n+1)/2)!} & \text{for odd } n \end{cases} \tag{8.33}$$

one can integrate this sum with $\exp\left(-\delta_T^2 x^2\right)$ and find that

$$J \asymp -\frac{1}{2\beta_T^2 k^2} \sum_{s=0}^{\infty} \frac{(-1)^s (2s+1)}{(s+1)!} \delta_T^{2s} - i\sqrt{\frac{\pi}{2}} \frac{k_0 - \beta k}{\beta_T^3 k^3} \sum_{s=0}^{\infty} \frac{(-1)^s}{s!} \delta_T^{2s}. \tag{8.34}$$

As it is easy to see, the case in question (small values of δ_T) corresponds to a large thermal spread of the beam particles. Under such conditions, plasma intrinsic oscillations are excited as a result of the kinetic instability development, which can be interpreted as negative Landau damping.

Really, substituting the main term of (8.34) in the general dispersion equation (8.6), one gets in the first approximation with respect to k_b^*:

$$\text{Im}\, k_0 = \frac{\sqrt{2\pi} k_b^{*2} \beta^4}{4 \beta_T^3 k_p^2} \kappa \quad \text{for} \quad |\kappa| \ll . \tag{8.35}$$

So, at one side of the Cherenkov resonance the kinetic increment is positive (Landau damping negative) and change the sign at the opposite side. In the resonance point itself, the increment vanishes because the distribution has the maximum there.

From this point of view the close vicinity of the resonance is always a domain of kinetic instability, but for not very hot beam the instability develops

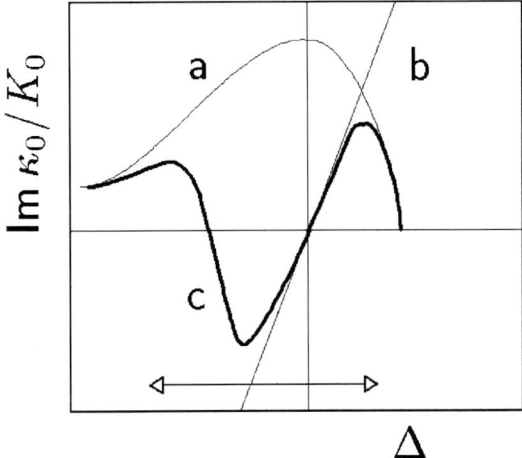

Fig. 8.5. The influence of kinetic effects on the fluid instability. (**a**) Fluid instability increment; (**b**) kinetic increment for a hot beam; (**c**) resulting curve. The arrows show the influence region

as a fluid one even at small detuning. The scheme of the effect is presented in Fig. 8.5.

To estimate the maximal increment (at least, in the low current approximation), one can use (8.18) with substitution

$$\Delta = \sqrt{2}\beta_\mathrm{T}\frac{k_s}{K}.$$

Of course, one cannot trust the numerical coefficient in this estimation, but an essential decrease in the gain for a warm beam is evident. Note, by the way, that Landau damping provokes a certain isolation of the Cherenkov instability from the electrostatic one at the long wave domain.

Of course, the brief sketch of theory above must be essentially supplemented to be applicable to more or less realistic devices. First of all, the problems of transverse plasma and beam nonuniformity as well as boundary conditions are of a great importance. We do not touch nonlinear effects, which require detailed computer simulations and are not typical for the book. Our aim was just basic physics of the involved processes, bearing in mind that many problems are still unsolved in this very young domain of electronics. Those who are interested in details can find them in monographs (i.e., [13]) which, unfortunately are rather few.

9
Cyclotron Resonance Masers (CRM)

9.1 General Principles

From the viewpoint of principles of the stimulated radiation emission, any of the systems that emit radiation and contain an electron beam may be called free–electron lasers or masers. This implies that emitting elements are not bound in atoms or in a crystal lattice.

We now accentuate the general properties, inherent in all beam systems with distributed parameters. Classifying these systems by the resonance conditions (the conditions of synchronism), one can divide them into the two large classes. The systems, the operation of which is based on Cherenkov mechanism for the field–particle interaction, belong to the first class. To the second one belong the beam systems where charged particles are oscillators. In this case, the oscillators, moving with a relativistic velocity, can emit radiation at very high frequencies due to Doppler effect.

For operation of Cherenkov systems, electrodynamic structures that can slow down electromagnetic waves are required. Therefore, the mechanism for Cherenkov radiation emission is effective only when oscillations are excited at relatively low frequencies. Surely, the quasi–Cherenkov effects (e.g., Smith–Parcel radiation emission) can be used for stimulating the radiation emission within the optical range. However, the transverse size of the active area of the field–particle interaction is very small ($l_\perp \approx \lambda$).

The beam systems of the second class are much better adapted for stimulating the short–range radiation emission (up to the x–ray band). In the systems of this type, no slow–wave structures are required. Electromagnetic waves are free (either completely or almost completely). Cyclotron resonance masers and free electron lasers (in the traditional sense of these terms) belong to this class.

In the given section, we are to investigate CRM. For providing the radiation emission in CRM, oscillators serve as an energy source. These oscillators are produced by injecting the beam electrons into an external constant homogeneous magnetic field at a certain angle. As it is known, the condition for the prolonged synchronism between the field and oscillators (the cyclotron resonance conditions) have the form: $\omega - kv \pm \Omega_0/\gamma = 0$, where $\Omega_0 = qB/mc$. This condition indicates that excitation and amplification of fast waves ($\beta_{\rm ph} > \beta$) is possible at the two frequencies:

$$\omega_{1,2} = \frac{\Omega_0}{\gamma\left(1 \pm \beta/\beta_{\rm ph}\right)}. \tag{9.1}$$

Here the sign "+" describes the low–frequency counterbeam propagating wave. The sign "−" determines the copropagating high–frequency wave. The energy of intrinsic waves of a passive electrodynamic structure is positive. At the same time, beam cyclotron waves can possess negative energy. It is the interaction between the positive–energy waves and the beam waves, characterized by negative energy, that is the origin of the development of cyclotron instabilities. In dispersion diagrams, parameters of the interacting waves are determined by the points of intersection of the corresponding partial branches. In Fig. 9.1, dispersion of the beam cyclotron waves and fast electromagnetic waves is plotted.

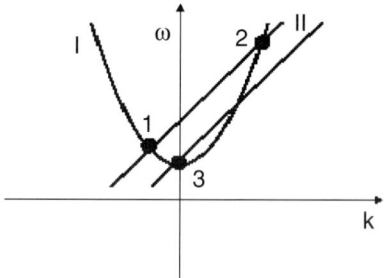

Fig. 9.1. Dispersion of the electromagnetic waves (I) and the beam cyclotron ones (II). Crossing point "1" corresponds to interaction of the beam waves with the backward low-frequency electromagnetic wave; the point "2" – with the high frequency wave; the point "3" – to the case of a gyrotron

There one can see the dispersion branch of fast electromagnetic waves $\omega^2 = k^2 c^2 + \omega_{\rm c.o.}^2$ propagating in a regular waveguide with a finite cutoff frequency $\omega_{\rm c.o.}$. The branch of the beam cyclotron wave, which corresponds to the Doppler normal effect ($\omega = kv + \omega_0/\gamma$), is depicted there as well.

The first point of the intersection corresponds to the excitation of the low–frequency wave moving opposite the beam. The second point corresponds to the excitation of a high–frequency wave, copropagating with the beam.

As regards more or less compact microwave devices, the relativism of the beam particles is rather low so that, in fact, there does not exist any essential Doppler heightening of the frequency. The case $k = 0$ is of a special interest. In the first approximation, the frequency of operation is independent of the particle longitudinal velocity. This makes an advantage of CRM over the systems with Cherenkov interaction, where the longitudinal velocity of the beam particles must be maintained to a high precision. This scheme of CRM, suggested in the early sixties, has been called *the gyrotron* (see [47]).

The choice of $k = 0$ means that the electric field of the high–frequency wave is homogeneous all over along the beam at any moment. The waveguide, where the beam particles interact with the field, operates at the standing mode as a cavity. The cavity output edge can be made in the form of a corrugated structure, which is a Bragg mirror. This construction of the electrodynamic structure provides the mode selection. It also permits to heighten essentially the system output power. It is worth mentioning that the internal field strength in the cavity is $\sqrt{Q}$ times higher than the output field strength (here Q is the cavity quality factor). This permits to increase the efficiency of CRM operation. Really, the CRM efficiency is inversely proportional to the number of revolutions N, performed by the particle within the interaction region. At the end of the interaction path saturation of the amplitude of the wave is desirable. Therefore, by increasing the field strength in the interaction area, one can reduce its geometrical size. Respectively, this provides for diminution of the number of the particle revolutions and increases the system efficiency.

Besides, the condition $k = 0$ permits to diminish essentially the Doppler broadening $\Delta(kv)$ of the cyclotron resonance line if there exists the spread in values of the forward velocity of electrons Δv.

The second point of intersection corresponds to the excitation of waves at higher frequencies. However, to realize this type of interaction between electromagnetic and cyclotron beam waves, one needs the values of γ, essentially higher than the ones mentioned above. It is worth mentioning that the wave autoresonance excitation ($\omega = \Omega_0/\gamma(1-\beta)$) is possible within this pattern. As it is known [28, 29], the conditions for the cyclotron autoresonance are independent of changes in the particle energy values. In principle, there arises the possibility of the unlimited resonance acceleration of charged particles and of the complete transfer of the particle energy to the wave field. The operation of cyclotron autoresonance masers (CARM) is based on this principle.

Stimulation of the radiation emission in the millimeter–submillimeter wave ranges (or in a band of shorter wavelengths) and on heightening the power level of the oscillations excited has an essential feature. The point is that in CRM charged particles can interact not only with a single spatial mode but with a large number of these modes as well. And what is more, the particles can interact with a single spatial mode when there simultaneously exist several cyclotron resonances. There occurs the interaction of this type when the field strength of the wave excited reaches a value, sufficient for overlap of nonlinear cyclotron resonances. In all these cases, the particle motion becomes chaotic. There are both drawbacks and advantages inherent in this regime of exciting oscillations. On the one hand, a disadvantage is that the level of fluctuations in the characteristics of the field excited becomes higher. There also arises a broadening of the field spectrum. On the other hand, one can control the oscillation spectrum width, which is a rather attractive point. Besides, in this regime, the particle motion is not restricted within a single isolated cyclotron resonance. In principle, this holds out prospects for heightening effectiveness of the wave–particle energy interchange. This mechanism for the microwave excitation has been investigated in [48].

9.2 CRM in Small–Signal Approximation

In fact, the theory of CRM in the small–signal approximation has been presented in Sect. 7. And what is more, it has been proved that the kinetic theory must be applied for describing the process of the instability development in CRM. Within the framework of the fluid dynamics approximation, it is impossible to correctly describe the process of the electron beam instability development in an external magnetic field. However, in Sect. 7, the principal attention has been paid to solving the paradox of the absence of the beam instability within the framework of the fluid description approximation. Besides, we have grounded the fact that the use of the kinetic theory is necessary. We have also dwelled on applicability of the notion of Landau damping. At the same time, as a matter of fact, the process of development of the radiative instability has not been investigated yet. In addition, in Sect. 7 an unlimited homogeneous beam of electrons was considered. In the given subsection, we are going in detail to dwell on development of the radiative instability in the limited beam. The simplest model of the beam of this type is to be studied. That is, we now investigate a two–dimensional model, where a ribbon beam of electrons is propagating in parallel to an external homogeneous constant magnetic field. The area of the beam–field interaction is restricted by two ideally conducting planes, located at $x = \pm a$. The point is that the parameters of the beam and those of the electrodynamic system are independent of y–coordinate. Hence, in the simplest case, we will regard all physical processes as independent of y–axis. Longitudinal and transverse velocities of all of the beam particles are regarded as equal to one another. This model of CRM has been examined in [49, 50].

9.2.1 Dispersion Equation for Ribbon Beams

In discussing the CRM theory, our starting point is Maxwell's equations for the fields and Vlasov kinetic equation for the distribution function of the beam electrons (e.g., see (8.42)). The distribution function f_0 describes stationary undisturbed states of the beam particles. In general, it is an arbitrary function of the characteristics of Vlasov equation. In the presence of just a constant external homogeneous magnetic field, these characteristics take the form:

$$\gamma = \text{const} ; \qquad p_z = \text{const} ;$$

9.2 CRM in Small–Signal Approximation

$$z - vt = \text{const} \; ; \qquad \Omega_0 x - p_x = \text{const} \; ; \tag{9.2}$$

$$\arctan(p_x/p_y) - \frac{\omega_0}{\gamma} t = \text{const} .$$

Here $p_{x,y,z}$ are corresponding momentum projections in mc units and γ is a particle energy in mc^2 units. It is easy to see that these characteristics are the solutions to the equation of motion of charged particles in a constant external homogeneous magnetic field. As we deal with a ribbon beam, the undisturbed distribution function is independent of the y coordinate. Consequently, it also does not depend on the azimuthal angle in the momentum space $\vartheta = \arctan(p_x/p_y)$, for example, $\partial f_0/\partial\vartheta = 0$. Besides, we consider the beam density to be low so that beam stationary fields (electric and magnetic) are negligible. We also neglect the beam permeability.

In the electrodynamic system, formed by two parallel perfectly conducting planes, the waves of both the E– and H–types can be excited. We now choose the temporal and longitudinal–coordinate dependencies of these waves in the form $\exp[\mathrm{i}(kz - \omega t)]$. As well as the field undisturbed distribution function, the fields are independent of y-coordinate. For simplicity, we limit ourselves to the case of excitation just of H–waves. The components of these waves are E_y, B_x, B_z. And what is more, by making use of Maxwell's equations, one can express the components via the wave electric field E_y. As a result, expressions for all the components can be written as

$$E_y = E(x) \exp[\mathrm{i}(kz - \omega t)] \; ;$$
$$B_z = -\frac{kc}{\omega} E(x) \exp[\mathrm{i}(kz - \omega t)] \; ; \tag{9.3}$$
$$B_z = -\mathrm{i}\frac{c}{\omega} \frac{\partial E}{\partial x} \exp[\mathrm{i}(kz - \omega t)] .$$

To determine E_y, one has to use the wave equation:

$$\frac{\mathrm{d}^2 E_y}{\mathrm{d}x^2} + \left(\frac{\omega^2}{c^2} - k^2\right) E_y = \frac{4\pi}{c} \frac{\partial j_y}{\partial t} . \tag{9.4}$$

On the RHS of (9.4), the current density j_y drives the field E_y. This variable can be found by solving the linearized Vlasov equation. For the function $f(\mathbf{r}, \mathbf{p}, t)$, which represents a small deviation from the stationary distribution function f_0, this equation can be presented as

$$\frac{\mathrm{d}f}{\mathrm{d}t} = \left[q\mathbf{E} + \frac{q}{c}[\mathbf{vB}]\right] \frac{\partial f_0}{\partial \mathbf{p}} . \tag{9.5}$$

Substituting the expression for the fields (9.3) into (9.5), we shall use a cylindrical coordinate system in the momentum space ($p_x = p_\perp \cos\vartheta$, $p_y = p_\perp \sin\vartheta$). Thus, (9.5) can be rewritten as

$$\frac{\mathrm{d}f}{\mathrm{d}t} = \frac{q}{mc\omega} \left[(\omega - kv)\frac{\partial f_0}{\partial p_\perp} + \frac{k p_\perp c}{\gamma} \frac{\partial f_0}{\partial p_z}\right]$$

$$\times E(x) \sin\vartheta \exp[\mathrm{i}(kz - \omega t)] . \tag{9.6}$$

The general solution of (9.6) can be found by direct integration:

$$f = \int dt \frac{q}{mc\omega}\left[(\omega - kv)\frac{\partial f_0}{\partial p_\perp} + \frac{kp_\perp c}{\gamma}\frac{\partial f_0}{\partial p_z}\right]$$

$$\times E(x)\sin\vartheta \exp[i(kz - \omega t)] . \qquad (9.7)$$

In (9.7), the integral must be taken along the characteristics (9.2). Therefore, one may make transition from integrating over t to integrating over another variable (e.g., over ϑ):

$$f = \int d\vartheta \frac{q\gamma}{mc\omega\Omega_0}\left[(\omega - kv)\frac{\partial f_0}{\partial p_\perp} + \frac{kp_\perp c}{\gamma}\frac{\partial f_0}{\partial p_z}\right]$$

$$\times E(x)\sin\vartheta \exp[i(kz - \omega t)] . \qquad (9.8)$$

We now suppose that Larmor radius and the ribbon beam thickness (a) are much smaller than the wavelength. Consequently, the electric field strength $E(x)$ in the integrand of (9.8) may be changed for the value $E(x_0)$ at the median plane of the beam and taken out from the integral. After that the integral in (9.8) can be easily calculated:

$$f = \frac{qE(x_0)}{2mc\omega}\left[(\omega - kv)\frac{\partial f_0}{\partial p_\perp} + \frac{kp_\perp c}{\gamma}\frac{\partial f_0}{\partial p_z}\right]$$

$$\times \left[\frac{\exp(-i\vartheta)}{\omega - kv - \Omega_0/\gamma} - \frac{\exp(i\vartheta)}{\omega - kv + \Omega_0/\gamma}\right]\exp[i(kz - \omega t)] . \qquad (9.9)$$

In the cylindrical system of axes, the expression for the current density in the momentum space can be submitted as

$$j_y = -\frac{q}{\gamma}\int f \sin\vartheta\, p_\perp\, dp_\perp dp_z d\vartheta . \qquad (9.10)$$

We now suppose that the equilibrium distribution function may be presented in the form of a product of several functions. One of them depends only on momenta, another one determines the beam structure along x-axis. Besides, we consider the beam to be cold and its thickness to be much shorter than the wavelength of the oscillations excited. In this case, the equilibrium distribution function may be presented in the form:

$$f_0 = n_0 2d\delta(x - x_0)\frac{\delta(p_z - p_{z,0})\delta(p_\perp - p_{\perp,0})}{2\pi p_{\perp,0}} , \qquad (9.11)$$

Here n_0 is the beam equilibrium density; d is the beam half–width; $x = x_0$ determine the location of the plane of the beam axis (see Fig 9.2).

Making use of the distribution function (9.11), we integrate over ϑ, p_z, and $p_\perp$. After simple but bulky calculations, one gets the following expression for the disturbed component of the beam current:

9.2 CRM in Small–Signal Approximation

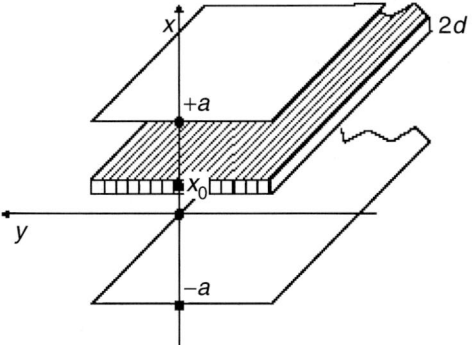

Fig. 9.2. Scheme of CRM with flat beam

$$j = \frac{\omega_b^2 d}{4\gamma} E(x_0) \delta(x - x_0) \exp\left[i(kz - \omega t)\right] G(\omega, k) , \qquad (9.12)$$

Here

$$G = -\frac{1}{\pi} \left\{ \left[\frac{1}{(\omega - kv_{z,0} - \Omega_0/\gamma)} + \frac{1}{(\omega - kv_{z,0} + \Omega_0/\gamma)} \right] \right. $$
$$\times (\omega - kv_{z,0})$$
$$\left. + \frac{v_\perp^2 k_\perp^2}{2} \left[\frac{1}{(\omega - kv_{z,0} - \Omega_0/\gamma)^2} + \frac{1}{(\omega - kv_{z,0} + \Omega_0/\gamma)^2} \right] \right\} .$$

We consider the beam narrow and its density sufficiently low. In a zeroth approximation over the beam density, one may not take into account the influence of this parameter on the field structure of the intrinsic wave of the electrodynamic system. All over the space where the beam is absent, we suppose that the field structure is the same as in the absence of the beam. The presence of the beam indicates itself by a jump of the microwave magnetic field component, tangential to the beam (H_z). There takes place this jump at the area where the beam is located. The magnitude of the jump can be found by integrating (9.4) over the beam small cross section:

$$\lim_{\varepsilon \to 0} \left[\left(\frac{\partial E_y}{\partial x}\right)_{x=x_0+\varepsilon} - \left(\frac{\partial E_y}{\partial x}\right)_{x=x_0-\varepsilon} \right]$$
$$= -2dk_\perp^2 E_y(x_0) - \frac{\pi \omega_b^2 d}{i\gamma c\omega} E_y(x_0) G . \qquad (9.13)$$

In addition, one must keep in mind that the electric field on conductive surfaces goes to zero: $E_y(x = \pm a) = 0$. As it follows from (9.13), there is a discontinuity in the derivative of the wave electric field component with respect to the transverse coordinate. Therefore, all over the areas where the beam is absent, one can look for the solution to (9.4) in the form:

$$E_y = \begin{cases} C \sin\left[k_\perp (x-a)\right], & a > x > x_0; \\ D \sin\left[k_\perp (x+a)\right], & -a < x < x_0, \end{cases} \quad (9.14)$$

Here C and D denote constants; $k_{\perp,n} = \pi n/2a$.

The solution (9.14) satisfies the boundary condition on metallic planes ($E_y(x = \pm a) = 0$). The constants C and D are related to one another by the continuity condition for the electric field (9.14) when $x = x_0$. Substituting the solution (9.14) into the boundary conditions (9.13) and excluding the constants, one gets the following dispersion relation:

$$L \equiv \frac{k_\perp}{2}\left[\frac{\sin(2k_\perp a)}{\sin k_\perp (x-a) \sin k_\perp (x+a)}\right] - k_\perp d$$
$$= -\frac{\omega_b^2 d}{2\gamma c^2}\left\{(\omega - kv_{z,0})\left[\frac{1}{(\omega - kv_{z,0} - \Omega_0/\gamma)}\right.\right.$$
$$+ \left.\frac{1}{(\omega - kv_{z,0} + \Omega_0/\gamma)}\right]$$
$$- \frac{k_\perp^2 v_{\perp,0}^2}{2}\left[\frac{1}{(\omega - kv_{z,0} - \Omega_0/\gamma)^2}\right.$$
$$+ \left.\left.\frac{1}{(\omega - kv_{z,0} + \omega_H/\gamma)^2}\right]\right\}. \quad (9.15)$$

If the thickness and density of the beam go to zero ($\omega_b \to 0$, $d \to 0$), (9.15) describes intrinsic waves of the electrodynamic structure, formed by two conductive parallel planes. In addition, $k_\perp = \pi n/2a$.

The beam exerts a substantial influence on the waveguide modes under the resonance conditions (the conditions of synchronism): $\omega - kv_{z,0} \pm \Omega_0/\gamma = 0$. This relation describes the beam cyclotron modes. Thus, one can see that there takes place an effective energy interchange between the beam and electromagnetic modes at the points of intersection of the dispersion branches (see Fig. 9.1). The frequency values, under which the branch intersection becomes possible, are prescribed by the expression:

$$\omega_{1,2} = \frac{\Omega_0}{\gamma\left(1 - v_{z,0}^2/c^2\right)}\left[1 \pm \frac{v_{z,0}}{c}\sqrt{1 - \frac{k_\perp^2 c^2 \gamma^2}{\Omega_0^2}\left(1 - \frac{v_{z,0}^2}{c^2}\right)}\right], \quad (9.16)$$

Here $k_\perp c$ is the minimum frequency that can propagate in the waveguide (the cutoff frequency).

As regards (9.15), it describes the relation of the system intrinsic frequencies to the intrinsic wave numbers. Generally speaking, this dispersion relation is transcendental, and it is difficult to give analysis to it. However, in the majority of the cases that are of practical interest, the analysis is possible. This equation can be solved with respect both to the frequency and to the wave numbers. Below we will investigate the three cases of the corresponding analysis. We consider them to be the most typical and interesting.

The First Case

Let us determine the conditions for the increase in the spatial disturbances when the waveguide intrinsic waves are synchronous with the beam cyclotron waves. Let us present the longitudinal wave number in the form $k = k_0 + h$ (here $k_0 = (\omega \pm \Omega_0)/v_{z,0} = \pm\omega/v_{\rm ph}$; $v_{\rm ph} = \pm\omega v_{z,0}/(\omega \pm \Omega_0)$ the magnitude h is a small disturbance of the wave number). In the expression for $v_{\rm ph}$, the signs "$\pm$" determine the copropagating and contrary waves, respectively. For simplicity, let us suppose that the beam central plane is located in $x_0 = 0$. We now substitute the expression for the wave number into the dispersion relation (9.15). Furthermore, we expand the terms in h. The beam density is regarded as small. We also take into account that the left–hand side of (9.15) goes to zero in a zeroth approximation with respect to the beam density ($L(k_0) = 0$). Thus, one gets the following algebraic equation of the third degree for determining amendments to the wave number:

$$L'\left(\frac{\partial k_\perp}{\partial h}\right)h^3 - \frac{\omega_b^2 d}{2\gamma c^2}\left[\left(\frac{\Omega_0}{\gamma v_{z,0}} - \frac{k^2 v_{\perp,0}^2}{v_{z,0}^2}\right)h + \frac{k_\perp^2 v_{\perp,0}^2}{2v_{z,0}^2}\right] = 0 . \qquad (9.17)$$

As regards the general case, in (9.17), we have preserved the terms of the same order of smallness. However, if the transverse velocity of the beam particles and the transverse wave number $k_\perp$ are high enough, the second term in the square brackets in (9.17) substantially exceeds the first one. Respectively, one gets the following equation for determining h:

$$h^3 = \frac{\omega_b^2 d k_\perp^2 v_{\perp,0}^2}{2ka\gamma c^2 v_{z,0}^2} . \qquad (9.18)$$

Here it is taken into account that if n is odd, $L'(\partial k_\perp/\partial k) \approx ku$.

As it follows from (9.18), under the conditions in question, there always occurs amplification of the microwave, the coefficient of amplification reaching its maximum:

$${\rm Im}\, h = -\frac{\sqrt{3}}{2}\left[\frac{\omega_b^2 d k_\perp^2 v_{\perp,0}^2}{2ka\gamma c^2 v_{z,0}^2}\right]^{1/3} . \qquad (9.19)$$

Otherwise, the beam transverse velocity and the transverse wave number can take values, not too high so that the first term in the square brackets in (9.17) exceeds the second one. At the same time, if the beam transverse velocity is high enough to provide fulfillment of the inequality $\left(v_\perp^2/v^2\right) > (\Omega_0/\gamma kv)$, there also takes place the microwave amplification. However, the coefficient of amplification is smaller:

$$h = -{\rm i}\sqrt{\frac{\omega_b^2 d}{4k\gamma ac^2}\left(\frac{kv_\perp^2}{v^2} - \frac{\Omega_0}{\gamma v^2}\right)} . \qquad (9.20)$$

On the other hand, if the beam transverse velocity is so low that the inverse inequality is true $\left(v_\perp^2/v^2\right) < (\Omega_0/\gamma kv)$, it is easy to see that there does not take place any amplification.

The Second Case

One can be interested in tracing out the field amplitude evolution not only in space but also in time. For this purpose, the dispersion relation (9.15) must be solved not with respect to the wave number but with respect to the frequency ω. On the analogy of solving this equation with respect to k, we also suppose that the beam density is a small parameter. In (9.15), it is convenient to present the frequency ω in the form: $\omega = \omega_0 + \delta$. The undisturbed value of the frequency ω_0 corresponds to the point of intersection of the waveguide beam modes with the cyclotron ones; i.e., it is one of the frequencies, determined by (9.16); δ is a small disturbance of the frequency. Let us substitute this expression for ω into (9.15). It should be taken into account that the function $L'(\omega_0)(\partial k_\perp / \partial \omega)$ is equal to $-ka$ if n is odd. The equation for δ can be written as

$$\delta^3 - \frac{\omega_b^2 \omega_0^2 d\Omega_0}{2a\gamma^2}\delta + \frac{\omega_b^2 \omega_0^2 dk_\perp^2 v_\perp^2}{4a\gamma} = 0. \tag{9.21}$$

Equation (9.21) indicates the following. If the beam transverse velocity is sufficiently high so that the inequality $v_\perp^2 \gg \delta\, 2\omega_H/\gamma k_\perp^2$ holds, the increment of the instability development reaches its maximum, equal to

$$\mathrm{Im}\,\delta = \frac{\sqrt{3}}{2}\left(\frac{\omega_b^2 dk_\perp^2 v_\perp^2}{4a\gamma\omega_0}\right)^{1/3}. \tag{9.22}$$

Otherwise, if the transverse velocity is low so that the opposite inequality $v_\perp^2 \ll \delta\, 2\Omega_0/\gamma k_\perp^2$ is true, there does not occur the instability development.

One could be interested in investigating the oscillations for which $k \to 0$. These waves are excited in gyrotrons. We now substitute $k = 0$ into (9.22). It should be taken into account that $k_\perp^2 = (\Omega_0^2/\gamma^2 c^2)$. Hence, one gets the following expression for the maximum increment:

$$\mathrm{Im}\,\delta = \frac{\sqrt{3}}{2}\left(\frac{\omega_b^2 d\Omega_0 v_\perp^2}{4a\gamma^2 c^2}\right)^{1/3}. \tag{9.23}$$

The Third Case

In the beam system examined, there can develop instabilities which are not related to the synchronism of waveguide and beam modes. This occurs if the dispersion equation (9.15) describes only the beam modes – that is, when, under the condition that $\omega_b \to 0$, $d \to 0$, the left–hand side of (9.15) does not go to zero, which means that $k_\perp \neq \pi n/2a$. In this case (under small values of the beam density), the solution to the dispersion equation (9.15) is located in the neighborhood of the point $\omega = kv + \Omega_0/\gamma \equiv \omega_0$. As above, let us substitute the solution in the form $\omega = \omega_0 + \delta$ into (9.15). For determining a small frequency addition δ, an algebraic equation of the second order has been derived. The solution to this equation can be written in the form:

$$\delta = -\frac{\omega_b^2 d\Omega_0}{8\gamma c^2 L} \pm \sqrt{\frac{\omega_b^2 d\Omega_0^2}{2\gamma^2 c^2 L}\left(\frac{\omega_b^2 d}{32 c^2 L} - \frac{k_\perp^2 v_\perp^2}{\Omega_o^2}\right)}. \quad (9.24)$$

As it follows from (9.24), there does not take place any instability if $L > 0$. Otherwise, if $L < 0$, the instability is developing under a sufficiently high value of the beam transverse velocity $v_\perp^2 > \left(\omega_b^2 \Omega_0^2 d / 32 k_\perp^2 c^2 |L|\right)$.

It is worth tracing back to the physical reasons that induce the first and second terms in the round brackets under the root of (9.24). The first one is conditioned by the presence of the power bunching of the particles. In its turn, there arises the second term due to the inertial bunching. As one can see, the two mechanisms for bunching act in antiphase. If the beam density is rather low, the inertial bunching prevails. The bunching of this type causes the instability development, the instability increment being proportional to the square root of the beam density. As the beam density is increasing, the instability increment value is becoming higher as well. At the same time, the influence of the power bunching becomes more and more essential in this process. When $\omega_b^2 = \left(32|L|c^2 k_\perp^2 v_\perp^2 / d\Omega_0^2\right)$, the two mechanisms for bunching start to compensate one another, which results in the instability derangement. If the beam density is $\omega_b^2 = \left(16|L|c^2 k_\perp^2 v_\perp^2 / d\Omega_0^2\right)$, the increment reaches its maximum. The instability in question has the same nature as "the negative – mass effect," known in the theory of accelerators [51, 52]. There arises this effect as a result of bunching of nonisochronous oscillators, interacting with one another via the microwave field of the beam mode.

9.2.2 Bunching of Particles in CRM

They usually call two mechanisms of bunching of particles during the beam instability development in CRM: they are the forced and inertial ones. However, these notions are rather conditional. Actually, bunching is always conditioned by the reaction of the excited field on dynamics of the particles. The notion of the forced bunching implies the field direct influence on particle phases with respect to the wave. Let us go back to the system (9.30). There, in the second equation, which describes dynamics of the wave phase, the forced bunching is due to the terms proportional to the wave strength parameter (g). The inertial bunching is stimulated by the wave field influence on the particle energy and its longitudinal and transverse momenta. Respectively, this causes changes in the resonance conditions. In the overwhelming majority of cases, the inertial bunching prevails over the forced one. So, one may retain only the terms that determine the inertial bunching in the third equation of the system (9.33) (in the resonance phase equation).

We now focus on the physical mechanism of the inertial bunching. It is conditioned by the nonisochronous motion of electrons in the homogeneous magnetic field. The notion of nonisochronous motion implies the dependence of the rotational frequency on the electron energy. Suppose that electron motion around the field lines of the external constant magnetic field is purely

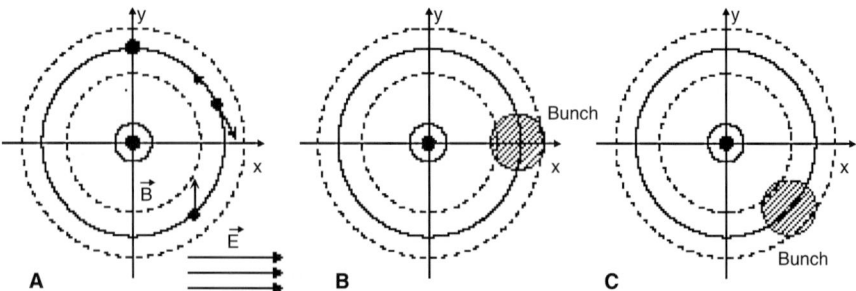

Fig. 9.3. Grouping of electrons into a bunch. (**A**) $t = 0$; $E_x = E_0 \cos \omega t$; (**B**) the case of the exact resonance $\omega = \Omega_0/\gamma$; (**C**) the case $\omega > \Omega_0/\gamma$

circular. We also suppose that the electrons are of the same energy (of the same Larmor radius), being uniformly distributed along the same circumference at the initial moment. The field of an external electromagnetic wave is switched on at this very moment. Besides, let Larmor radius be much smaller than the wavelength. In addition, we suppose that the electric field strength of the wave is directed along x–axis (see Fig. 9.3a).

As this graph indicates, the electrons, located in the half–space $y < 0$, are being decelerated by the wave field. At the same time, the electrons, located above the x–axis (i.e., where $y > 0$), are being accelerated by the same field. As the magnitude of Larmor radius depends on the particle energy ($r_L = \gamma v_\perp / \Omega_0$), the particles under deceleration pass over to a circumference of a smaller radius. The rotational frequency of these electrons is increasing. Respectively, the electrons under acceleration pass to a circumference of a larger radius, and their rotational frequency decreases. Thus, the electrons under deceleration come on in azimuthal direction to the electrons that have gained in energy. As a result, a bunch of electrons is formed. If $\omega = \Omega_0/\gamma$, the bunch is rotating synchronously with the electric field of the external electromagnetic wave. Under the condition $\omega = \Omega_0/\gamma$, the number of the decelerated electrons is approximately equal to the number of the accelerated ones. If $\omega > \Omega_0/\gamma$, the wave slips with respect to the particles. So the bunch gets into the wave decelerating phase; that is, the particles of the bunch transfer in the average their energy to the wave. This mechanism of bunching typical for cyclic accelerators is illustrated in Fig. 9.3.

As regards beams used in microwave devices, most often a somewhat different pattern of the particle bunching is realized. It is rather not bunching but phasing of rotation. Actually, the bunch geometrical sizes usually exceed Larmor radii of the electron rotation in the external magnetic field. In this case, each electron rotates with respect to its driving center. If no special conditions are prescribed, phases of rotation of the electrons are arbitrary (see Fig. 7.4). Therefore, the total current equals to zero because there always exists an electron rotating in antiphase.

We now suppose that, in addition to the external constant magnetic field, in the system also exists the field of an electromagnetic wave. In this case, electrons behave as it has been described above. That is, the rotation of the electrons that got into the accelerating phase is slowing down. On the contrary, the electrons in the decelerating phase rotate faster. This results in phasing of rotation of all electrons. As a consequence, there arises some nonzero total current (see Fig. 7.4). Exactly as in the previous case, to provide the total current energy transfer to the wave, the wave frequency has to be somewhat higher than the relativistic frequency of electron rotation in the magnetic field.

Generally speaking, electrons in real microwave devices, in addition to the rotational velocity, possess a longitudinal velocity. The resonance frequency ($\omega = kv + \Omega_0/\gamma$) depends not only on the energy and transverse velocity of the particles but on their longitudinal velocity as well. Under such conditions, there occurs the particle bunching both in the azimuthal direction, examined above, and in the longitudinal one. The process of bunching just slightly differs from the corresponding physical pattern, already described. However, one should take into account the existence of the longitudinal bunching, which can change the sign of the phase relations of the wave to the particle. In particular, it is evident that a certain condition exists, under which the wave frequency must be smaller than the particle rotational frequency in the magnetic field ($\omega < \Omega_0/\gamma$), which would provide the particle energy extraction. That is, the condition of particles bunching can change its sign (see Chap. 7).

9.3 Particle Interaction with Large Amplitude Wave

In what follows we discuss certain general features of the wave–particle interaction in a uniform magnetic field which are of importance for different types of CRM as well as for particle acceleration by a high-frequency field.

9.3.1 Averaged Equations of Motion

Let us consider a charged particle moving in an external constant magnetic field $\mathbf{B}_0$ directed along the z–axis. In addition, the particle is influenced by the wave field of an arbitrary polarization:

$$\mathbf{E}\exp\left(i\mathbf{k}\mathbf{r} - i\omega t\right) \; ; \quad \mathbf{B} = \frac{c}{\omega}\left[\mathbf{k}\mathbf{E}\right]\exp\left(i\mathbf{k}\mathbf{r} - i\omega t\right) \; . \tag{9.25}$$

Not losing generality, one can suppose that only two components of the vector $\mathbf{k}$ (k_x and $k \equiv k_z$) are nonzero. In what follows we measure time in units of ω^{-1}, the velocity in units of c, the wave numbers in units of ω/c, and the momentum in units of mc. We also introduce the dimensionless field $\mathbf{g} = q\mathbf{E}/mc\omega$. Respectively, the equations of the particle motion are reduced to

$$\dot{\mathbf{p}} = \left(1 - \frac{\mathbf{kp}}{\gamma}\right) \operatorname{Re}\left(\mathbf{g} \exp i\psi\right) + \frac{\Omega_0}{\gamma}\left[\mathbf{pe}\right] + \frac{\mathbf{k}}{\gamma} \operatorname{Re}\left(\mathbf{pg}\right) \exp i\psi \ ;$$
$$\dot{\mathbf{r}} = \mathbf{p}/\gamma \ ; \qquad\qquad\qquad (9.26)$$
$$\dot{\psi} = \mathbf{kp}/\gamma - 1 \ .$$

Here $\mathbf{e} \equiv \mathbf{B}_0/B_0$, $\Omega_0 \equiv qB_0/mc\omega$, $\psi = \mathbf{kr} - t$.

The point to be made is that the field dimensionless amplitude g coincides with wave strength parameter [53]. They also call this value "the parameter of nonlinearity" or "the wave acceleration parameter." By the order of magnitude, this parameter is equal to the ratio of the work performed by the wave on the particle within the distance equal to the wavelength to the particle rest energy. Being small, it is also equal to the ratio of the particle oscillation velocity in the wave field to the velocity of light.

We now multiply the first equation in (9.26) by $\mathbf{p}$, also taking into account that $p^2 = \gamma^2 - 1$. Thus, one gets the following equation for the particle energy:

$$\dot{\gamma} = \operatorname{Re}\left(\mathbf{vg}\right) \exp i\psi \ . \qquad (9.27)$$

Then (9.26) yields the integral of motion:

$$\mathbf{p} - \operatorname{Re}\left(i\mathbf{g} \exp i\psi\right) + \Omega_0 \left[\mathbf{re}\right] - \mathbf{k}\gamma = \operatorname{const} \ . \qquad (9.28)$$

The integral of motion (9.28) represents the generalized form of the integral, derived in [28, 29]. the direction between $\mathbf{k}$ and the external magnetic field is arbitrary and the field strength parameter g is taken into account.

For the further calculations, it is convenient to make transition to the new variables $p_\perp$, $p_\parallel$, ϑ, ξ, and, η. They are related to the former ones as

$$\begin{aligned} p_x &= p_\perp \cos\vartheta \ ; \\ p_y &= p_\perp \sin\vartheta \ ; \\ p_z &= p_\parallel \ ; \\ x &= \xi - \frac{p_\perp}{\Omega_0} \sin\vartheta \ ; \\ y &= \eta - \frac{p_\perp}{\Omega_0} \cos\vartheta \ . \end{aligned} \qquad (9.29)$$

Taking into account the integral (9.26), one can rewrite (9.28) in the new variables:

$$\begin{aligned} \dot{p}_\perp &= (1 - kv) \sum_n \left(g_x \frac{n}{\mu} J_n - g_y J'_n\right) \cos\vartheta_n + k_x v g_z \sum_n \frac{n}{\mu} J_n \cos\vartheta_n \ ; \\ \dot{\vartheta} &= -\frac{\Omega_0}{\gamma} + \frac{(1 - kv)}{p_\perp} \sum_n \left(g_x J'_n - g_y \frac{n}{\mu} J_n\right) \sin\vartheta_n \\ &\quad + \frac{k_x v_\perp}{p_\perp} g_y \sum_n J_n \sin\vartheta_n + \frac{k_x v}{p_\perp} g_z \sum_n J'_n \sin\vartheta_n \ ; \end{aligned} \qquad (9.30)$$

9.3 Particle Interaction with Large Amplitude Wave

$$\dot{p}_{\|} = \sum_n \cos\vartheta_n \left[g_z J_n + (kv_\perp g_x - k_x v_\perp g_z)\right] \frac{n}{\mu} J_n - kv_\perp g_y J'_n \; ;$$

$$\dot{\xi} = -\frac{1}{\Omega_0} \sum_n J_n \sin\vartheta_n \left[g_y(1-kv) + \frac{n}{\mu} k_x v_\perp g_y\right] \; ;$$

$$\dot{\gamma} = \sum_n \cos\vartheta_n \left[J_n \left(v_\perp g_x \frac{n}{\mu} + vg_z\right) - v_\perp g_y J'_n\right] \; ;$$

$$\dot{z} = v \; .$$

In deriving (9.30), the use is made of the expansion:

$$\cos(x - \mu\sin\vartheta) = \sum_{n=-\infty}^{\infty} J_n(\mu)\cos(x - n\vartheta) \; . \tag{9.31}$$

Let us investigate the case of small amplitudes of the electromagnetic wave ($g \ll 1$). Respectively, the particle effectively interacts with the wave if one of the resonance conditions takes place:

$$\Delta_s(\gamma) \equiv kv + s\frac{\omega_0}{\gamma} - 1 = 0 \; . \tag{9.32}$$

Regarding (9.32) as fulfilled, we also introduce the resonance phase $\vartheta_s = s\vartheta - t$. After averaging, the system (9.30) yields the following equations of motion:

$$\dot{p}_\perp = \frac{1}{p_\perp}(1-kv)W_s g\cos\vartheta_s \; ;$$

$$\dot{p}_z = \frac{1}{\gamma} kW_s g\cos\vartheta_s \; ;$$

$$\dot{\vartheta}_s = \Delta_s \equiv kv + s\frac{\Omega_0}{\gamma} - 1 \; ; \tag{9.33}$$

$$\dot{\gamma} = \frac{g}{\gamma} W_s \cos\vartheta_s \; .$$

Here

$$W_s \equiv \alpha_x p_\perp \frac{s}{\mu} J_s - \alpha_y p_\perp J'_s + \alpha_z p_z J_s$$

where $\alpha_{x,y,z}$ are the components of the wave polarization unit vector. In (9.33), the last equation follows from the other ones. In deriving (9.33)), the terms proportional to $\Delta_s g$ have been neglected.

It is worth to note that the system (9.33) is derived after averaging over varying quickly phases (nonresonant ones). Resonances at various harmonics of the cyclotron frequency can take place depending on the wave and particle parameters. That is, generally speaking, s can be an arbitrary integer number. However, if the wave is propagating strictly along the constant external magnetic field, one can neglect its transverse structure and one should put $k_x \to k_y \to 0$. Respectively, $\mu \to 0$. Consequently, only the terms that describe the resonances $s = 0, \pm 1$ remain nonzero. They correspond to Cherenkov resonance and also to resonances with normal and anomalous Doppler effect. Thus, the cyclotron frequency harmonics are driven by the transverse inhomogeneity of the wave.

9.3.2 Qualitative Analysis

Giving analysis to the above–derived equations, which describe dynamics of the particle motion even in the simplified form (see (9.33)), is rather hampered. However, some information can be obtained by examining the integrals of motion (9.28) and the resonance conditions (9.32). Besides, in practice, the dimensionless amplitudes of the waves excited are usually small. Therefore, there exists the possibility of some substantial energy interchange between the wave and particles only under the conditions of their rather prolonged synchronous interaction. In this case, a particle phase $\vartheta_s = s\vartheta - t$ with respect to the wave is of the main interest. Actually, the phase relations, integrals of motion and resonance conditions are depicted by rather simple algebraic expressions.

The starting point is that in the space $(\gamma, p_\parallel, p_\perp)$ the particle can move only in the surface

$$\gamma^2 = p_\parallel^2 + p_\perp^2 + 1 , \qquad (9.34)$$

which is a rotational hyperboloid. One should keep in mind that the particles cannot get into all areas of the surface and stay out of the areas, limited by the inequalities $\gamma < 0$ and $p_\perp < 0$.

Integrals of Motion

The integral (9.28) is presented in the vector form. In reality, one deals with three algebraic relations, that is, with the projections of the integral (9.28) on to the axes of Cartesian system (x, y, z). During the wave–particle interaction, these projections keep on being constant (i.e, they are integrals as well). As regards these integrals, the third one is of especial importance (it is the projection of the integral (9.28) on to z–axis). It can be essentially simplified if we consider an electromagnetic wave propagating strictly along z–axis ($k_x = k_y = 0$, $k_\perp = 0$). Besides, averaging over the fast phase $\psi = \mathbf{kr} - t$, simplifies this integral as well. In both the cases, the integral takes the form:

$$p_\parallel - k = p_{\parallel,0} - k\gamma_0 \equiv C = \text{const} . \qquad (9.35)$$

In (9.35) the subscript "0" designates the initial values of the longitudinal momentum and energy of the particle.

It is worth mentioning that (9.35) follows from the laws of conservation of energy and momentum at emission of a wave quantum. Really, these laws may be presented in the form:

$$\Delta\gamma = \gamma_0 - \gamma = \hbar\omega/mc^2 \;;$$
$$\Delta\mathbf{p} = \mathbf{p}_0 - \mathbf{p} = \mathbf{e}\hbar\omega/mcv_{\rm ph} \;.$$

As it is easy to see, if the quantum is emitted along z–axis, one can derive the integral (9.35) by substituting $\hbar\omega$ from the first equation of this system into the second one. Note that the relation obtained does not contain Planck constant, that is, it is classic.

On the plane $(\gamma, p_\|)$, the integral (9.35) takes the form of an equation of parallel straight lines. They differ from one another in values of the constant C. Several of these lines are plotted in Fig. 9.4.

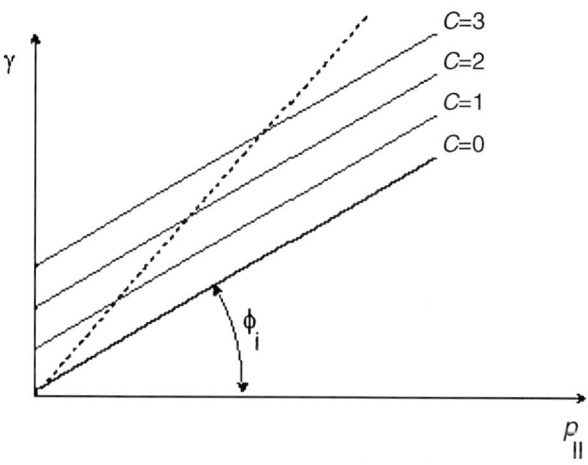

Fig. 9.4. Integral's straight lines in the plane $(\gamma, p_\|)$ for the case $k > 0$

Inclination of these lines of integrals with respect to $p_\|$–axis, prescribed by the longitudinal wave number k, is equal to $\arctan k^{-1}$. Running ahead, the resonances in the plane $(\gamma, p_\|)$ also are represented by straight lines. The angle of their inclination with respect to $p_\|$–axis is equal to $\arctan k$. It is easy to see that for a wave propagating strictly along a constant external magnetic field in vacuum, $k = 1$. Respectively, the straight lines that depict these integrals are parallel to the resonance straight lines. In addition, if $C = s\Omega_0$, these straight lines coincide. These particular specificities make the conditions for autoresonance, which is of considerable independent interest.

It is worth depicting the integrals in the space $(\gamma, p_\|, p_\perp)$. One should take into account that in this space the particle can move only over the surface of the rotational hyperboloid. Therefore, the integrals (9.35) can be presented in the form of a line of intersection of this plane with the hyperboloid. If the wave phase velocity along z–axis is lower than the velocity of light ($k > 1$), this curve of intersection takes the form of an ellipse (see Fig. 9.5).

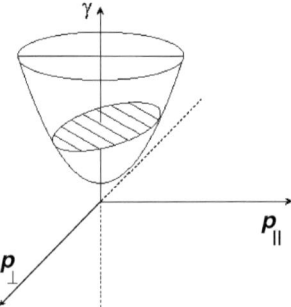

Fig. 9.5. Specific section of the hyperboloid $\gamma^2 = 1 + p_\parallel^2 + p_\perp^2$ by the resonant condition plane for the case $k > 1$. The same shape has the section of the hyperboloid by the integral plane for a fast wave ($k < 1$)

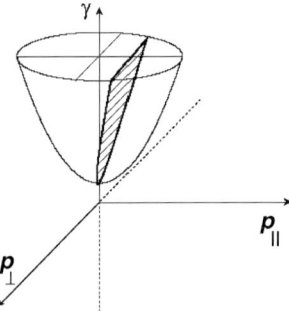

Fig. 9.6. Specific intersection of the hyperboloid (9.34) with a resonant condition plane for the case of particles interacting with a fast wave ($k < 1$). The same shape has the intersection of the hyperboloid with the integral planes for the case $k > 1$ (slow wave)

Otherwise ($k < 1$) the curve of intersection takes the form of a hyperbola (see Fig. 9.6). Even by examining the two plots, one can come to important physical conclusions. In particular, if the particle interacts with the fast wave ($k < 1$), the particle energy, not restricted by the integrals, can reach arbitrary positive values. This is the case when there principally exists the possibility of unlimited acceleration of charged particles. Transfer of a substantial amount of the particle energy to the wave also becomes possible. Surely, the problem of realization of the energy interchange of this type remains open. Below we will consider certain methods to do that. It should be noted that if the slow wave ($k > 1$) is interacting with the particle, the energy interchange is limited by the ellipse characteristics.

Let us examine the projections of the curves, located in the hyperboloid surface, on to the plane ($\gamma, p_\perp$). As it is easy to see, such projection is a second-order curve. Is is presented by the equation:

9.3 Particle Interaction with Large Amplitude Wave

$$\frac{p_\perp^2}{A^2} + \frac{(\gamma - \gamma_*)^2}{B^2} = 1 \,. \tag{9.36}$$

Here

$$A^2 = \frac{C^2}{k^2 - 1} - 1 \,; \quad B^2 = \frac{C^2 - k^2 + 1}{(k^2 - 1)^2} \,; \quad \gamma_* = C \frac{k}{1 - k^2} \,.$$

If $A^2 > 0$ and $B^2 > 0$, (9.36) is an equation of an ellipse with its center being located at the point $p_\perp = 0$, $\gamma = \gamma_*$. In particular, this case is realized if the particle interacts with a slow wave ($k > 1$). One can be interested in determining the conditions under which the particle could transfer its total energy to the wave. If so, $\gamma \to 1$, $p_\parallel \to 0$, and $C = -k$. In this case, the following relation of the particle longitudinal momentum to the particle initial energy has to take place: $p_{\parallel,0} = k(\gamma_0 - 1)$. While the particle is transferring its energy to the wave, the particle transverse momentum increasing at the beginning then reaches its maximum, equal to $1/\sqrt{k^2 - 1}$. After that, it goes to zero.

If the particle interacts with a fast wave ($k < 1$), the curve of intersection of the integral with the hyperboloid is a hyperbola. In this case, the parameter A^2 in (9.36) is negative. The particle energy transfer to the wave is accompanied by the monotonous decrease in the particle transverse momentum. There occurs the total transfer of the particle energy to the wave ($\gamma \to 1$) under the same initial conditions, under which the particle interacts with the slow wave; that is, when $C = -k$ ($p_{\parallel,0} = (\gamma_0 - 1)k$).

As regards the wave–particle energy interchange, the dependence of the particle longitudinal momentum on the transverse one is of interest. Considering the case $k \neq 1$, one can use the equation of the hyperboloid (9.34) and the expression (9.35) for the integral. Correspondingly, this dependence can be presented as

$$\frac{p_\perp^2}{A^2} + \frac{\left(p_\parallel - p_*\right)^2}{B^2} = 1 \,, \tag{9.37}$$

where

$$A^2 = \frac{C^2}{k^2 - 1} - 1 \,; \quad B^2 = k^2 \frac{C^2 - (k^2 - 1)}{(k^2 - 1)^2} \,; \quad p_* = -\frac{C}{k^2 - 1} \,.$$

If $k < 1$, (9.37) is the equation of a hyperbola. In this case the particle energy transfer to the wave is accompanied by the simultaneous and monotonous decrease in both the longitudinal and transverse momenta of the particle.

If the particle interacts with a slow wave ($k > 1$), (9.37) takes the form of an ellipse. In this case, the particle energy transfer to the wave can be accompanied by an initial increase in the particle transverse momentum, but after that its magnitude goes to zero.

It is also worth mentioning that the integral (9.35) takes the form of unlimited rays in the plane $(\gamma, p_\parallel)$ if the particle interacts with a fast wave ($k < 1$). Otherwise, if the particle interacts with a slow wave ($k > 1$), this process is depicted by limited segments of straight lines.

Resonances

In the plane $(\gamma, p_\parallel)$, the resonance conditions (9.32) as well as the integrals (9.34), take the form of the equations of straight lines (see Fig. 9.4). In this plot, the inclination angle prescribed by the longitudinal wave number k is equal to $\arctan(k)$. In contrast to the integral straight lines, the resonance lines take the form of rays if $k > 1$. If the particle is interacting with the fast wave $(k < 1)$, the resonance lines take the form of limited segments of straight lines.

It is worth investigating the resonance conditions in the space $(\gamma, p_\parallel, p_\perp)$. There the resonance lines are the curves of intersection of the hyperboloid (9.34) with the planes of the resonances (9.35). In their typical form, these curves are analogous with the curves of intersection of the hyperboloid with the integrals. The difference is the following: if the particle interacts with the fast wave, they take the form of ellipses (it should be noted that if the hyperboloid intersects the integrals, they are hyperbolas). Otherwise, if the particle interacts with a slow wave, the curves are hyperbolas (if the hyperboloid intersects with the integrals, they are ellipses).

The analytical expression for the projection of these curves of intersection on to the plane $(p_\perp, p_\parallel)$ is analogous with (9.37):

$$\frac{p_\perp^2}{A^2} + \frac{(p_\parallel - p_*)^2}{B^2} = 1, \qquad k \neq 1, \qquad s^2 \Omega_0^2 \neq 1 - k^2;$$

$$p_\perp^2 + n^2 \Omega_0^2 \left[p_\parallel - \frac{ks\Omega_0}{1-k^2} \right]^2 = 0, \qquad s^2\Omega_0^2 = 1 - k^2. \qquad (9.38)$$

Here

$$A^2 = \frac{n^2 \Omega_0^2}{k^2 - 1} - 1;$$

$$B^2 = \left(\frac{n^2 \Omega_0^2}{k^2 - 1} - 1\right) \Big/ (1 - k^2);$$

$$p_* = \frac{kn\Omega_0}{1 - k^2}.$$

For $k < 1$, the first equation in the system (9.38) represents an ellipse. If $k > 1$, this is a hyperbola. If the particle interacts with a fast wave $(k < 1)$, the frequency of which exceeds the particle Larmor frequency $(\Omega_0^2 < 1)$, there exists a certain resonance number $s < s_c = \sqrt{1 - k^2}/\Omega_0$, which would correspond to a negative value of the denominator of the first term in left-hand side of (9.38). That is, the resonance conditions (9.32) cannot be satisfied under any values of the particle momentum. One can choose certain values of the cyclotron frequency and the wave vector longitudinal component so that the parameter s_c would be an integer number. In this case, the plane of the resonance conditions (9.32) becomes tangential with respect to the hyperboloid if $s = s_c$. This case is described by the second equation in the system (9.38).

The analytical expression for the projection of the lines of the hyperboloid intersection with the resonances in the plane $(\gamma, p_\perp)$ can be presented in the form:

$$\frac{(\gamma - \gamma_*)^2}{A^2} + \frac{p_\perp^2}{B^2} = 1, \qquad (9.39)$$

where

$$B^2 = A^2 \frac{(1 - k^2)}{k^2} \ ; \qquad A^2 = \frac{\left(s^2 \Omega_0^2 + k^2 - 1\right) k^2}{(1 - k^2)^2}.$$

The particle moves along the integral curves. As it is known, if the amplitude of the wave that interacts with the particle is small ($g \ll 1$), there takes place an effective energy interchange between the wave and the particle under the condition of synchronism, that is, under the resonance conditions (9.32). Therefore, it is worth examining a graph where the resonance and integral curves are presented simultaneously (see Fig. 9.7).

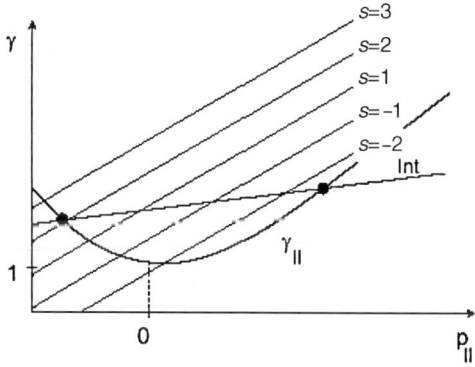

Fig. 9.7. Resonant conditions and the integral in plane $(\gamma, p_\parallel)$ for the case $k > 1$

For distinctness, this graph illustrates the case of the particle interaction with a fast wave. In this plot are the resonance curves, calculated for $s = 0, \pm 1, \pm 2, \pm 3$, and an integral of the system (9.28). In this very figure, the hyperboloid (9.34) is projected on to the plane $(\gamma, p_\parallel)$ under the supposition that $p_\perp = 0$ (the corresponding curve is denoted by $\gamma_\parallel$). In particular, Fig. 9.7 indicates that under the given conditions the wave–particle resonance interaction is possible just in the case when $s = 0, \pm 1, \pm 2$.

Suppose that at an initial moment of time the particle is located at the point 0 and then moves along the integral curve (9.35). As one can see, the area of the wave–particle resonance interaction (i.e., the area where the integral intersects the resonances) is small within the framework of the approximation of an isolated resonance. And what is more, Fig. 9.7 does not permit determining sizes of this area because there takes place intersection of the integral with the resonances at one point. However, in fact, if one takes into account nonlinearity, the resonances are characterized by a certain width (see below) proportional to $\sqrt{g}$. It should be mentioned that in the averaged integral (9.35) we have neglected the term proportional to g. The nonlinear resonance width is much larger ($\sqrt{g} \gg g$). Therefore, to determine this width, one can make use of the averaged integral (9.35). These simple qualitative arguments indicate that the synchronous (resonance) interaction between charged particles and the wave is possible only in a limited area of the plane $(\gamma, p_\parallel)$. Suppose that at the initial moment the particle is in a cyclotron resonance with the wave. As a result of the particle–wave interaction, the particle, moving along the integral curve, quickly leaves this resonance. Under such conditions, any substantial energy interchange between the wave and the particle is hardly possible. However, there exist exceptions: these are the cases of the autoresonance and of the stochastic wave–particle energy interchange. These cases will be investigated below.

Phase Relations

Let us suppose that particle dynamics is restricted by an isolated nonlinear resonance. As it has been demonstrated above, there occurs the synchronous wave–particle interaction in a relatively small area where the integral intersects the resonance. Here we are going to estimate qualitatively effectiveness of this interaction.

To provide the long–term synchronism between the wave and the particle, the resonance condition (9.35) has to be fulfilled. In fact, there occurs the wave–particle interaction in a limited area of the characteristic size L during a limited time interval $T \approx L/v$. The sign of the particle energy transfer to the wave (and v.v.) keeps on being the same all over the interaction area if the phase shift of the rotating electron with respect to the wave $\Delta = (\omega - kv - s\Omega_0/\gamma)T$ is smaller than 2π:

$$|\Delta| \leq 2\pi . \tag{9.40}$$

One can single out the two factors that cause the phase shift.

1. The phase shift can be conditioned by some initial deviation of the wave frequency from the frequency of the precise synchronism $\omega_r = kv + s\,\Omega_0/\gamma$. They call this a kinematic phase shift. It is easy to determine its value:

$$\Delta_k \approx 2\pi N s \frac{(\omega - \omega_r)}{\omega_r} . \tag{9.41}$$

In (9.41), the parameter $N = \Omega_0 T\gamma/2\pi$ is introduced. It determines the number of revolutions, performed by the particle during its interaction with the field. Naturally, under the condition of the long–term synchronism this magnitude is large ($N \gg 1$).

9.3 Particle Interaction with Large Amplitude Wave

2. Besides, the phase shift can be stimulated by the wave influence on the particle motion. The phase shift, conditioned by this effect, is called a dynamic phase shift. Under the influence of the wave, the velocity and energy of the particle change their values. Thus, there takes place deviation from the resonance conditions. As one can readily see, the shift magnitude is

$$\Delta_{\mathrm{d}} = k\left(v_{\mathrm{r}} - v\right) + s\Omega_0\left(\gamma_{\mathrm{r}}^{-1} - \gamma^{-1}\right). \tag{9.42}$$

Making use of (9.35), one can relate the deviation of the particle velocity to the deviation of its energy. Respectively, the dynamic phase shift (9.42) may be rewritten as

$$\Delta_{\mathrm{d}} = 2\pi N s \frac{\Delta\gamma}{\gamma v_{\mathrm{ph}}} \frac{c^2 - v_{\mathrm{ph}}^2}{v_{\mathrm{ph}} - v_{\mathrm{r}}}. \tag{9.43}$$

In particular, if the wave phase velocity is equal to the velocity of light ($v_{\mathrm{ph}} = c$), then $\Delta_{\mathrm{d}} = 0$, that is, there takes place a total compensation of the phase shift. This is the case of the autoresonance. If at the initial moment the resonance conditions are precisely satisfied (i.e., there does not occur any kinematic phase shift), under the condition of the autoresonance there takes place no phase shift at all.

Knowing the magnitude of the dynamic phase shift and making use of the inequality (9.40), it is easy to determine the admissible change in the particle energy:

$$\frac{\Delta\gamma}{\gamma} = \frac{v_{\mathrm{ph}}}{sN} \frac{v_{\mathrm{ph}} - v_{\mathrm{r}}}{c^2 - v_{\mathrm{ph}}^2}. \tag{9.44}$$

It is worth mentioning that in some devices (e.g., in a gyrotron), the wave phase velocity substantially exceeds the velocity of light. Respectively, as it follows from (9.44), just small changes in the particle energy are possible [54]. At the same time, the wave phase velocity can be close to the velocity of light (i.e., the conditions are close to the conditions of the autoresonance). In this case, this very formula (9.44) indicates that one can essentially change the particle energy ($\Delta\gamma \approx \gamma$).

To characterize effectiveness of the wave–particle energy interchange, let us introduce a parameter of efficiency:

$$\eta = \frac{\Delta\gamma}{\gamma - 1}, \tag{9.45}$$

which is equal to the relative change in the electron kinetic energy. In practice, field interacts with a large number of charged particles, so one has to substitute $\langle\Delta\gamma\rangle$ for $\Delta\gamma$ in (9.45). Here the angular brackets designate averaging over the initial phases of the particles at the entrance to the interaction space. The magnitude η, determined in this way, is called a single–particle efficiency.

Making use of the equation of the rotational hyperboloid (9.34) and of the integrals (9.35), one can derive the following relation of the change in the particle energy to the change in the particle transverse momentum:

$$\Delta\gamma = \frac{1}{2}\frac{v_{\text{ph}}\Delta p_\perp^2}{v_{\text{ph}} - v}. \qquad (9.46)$$

In the optimal regime of the wave–particle energy interchange, a particle should lose completely its transverse momentum, that is, $p_\perp \to 0$. Then, by substituting (9.46) into (9.45), one gets the following expression for the maximum available efficiency of a microwave device based on the emission of magnetic bremsstrahlung:

$$\eta_{\max} = \frac{v_\perp^2 v_{\text{ph}} \gamma}{(v_{\text{ph}} - v)}. \qquad (9.47)$$

The maximum effectiveness of the particle energy transfer to the wave requires a certain value of the wave field strength. Really, passing through the region of interaction with the field, the particle gives away the maximum amount of its energy just when its dynamic shift (9.43) is of the order of 2π. If the field strength is lower than the optimal value, the dynamic shift is smaller than 2π and the energy transfer is small. Otherwise, if the field strength exceeds the optimal value, the particle is shifted from the phase of deceleration to the phase of acceleration. Thus, it starts to absorb the wave energy. To estimate the field strength optimal value, one can make use of the fact that the work performed by the wave field on the particle in the interaction region has to be equal to the optimal losses of the particle energy: $A \approx eE(2\pi r_\text{L} N) \approx mc^2 \Delta\gamma$, where r_L is Larmor radius. Substituting $\Delta\gamma$ from (9.44) into this relation, one can evaluate the optimal field strength:

$$g_{\text{opt}} = \frac{\gamma\beta_{\text{ph}}(\beta_{\text{ph}} - \beta)}{2\pi N^2 \beta_\perp (\beta_{\text{ph}} - 1)}. \qquad (9.48)$$

If there takes place the wave–particle interaction in vacuum and the wave propagates strictly along the external magnetic field, then $v_{\text{ph}} \to c$. In this case, the straight lines of the integrals (9.35) are parallel to the straight lines of the resonances (9.32). Under certain initial conditions, these straight lines coincide. Thus, infinitely long synchronous resonance interaction of the particles with the field becomes possible. Really, as (9.43) indicates, the magnitude of the dynamic phase shift goes to zero. Therefore, if the initial conditions are chosen to keep the kinematic phase shift, the total phase shift during the whole time of the wave–particle interaction is equal to zero as well.

9.3.3 Stochastic Regime

At present, microwave electronics tends toward heightening the power of the oscillations excited and toward shortening the wavelength of the waves generated. The two tendencies inevitably result in certain complications of the described processes. Particles interact not with a singled–out mode of the electromagnetic field but with a large number of modes. Really, the equation of

9.3 Particle Interaction with Large Amplitude Wave

balance between the power of the microwave losses in the cavity and the power transferred to the field in the cavity by the electron beam can be written as

$$E^2 V \frac{\omega}{Q} = \eta j U S \,. \tag{9.49}$$

Here E represents the microwave electric field strength in the cavity; Q and V denote the quality factor and volume of the cavity, respectively; j is the beam current density; η is the electron efficiency; the voltage is labeled by U; S reads the beam cross section. As this expression indicates, the increase in the microwave power is achievable in the two ways: either by heightening the current density of the beam or by enlarging the beam cross section.

As we have seen above (see (9.48)), the maximum efficiency is achievable only under certain optimal values of the field strength. Therefore, if one intends to rise the microwave oscillation power by heightening either the current density or the voltage of the beam, this inevitably causes an increase in the wave field strength in the area of the wave–particle interaction. However, this inevitably results in regrouping of the particles and in diminution of the electron efficiency. To avoid these phenomena, one has to heighten the microwave power by enlarging the geometrical sizes of the interaction area (V and S). However, if one enlarges the transverse sizes of the interaction area under a fixed value of the wave frequency, the beam particles interact with high spatial modes of the electromagnetic field. The field structure of the spatial modes, located close to one another, hardly distinguish from the field structure of the desirable wave. Therefore, these modes can be excited as well. Dynamics of the particle in the field of several waves essentially differ from the particle dynamics in the field of a single wave.

There also exists a qualitative difference in dynamics of charged particles if the conditions for several cyclotron resonances can be realized simultaneously. This happens under conditions when the field strength of the wave excited reaches a certain value high enough. As (9.48) indicates, the field strength can be heightened, for instance, when the wave phase velocity is approaching the velocity of light ($\beta_{\rm ph} \to 1$). Below it will be proved that an increase in the field strength value can cause overlapping of nonlinear cyclotron resonances. Then in the field of a single regular wave, the particle dynamics is determined by a large number of the resonances and becomes chaotic. This fact can play either a positive role or negative role. The advantage is that changes in the particle energy become unrestricted by the width of one resonance. Thus, in principle, there arises the possibility of a substantial heightening of the amount of the particle energy that could be transferred to the wave (in comparison with the case of interaction with an isolated resonance). However, chaotic particle dynamics causes phase scattering with respect to the wave. In its turn, this circumstance can be an additional factor of power stabilization of the excited wave. It is a negative aspect of the interaction with a large number of resonances. Besides, the spectrum of the excited oscillations becomes broader.

9 Cyclotron Resonance Masers (CRM)

Now, we consider the conditions of overlapping of cyclotron resonances. Suppose for the moment that the wave–particle interaction does not influence essentially the particle energy: $\gamma = \gamma_0 + \tilde{\gamma}$, $\tilde{\gamma} \ll 1$. Besides, the resonance condition (9.32) is considered to be precisely satisfied for the particles of the energy γ_0. In this case, after expanding $\Delta_s(\gamma)$ in vicinity to γ_0, the last two equations in the system (9.33) yield a closed system of two equations for $\tilde{\gamma}$ and ϑ_s:

$$\frac{d\tilde{\gamma}}{dt} = \frac{g}{\gamma_0} W_s \cos \vartheta_s \; ; \qquad \frac{d\vartheta_s}{dt} = \frac{k^2 - 1}{\gamma_0} \tilde{\gamma} \; . \qquad (9.50)$$

The system (9.34) describes a nonlinear mathematical pendulum. It yields the nonlinear resonance width:

$$\Delta \dot{\vartheta}_s = 4\sqrt{(k^2 - 1) g W_s / \gamma_0^2} \; . \qquad (9.51)$$

It is handy to present this parameter in energy units:

$$\Delta \tilde{\gamma}_s = 4\sqrt{g W_s / (k^2 - 1)} \; . \qquad (9.52)$$

To determine a distance between resonances, let us write the resonance conditions (9.32) and the averaged energy conservation law (9.28) for two neighboring resonances:

$$kp_{s+1} + (s+1)\Omega_0 - \gamma_{s+1} = 0 \; , \qquad \gamma_{s+1} - p_{s+1}/k = C \; ;$$
$$kp_s + s\Omega_0 - \gamma_s = 0 \; , \qquad \gamma_s - p_s/k = C \; .$$

One should keep in mind that the value of the constant C is the same for both resonances. Making use of these conditions, one gets the following value of the distance between the resonances:

$$\delta\gamma = \Omega_0 / (1 - k^2) \; . \qquad (9.53)$$

The expressions (9.52) and (9.53) indicate the following. If the inequality

$$g > \frac{\Omega_0^2}{4\left(\sqrt{W_s} + \sqrt{W_{s+1}}\right)^2 (1 - k^2)} \qquad (9.54)$$

holds, the sum of half–widths of the nonlinear resonances $(\Delta\tilde{\gamma}_s + \Delta\tilde{\gamma}_{s+1})/2$ is larger than the distance between the resonances $\delta\tilde{\gamma}$. In this case, there occurs overlapping of the resonances.

For practical applications, it could be convenient to rewrite (9.52), (9.53), and (9.54) in dimensional units:

$$\Delta\tilde{\gamma}_s = 4\sqrt{\frac{qEW_s}{mc\left(\omega^2 - k^2c^2\right)}} \; ;$$
$$\delta\gamma = \omega\Omega_0 / \left(\omega^2 - k^2c^2\right) \; ;$$
$$E > \frac{mc\omega\Omega_0^2}{4q\left(\omega^2 - k^2c^2\right)\left(\sqrt{W_s} + \sqrt{W_{s+1}}\right)^2} \; ,$$

9.3 Particle Interaction with Large Amplitude Wave

where

$$W_s \equiv \left(\frac{\alpha_x s p_\perp}{mc\mu + \alpha_z p_z/mc}\right) J_s(\mu) - \frac{\alpha_y p_\perp}{mc} J'_s(\mu) ; \quad \mu \equiv \frac{k p_\perp}{m\Omega_0}.$$

The expression (9.52) for the nonlinear resonance width and the condition (9.54) for the emergence of the particle motion stochastic instability are rather general. They describe the most important cases of the wave–particle resonant interaction. Really, (9.52) yields the nonlinear resonance width in the cases of the field–particle Cherenkov interaction ($s = 0$), for the cyclotron resonances ($k_z = 0$) and for nonlinear resonances Doppler normal ($s > 0$) and anomalous ($s < 0$) effect taken into account. Respectively, (9.54) describes the condition for the emergence of the stochastic instability, conditioned by overlapping of the corresponding nonlinear resonances.

We dwell now on certain specific cases.

1. Let us consider first of all the charged particle interaction with the longitudinal wave in a constant magnetic field. Under such conditions, a criterion of emergence of chaotic motion is determined in [55, 56, 57]. Corresponding expressions can be derived from (9.54). Surely, using (9.33) in the case of the longitudinal wave ($\alpha_x = k_x/k$, $\alpha_y = 0$, $\alpha_z = k_x/k$) and taking into account the resonance conditions ($s\Omega_0 + kp_z = \gamma$) one gets $W_s = \gamma J_s(\mu)/k$. Under the supposition that $\mu \gg 1$, (9.54) yields the following condition for emergence of the stochastic instability, stimulated by overlapping of the Cherenkov resonance ($s = 0$) with the neighboring Doppler-shifted resonances:

$$g > \frac{\Omega_0^2 \sqrt{\mu}}{\gamma(1-k^2)} \frac{1}{16} \sqrt{\frac{\mu}{2}}. \tag{9.55}$$

2. Consider now a transverse electromagnetic wave propagating perpendicularly to the external magnetic field. In this case, overlapping of resonances is conditioned by relativistic effects only.

For an E–wave with polarization $\{\alpha\} = (0, 1, 0))$, the criterion of overlapping is

$$g > \frac{\Omega_0^2}{16 p_\perp J'_s(\mu)}. \tag{9.56}$$

Note that it is independent of the longitudinal velocity.

For an H–wave ($\{\alpha\} = (0, 0, 1)$), (9.54) takes the form:

$$g > \frac{\Omega_0^2}{16 p_z J_s(\mu)}. \tag{9.57}$$

In contrast to the case of the E–wave, the value of the amplitude, required for the development of the stochastic instability, essentially depends on the magnitude of the particle longitudinal momentum.

3. Let us stay now on the condition (9.54) considering the particle motion in the field of a plane–polarized wave, propagating at the angle φ with respect to the external magnetic field in a medium, characterized by a dielectric constant $\varepsilon > 1$.

As regards overlapping of Cherenkov resonance ($s = 0$) with neighboring cyclotron resonances in the E–wave field $\{\alpha\} = (\cos\varphi, 0, \sin\varphi)$, the condition (9.54) takes the form:

$$g > \frac{\Omega_0 v_z}{16 J_0(\mu)\gamma(1 - v_z^2)\sin\varphi} . \tag{9.58}$$

In the case of the H–wave field (9.54) looks as

$$g > \frac{\Omega_0 v_z^2}{16 J_1(\mu) p_\perp (1 - v_z^2)} . \tag{9.59}$$

The formulae (9.58) and (9.59) indicate that an increase in the particle longitudinal velocity heightens the amplitude sufficient for overlapping of resonances.

4. Particular attention should be given to the case of a longitudinal wave propagating in vacuum. Here $k = 1$, and there is no stochastic instability within the framework of the given approximation. The resonance condition now coincides with the integral of motion (see (9.53b)). Changes in the particle energy, which result from the wave–particle interaction, do not cause any violation of the resonance condition. That is, the conditions of autoresonance [48, 49] are realized. So, one may state that the stochastic instability of the particle motion does not develop under the conditions of autoresonance.

5. From the viewpoint of stochastic acceleration, one could be interested in the case of a high–energy particle ($\gamma \gg 1$) interacting with a plane E–wave ($\{\alpha\} = (0, 1, 0)$) propagating perpendicularly to the external magnetic field ($k = 0$). For simplicity, the particle longitudinal velocity is considered zero ($p_z = 0$). Besides, we suppose that there takes place the wave–particle interaction at high cyclotron resonances ($s \gg 1$). The last condition corresponds to the case of the particle stochastic acceleration in the wave field, the frequency of which substantially exceeds the cyclotron one ($\omega \gg \Omega_0$). Here the resonance condition has the form: $\Omega_0 = s/\gamma$. As $p_{\perp s} \approx \gamma$, one gets $\mu \approx s \gg 1$. Consequently, the use can be made of the Bessel function asymptotic $J_s(\mu) \approx 0.44/s^{1/3}$. Then (9.54) yields

$$g > 0.28\, \Omega_0 s^{1/3} . \tag{9.60}$$

As it follows from (9.60), the wave amplitude, necessary for overlapping of resonances, increases with an increase in the resonance number.

9.4 Nonlinear Regime of Operation

In the previous section, we have demonstrated that if the wave amplitude is high enough so that there takes place overlapping of nonlinear resonances, the motion of charged particles becomes stochastic. One can expect that the amplitude of the excited field can reach this level as a result of the collective instability development. In this case, the motion of charged particles becomes stochastically unstable. Consequently, the system passes on to the regime of exciting stochastic oscillations. Besides, it is quite possible that the noncorrelated chaotic motion of the particles could hamper the instability development. That is, the mechanism for the stochastic instability development could play the role of a mechanism stabilizing the output power level.

To consider the subject, our starting point will be the self–consistent nonlinear problem of exciting microwave oscillations by a system of "cold" in-phase rotators in the coordinate frame where their longitudinal momentum is equal to zero. Besides, at the initial moment, the oscillators possess just the transverse component of their momentum. As above, z–axis is directed along the strength line of a homogeneous constant external magnetic field.

So, the distribution function may be presented as

$$f = \frac{\varrho}{p_\perp} \delta\left(p_\perp - p_{\perp,0}\right) \delta\left(p_z\right) \delta\left(\vartheta - \vartheta_0 + \Omega_0 t/\gamma\right), \qquad (9.61)$$

where ϱ denotes the density.

The complete self–consistent system of Eq. (9.61) describes the electromagnetic radiation emission by the particles. It contains the equations of particles motion and Maxwell's equations for the electromagnetic field proportional to $\exp(ikz)$:

$$\frac{d\mathbf{p}}{dt} = q\mathbf{E} + \frac{q}{mc\gamma}\left[\mathbf{p}\left(\mathbf{B} + \mathbf{B}_0\right)\right] ; \quad \frac{d\mathbf{r}}{dt} = \frac{\mathbf{p}}{mc\gamma} ;$$

$$\frac{\partial \mathbf{B}}{\partial t} = -ic\left[\mathbf{k} \times \mathbf{E}\right] ; \quad \frac{\partial \mathbf{E}}{\partial t} = ic\left[\mathbf{k} \times \mathbf{B}\right] ; \quad kE_z = -4\pi\varrho . \qquad (9.62)$$

Here $\mathbf{E}$ and $\mathbf{B}$ denote the electric and magnetic field strengths; ϱ is the charge density; q and m designate the charge and rest mass of particles, respectively. Note that the temporal dependence of the field is not singled out as a harmonic one in (9.62). This approach permits to describe temporal evolution of the fields and the motion of charged particles in the stochastic regime (i.e., when the excited fields are characterized by a broad frequency spectrum). The system (9.62) takes into account excitation of a longitudinal electric field, that is, the collective Coulomb field of charged particles.

Regarding the field strengths as harmonic functions of time, one deals with the problem of motion in a prescribed electromagnetic field. If $\varrho \to 0$, there arises the problem of motion of a single charged particle in the external constant magnetic field and in the field of an electromagnetic wave of a prescribed amplitude. This problem has been considered in the previous subsection. In particular, we have determined the conditions of appearance of a stochastic instability in the particles motion (9.54). The complete self–consistent system of Eq. (9.62) can be investigated only by numerical simulations. That has been carried out in [48]. Below we will briefly describe the most important results of this analysis.

The system (9.62) has been analyzed by numerical simulations under various values of the plasma frequency and fixed values of the cyclotron frequency $\Omega_0/\gamma = 0.5$. In this case, the resonance condition is fulfilled for $s = 4$. The field evolution in time, spectra of the excited fields, correlation functions, and evolution of the particles energy distribution have been displayed.

The result of the numerical analysis shows that the most important characteristics of particle dynamics and fields in a self–consistent system can be forecasted analyzing the single-particle dynamics in external electromagnetic fields.

If the density of charged particles is low ($\omega_b = 0.1$), the transverse component of the electric field is mainly excited. At the initial stage of the instability development, there takes place an exponential increase in the transverse electric field amplitude. Further, the amplitude heightening gives way to slow oscillations. These oscillations are stimulated by phase oscillations of particle bunches trapped in the wave field (see Fig. 9.8).

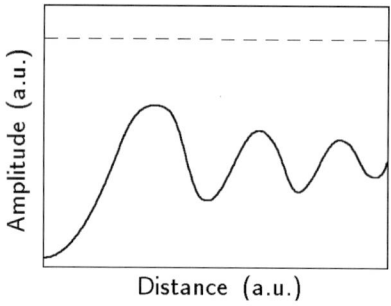

Fig. 9.8. Evolution of the envelope of the microwave transverse field amplitude. The resonances are not overlapping. The dashed line shows the threshold of stochastic instability

In this graph, the dash line designates the field power level, necessary for overlapping of the nonlinear resonances. In the electric field transverse component spectrum, there is a narrow peak at the basic frequency of the oscillations. In addition, there exist two satellites located on both sides of the peak (Fig. 9.9).

The presence of the satellites is conditioned by the wave modulation by phase oscillations of the bunches in the wave field. The correlation function of the transverse electric field oscillates with a slowly decreasing amplitude.

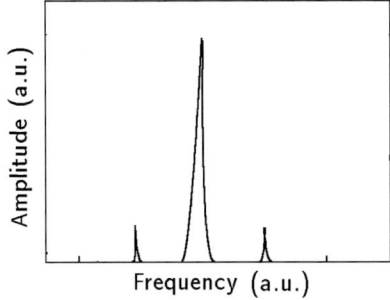

Fig. 9.9. Spectrum of the excited field. The case of an isolated resonance

The temporal dependence of the longitudinal electric field is somewhat more complicated. This is conditioned by superposition of the gyrofrequency harmonics. However, even in this case, there occurs an exponential increase in the wave field amplitude at the initial stage of the instability. Later on, the amplitude starts to oscillate at the frequency of the phase oscillations in the transverse wave field. In the longitudinal electric field spectrum, there exist several narrow peaks at the cyclotron frequency harmonics. The correlation function of the longitudinal field is a slowly decreasing periodic function of the frequency.

Thus, a beam of low density excites regular oscillations, characterized by a discrete spectrum. It is easy to see that the maximal amplitude of the transverse field is smaller than the field strength, necessary for overlapping of the resonances (9.54). Therefore, the particles are locked in an isolated resonance with the wave and their motion is practically regular. Analysis of the function of the particle distribution in energy indicates that the excitation of oscillations is accompanied by a broadening in the distribution function. However, this broadening remains within the limits of the nonlinear resonance width; that is, the particles keep on moving in an isolated cyclotron resonance, not passing on to the neighboring ones.

It is worth mentioning that the system efficiency, determined by the relation

$$\eta = \left(|E|^2/4\pi\right) q \Big/ \varrho_0 mc^2 \left(\gamma_0 - 1\right),$$

turns out to be rather high. Under conditions above, it reaches 37%.

If one heightens the particle density up to the values that correspond to the condition of nonlinear resonances overlapping, the instability development substantially changes. At the beginning of the process occurs an exponential increase in the transverse electric field amplitude as in the case of an isolated resonance. This increase is limited by trapping of particles by the field of the excited field (see Fig. 9.10; $t \leq 50$).

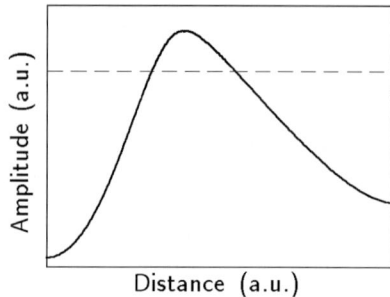

Fig. 9.10. Evolution of the transverse field amplitude envelope. The resonances are overlapping

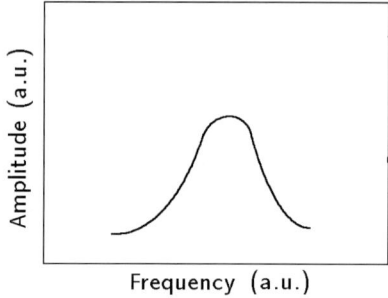

Fig. 9.11. Spectrum of the excited transverse field. The resonances are overlapping

The field power level is approximately twice as high as the level necessary for resonances overlapping. In consequence, the motion becomes chaotic. In its turn, the chaotic motion results in a chaotic modulation of the transverse field amplitude ($50 \leq t \leq 200$) and in appearance of a chaotic longitudinal field. The difference in a degree of their chaos can be explained in the following way. According to (9.62), the temporal evolution of the longitudinal Coulomb field is completely determined by the motion of charged particles. Respectively, the chaotic character of motion causes the self–consistent Coulomb field. The transverse electromagnetic field evolution is described by the inhomogeneous wave equation. Therefore, the beam chaotic current on the right–hand side of this equation can cause only irregular modulations of the transverse field complex amplitude.

The spectrum of the excited oscillations and the evolution of the distribution function correspond to this scenario of the instability development. Although the transverse field spectrum has a maximum at the basic frequency $\omega = 1$, it is substantially broadened. In contrast to the case of a low–density beam, the field correlation function quickly decreases in time. The longitudinal field spectrum is continuous, and it is much broader than the transverse field spectrum. The form of the distribution function indicates that until the moment $t \approx 40$ the instability behaves as in the case of an isolated resonance. However, starting from $t \approx 80$, the particle distribution function seizes several resonances. In addition to the slowed–down particles, there appears a group of stochastically scattered ones. Then the distribution function becomes more and more fuzzy. A number of the accelerated particles increases. However, generally speaking, the decelerated particles predominate over accelerated ones. This chaotic motion of oscillators accompanied by smearing of the energy distribution function causes a limitation of the self–consistent field. And what is more, starting from the moment $t \approx 225$, the field average amplitude corresponds to the value necessary for overlapping of the resonances. This means, in the long run, that the level of the field saturation turns out to be prescribed by the condition of overlapping rather then by trapping of the particles.

It is worth mentioning that even if the beam density is low so that the particles are under action of a single isolated resonance, their dynamics can also become chaotic. Really, the resulting self–consistent field is the wave field, the amplitude of which varies periodically in time. The particle motion in the field of this kind is equivalent to the motion in the field of three waves, the frequencies of which differ by frequency value of the bounce oscillations of the trapped particles. The wave amplitude values are large enough to provide nonlinear resonances of the three waves overlapping. Under these conditions, dynamics of the particle motion has to be chaotic. In its turn, the chaotic character of the particle motion has to cause smoothing down of the amplitude of modulation of the wave excited. However, numerical simulations indicate that the period of the wave modulation is much larger than the period of bounce oscillations. The reason of such prolonged maintenance of the regular modulation of the amplitude of the wave is the following. During the process of bunching, the particles mainly become bunched in the area of the phase space which corresponds to "an island" of stochastic stability of the particle motion. Finally, their motion does become chaotic but the amplitude of the excited oscillations decreases and is subjected just to small incidental modulations.

The principal attention has been paid above to the description of the interaction of charged particles with electromagnetic waves under the conditions of their cyclotron synchronism. No great attention has been paid to the operation of real microwave devices, which is a subject of extensive literature. Among them, gyrotrons and cyclotron autoresonance masers (CARMs) are of a special interest. As regards gyrotrons, the straight lines of the integrals are perpendicular to the resonance straight lines. As the particles move along the integral curves, there does not take place any energy interchange between the particles and waves within the framework of the small–signal approximation. The energy interchange is possible only when one regards finiteness of amplitudes of the waves. Under the conditions of an isolated resonance, the maximum amount of the particle energy transferred to the wave is of the order of magnitude of the nonlinear resonance width.

In CARMs, the resonance straight lines are parallel to the straight lines of the integrals. There can take place the infinitely long resonant field–particle interaction. Limitations on the magnitude of the energy transferred (either from the particles to the wave field or v.v.) are prescribed by the two reasons. It could be either depletion of an energy source or geometrical sizes of electrodynamic structures, where the field–particle interaction occurs. Notwithstanding the circumstance mentioned above, effectiveness of the gyrotron operation is all the same rather high reaching in practice tens of percent.

The above–studied physical mechanism of the field–particle cyclotron interaction permits to describe qualitatively new modes of CRM operation, that is, the stochastic regimes. The stochastic mechanisms indicate themselves more and more often while the power of the oscillations excited increases and one is advancing into a range of shorter wavelengths. Besides, this very approach can be used for deeper understanding of various processes, that is, particle acceleration and mechanisms of stochastic heating of an ensemble of charged particles [48]. In particular, not long ago, by making use of the above-described mechanisms, plasma heating up to high temperatures ($\approx 1.5\,\mathrm{MeV}$) has become possible, the effectiveness being rather high ($\approx 50\,\%$) [58, 59]. Probably, there exists no alternative to the described mechanism of stochastic heating. Really, there occurs a direct transformation of the regular wave energy to the energy of the particles chaotic motion without any intermediate stages.

An important conclusion, which can be made from the above-presented results, consists in the fact that, considering the dynamics of a single particle, one can describe correctly the entire physical picture of interaction of a flow of charged particles with the electromagnetic waves. That is, within the framework of the single–particle model, the levels of excited oscillations can be determined as well as thresholds for appearing the chaotic particle dynamics. Determined in this way the thresholds and levels are in fairy good agreement with the results of numerical simulation. In addition, having considered the case of an isolated resonance shows certain modulation at the bounce frequency. This modulation taken in mind, one can conclude that the analysis of single–particle dynamics produces not only correct qualitative results but also quantitative esteems of transition from the regular particle dynamics into the chaotic one. This enables one to determine the amplitude saturation level of the wave, the shape of the energy distribution function, and the main statistical characteristics of the excited field (spectra, correlation functions, dispersion, etc.).

10

Free Electron Lasers (FELs)

In the broad sense of the word, the term "Free Electron Laser"(FEL) implies a device the operation of which is based on the stimulated emission of undulator and/or Cherenkov radiation of relatively short wavelengths. However, Cherenkov mechanism, based on applying dielectric channels and diffraction lattices, is limited by the above–mentioned difficulties in slowing the wave velocity down to the velocity of the beam. The corresponding devices rather should be called the free electron masers.

Below we will use the term "FEL" in a narrower sense. That is, it refers to the devices the operation of which is based on applying straight beams under conditions of strong Doppler effect. The existence of this effect permits stimulating the radiation emission in the infrared, ultraviolet, and even in the x–ray ranges. Properly speaking, the very name of this device issues from the phenomenon of light amplification. At the same time, the term "FEL" does not depict the essence of the process so perfectly. The point is that electrons in FEL are not completely free because they move in external electromagnetic fields. However, introduction of this term has emphasized the general character of the processes that take place in quantum oscillators and purely classical devices, operating on the principles analogous with the work of a well–known TWT.[1]

The above–submitted basic principles permit us to clearly understand the price we pay for the shift toward the short wavelength range:

1. In practice, slowing down of short waves is possible only in dielectrics. An intense electron beam has to interact with the transverse electromagnetic waves propagating in vacuum at the velocity of light. To provide this interaction, electrons have to perform transverse oscillations of rather high frequencies. Besides, the Doppler transformation of the radiation frequency must be large. At present, magnetic undulators are the only practical means

---

[1] It is worth mentioning that the usual abbreviation "laser" (Light Amplification by Stimulated Emission of Radiation) implies the quantum nature of this process. At the same time, this term bears no direct references to it.

of generating oscillations of this type. However, generally speaking, the radiation emission in the electrostatic field of a crystal lattice and the backward induced scattering of an external counterpropagating wave are of interest as well.

2. For providing the large Doppler frequency transformation, the particle longitudinal velocity must be close to the velocity of light: $\beta_\| \approx 1$. Thus, the admissible amplitude of the transverse oscillations at the given frequency is small

$$\beta_\perp^2 = 1 - \beta_\|^2 - \gamma^{-2} \ll \gamma^{-2}.$$

Effectiveness of the energy transfer from the wave to the particle (and v.v.) due to the transverse electric field is in direct proportion to $\beta_\perp^2$. Consequently, as regards shortwave–range FELs, where the use of ultrarelativistic beams is inevitable, their length must be large – about scores or even hundreds of meters. Besides (or otherwise), rather heavy–current high–power beams must be used.

3. The last but not the least, shortness of the wavelength and the corresponding finesse of particles phasing, on which the mechanism of the stimulated radiation emission is based, require a high-beam quality. This implies, at least, that the beam must be almost monoenergetic. Besides, its transverse phase volume must be small and stable in time. All these requirements must be met under the conditions of a rather high current and high energy of the beam particles.

On the other hand, as regards stimulation of emission of the coherent shortwave radiation, potentials of FEL are much higher than those of other sources. In the first place, the following advantages should be enumerated.

1. The broadband frequency tuning can be realized both by varying the undulator controllable parameters and by changing the beam energy. In this respect, FEL does not meet competition from other sources, quantum oscillators included.

2. The spectrum of the radiation emitted by FELs rather easily overlaps the infrared range. The beam devices of other types do not fit this band. Quantum oscillators cannot operate in this range because there are no corresponding quantum transitions. This is the wavelength range (10–500 microns) where FEL potentialities can be used for important practical purposes, that is, the surface spectroscopy, physics of superconductivity, physics of narrowband semiconductors, chemistry of free radicals, etc. In the optical range, FEL can hardly concur with dye lasers, compact and inexpensive. At the same time, FELs are the only sources emitting the coherent radiation within UV and far UV ranges. As regards the spectral brightness and intensity, FEL exceeds synchrotrons by many orders of magnitudes (see Fig. 10.1).

3. If one compares FEL with ordinary lasers, the following fact could strike. There does not exist working body, subjected to the overheating at a high

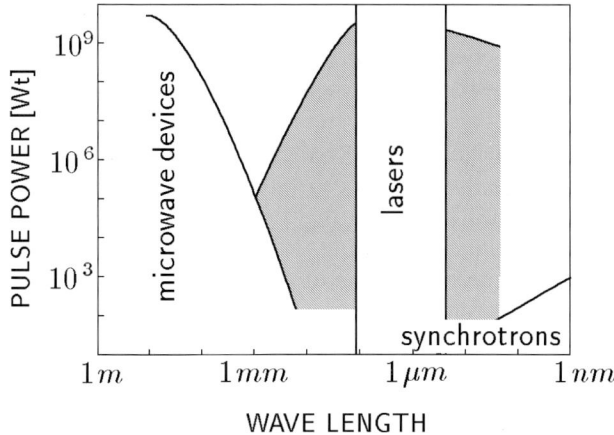

Fig. 10.1. Up-to-date potentialities of free electron lasers

level of the radiation average power. As a matter of fact, the notion of the working body may also imply the electron beam. However, the beam skips all over through FEL at the velocity of light. As a rule, overheating of the beam is not essential. Thus, there arises an opportunity of generating very high powers, not only of the pulse nature, but average as well. The prospects for the FEL applications are very teasing – to start from the stimulated pulse fusion and up to the directed energy weapons.

In all fairness, it should be mentioned that all these advantages could hardly be combined in the same device. There exists inexorable contradiction between obtaining the maximum gain in a narrowband system, and the high efficiency and output power in the broadband system. Because of this problem, FELs have been divided into the two classes: oscillators and amplifiers. It is worth considering the devices of these types separately.

10.1 FEL–oscillators. Low-Gain Regime

As it has been already mentioned above that, notwithstanding the common principles of their operation, the devices based on the emission of the stimulated undulator radiation are divided into the two classes: oscillators and amplifiers. To a large extent, they differ in the currents available and in the beam quality. These differences condition not only technical but also physical peculiarities of each system.

Surely, this difference originates from the coefficient of amplification (gain) obtainable or from the characteristic length of the field spatial increase. The last notion is applicable only when one deals with the systems sufficiently long so that the increase in the field strength establishes in the spatially exponential form at the distances, rather remote from the input. At the short distances, the field spatial increase is much slower. Hence, if the system length admissible is limited, the spatial coefficient of the field amplification can be very close to unit. All the same, if the gain exceeds the unavoidable losses, by providing deep and positive feedback, this system can be transferred to the regime of generation with a rather high level of the output power.

Generally speaking, both the beam and field feedbacks can be realized. However, it is rather difficult to bring the beam back to the system input without losing the obtained bunching. It is much simpler to realize the field feedback. For this purpose, surfaces of a rather good reflectivity can be inserted at the input and output of the system. At the output, the mirror must be semitransparent for extracting a part of the circulating power from the system. The input mirror is necessary for the excitation of the copropagating reflected wave. Under the condition of large Doppler effect, the beam can interact only with such wave.

10.1.1 Optical Cavity

We are interested in generating high frequencies so that the wavelength is supposed to be much shorter than any of the system geometrical sizes. So the wave is a superposition of almost free plane waves, their angular deviation from the beam being very small. Thus, we logically come to the scheme of an open optical cavity operating at a very high mode. Apropos, in such an "oversized" system, the problem of the feedback sign becomes insignificant; that is, there is no need to control the reflected wave phase. If positions of the mirrors are strictly fixed, among multitude of the almost identical possible modes always exist the ones characterized by the positive feedback. Those very modes are self–excited. As far the authors know, even the very question about design and construction of a single-mode short-wave FEL-oscillator has not yet been brought up.

As regards the scheme depicted in Fig. 10.2, the electron beam is moving in the standing–wave field.[2] However, the beam interacts only with the copropagating component of this standing wave because the condition for synchronism is not satisfied for the counterpropagating component. Consequently, it is perfectly legitimate to apply the analytical conclusions elaborated in Sect. 6. True, in Sect. 6, the mechanism for the induced phasing has been investigated for the Cherenkov radiation emission. All the same, it is easy to see that the equations that describe the process of phasing are isomorphic with those derived for the undulator radiation stimulated emission under the conditions of the Doppler normal effect.[3]

---

[2] To be more precise, the wave is quasi standing – neglecting to the output power, respectively low in comparison with the power that is circulating within the cavity.

[3] Generally speaking, the regime of the Doppler anomalous effect could be used as well. But, naturally, then a structure or a medium is necessary to provide slow waves propagation.

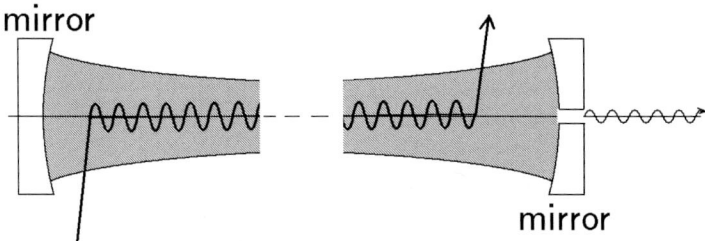

Fig. 10.2. The scheme of the standing–wave FEL–oscillator. The dashed area is occupied by the beam of light

However, as regards the oscillator, some other aspects should be emphasized. Above we have dwelled on the response of the particles of different energies to the wave of a prescribed frequency, propagating along z–axis. But now we consider evolution of various modes in the presence of the beam of a prescribed energy. In FEL–oscillators of the short–wave range, the open optical cavity is used as a resonant element (perhaps, there could be no alternative to it). Hence, it is worthwhile to dwell on the cavity intrinsic oscillations themselves.

First of all, the open FEL cavity length reaches scores of meters, while the wavelength is shorter than hundred of microns. So, it operates under very high numbers of the longitudinal harmonics. As a matter of fact, their spectrum can be regarded as continuous. As a rule, this is not justified for the spectrum of the transverse wave numbers, notwithstanding the fact that the cavity transverse sizes are also large in comparison with the wavelength excited. The matter is that an open system always emits radiation. Consequently, generally speaking, its quality value is low. The modes characterized by relatively small transverse wave numbers make an exception when they are excited with the help of slightly concave mirrors and are limited by caustics, typical of quasi–optics. Totality of these modes form so–called light beams of finite transverse sizes. Inside these beams, the wave is quasi plane in the sense that its wave front is close to the plane perpendicular to the beam. At the same time, outside caustics practically does not exist any field at all.

The behavior of beams of light is described by particular solutions of the wave equation.[4] These beams are characterized also by two transverse wave numbers. In contrast to the case of proper oscillations in a closed volume, these solutions do not form a complete system of eigenfunctions of the wave equation. The fields, not taken into account by these solutions, generate a certain nonresonant background. A part of the source power is spent on exciting this background (it is especially characteristic of a localized power source). However, the following fact is predictable quite easily. Generally speaking, these are the beams that have to be excited by the sharply directed paraxial

---

[4] To be more precise, one deals with the wave equation in the so–called parabolic approximation.

radiation emitted by a relativistic electron flow of small transverse dimensions. Therefore, the nonresonant background generates fields, small in comparison with the field induced by the beams, localized relatively sharply and subjected to multiple reflections from the mirrors. To some extent, Fig. 10.2 depicts caustic and the beam behavior in the open cavity. One can find more detailed information on this subject in the special literature, for example, [60]. There are rather many works dedicated to this theme because open cavities long ago have become an integral part of laser technology.

Below we will limit ourselves to the most illustrative case of the cavity, characterized by the rotational symmetry, (as regards a plane undulator, the presentation of the field in the form of plane beams of light would be more adequate). Besides, we examine the beams without azimuthal variations, which corresponds to geometry of a helical undulator (however, if the electron trajectory is as a whole shifted with respect to the undulator axis, this presentation is somewhat incomplete). Under these conditions, the frequencies of the cavity of the length L_c can be obtained from the approximate expression:

$$k_0 L_c = s\pi + 2(2n+1)\sigma; \qquad s, n = 0, 1, 2, \ldots . \tag{10.1}$$

Here the parameter σ is prescribed by the radius of curvature of the mirrors $\mathcal{R}$:

$$\sin \sigma = \sqrt{L_c/2\mathcal{R}} . \tag{10.2}$$

In particular, this formula indicates that the radius of curvature of the mirrors must exceed the half of the cavity length; otherwise, there cannot exist any stable beam. The parameters L_c and $\mathcal{R}$ also prescribe the size of the beam of light at the caustics crossover a and on the mirror surfaces a_0:

$$2k_0 a^2 = \sqrt{L_c(2\mathcal{R} - L_c)}; \quad k_0 a_0^2 = 2(2n+1)\sqrt{2\mathcal{R}L_c} . \tag{10.3}$$

For providing the existence of the beam of light, it is evident that the parameter a_0 must be smaller than the mirror radius a_m. This inequality determines the maximal number of the beams that can be excited. Thus, the mirror minimal radius admissible is by the order of magnitude equal to the geometrical average of the cavity length and the length of the wave excited. In the crossover, the radius of the beam of light can be several times smaller than the geometrical average. This radius becomes larger as the parameter a_m increases; that is, as the number of the beams of light, which could be excited, increases.

Inside the beam of light, the field amplitude distribution depends on the beam number n:

$$E_0 \propto \cos \zeta \, L_n \left(\frac{r}{a} \cos \zeta \right)$$
$$\times \exp \left[-i(2n+1)\zeta - \frac{r^2}{2a^2} \cos \zeta \, \exp(-i\zeta) \right]. \tag{10.4}$$

Here the distance

10.1 FEL–oscillators. Low-Gain Regime

$$\zeta = \arctan \frac{z}{k_0 a^2}$$

is counted out from the cavity middle; L_n are Laguerre polynomials:

$$L_0(x) = 1; \quad L_1(x) = x^2 - 1; \quad L_2(x) = x^4 - 4x^2 + 2\ldots .$$

Because of this fact, the particular solutions of (10.4) are called Gauss–Laguerre beams.

In the cavity cross section, the principal beam field obeys Gauss distribution. If the electron beam transverse size is much smaller than a, the field can be regarded as homogeneous. The fields induced by the beams of greater numbers go to zero at certain radii. Therefore, the estimations that are submitted below must be somewhat corrected.

10.1.2 Gain

Proceeding from the above–presented reasoning, let us consider the standing transverse plane wave that occupies the area $r < a$. We also neglect the wave front distortions, caused by finiteness of the beam radius:

$$E_x = E_0 \cos(k_0 z) \cos k_0 c t.$$

Let us consider an electron of the energy γ moving along z–axis with the velocity $\beta_z c$. Besides, the electron is supposed simultaneously to perform small transverse oscillations along the x–axis with the velocity $c\beta_x = c\beta_\perp \cos k_u z$ ($l_u = 2\pi/k_u$ is the undulator period). The longitudinal component of the electron velocity is also modulated because of these oscillations. Respectively, the moment at which the electron reaches a point z in the axis is a nonlinear function of z:

$$\beta_z = \sqrt{\beta^2 - \beta_\perp^2 \cos^2 k_u z} \approx \beta \left(1 - \frac{\beta_\perp^2}{2\beta^2} \cos^2 k_u z\right); \quad (10.5)$$

$$\int \frac{dz}{\beta_z} \approx \int \left(\frac{1}{\beta} + \frac{\beta_\perp^2}{2\beta^3} \cos^2 k_u z\right) dz.$$

The wave performs the work on the electron per a unit of length:

$$w = qE_0 \frac{\beta_\perp}{\beta} \cos k_u z \cos(k_0 z) \cos \int k_0 \left(\frac{1}{\beta} + \frac{\beta_\perp^2}{2\beta^3} \cos^2 k_u z\right) dz . \quad (10.6)$$

On the right–hand side of the expression for the performed work, we have neglected small corrections, caused by oscillations of the electron longitudinal velocity.[5] However, in phase relations, where one deals with a difference between two great magnitudes, these corrections must be taken into account.

---

[5] These corrections can be essential in the case of high harmonics, for which the condition for synchronism takes the form $k_0/\beta - k_0 \approx sk_u$.

Really, let us single out in (10.6) the quasi–synchronous component in the product of the two cosine functions. Neglecting the terms oscillating at a high frequency of the order of the radiation field frequency, we get

$$w \approx \frac{qE_0\beta_\perp}{2\beta} \cos k_u z \cos \int \left(\frac{1}{\beta} + \frac{\beta_\perp^2}{2\beta^3} \cos^2 k_u z - 1 \right) k_0 \, dz . \qquad (10.7)$$

We now average this expression over the oscillator period, considering the particle energy as constant within this period. Making use of the well–known presentation of Bessel functions,

$$\overline{\cos(Y \sin 2z)} = J_0(Y) ; \quad \overline{\sin 2z \sin(Y \sin 2z)} = J_1(Y) ,$$

one gets

$$\overline{w} = \frac{qE_0\beta_\perp}{4\beta} \left[J_0 \left(\frac{\beta_\perp^2 k_0}{8\beta^3 k_u} \right) - J_1 \left(\frac{\beta_\perp^2 k_0}{8\beta^3 k_u} \right) \right]$$
$$\times \cos \int \left(k_u - \frac{k_0}{\beta} - \frac{\beta_\perp^2 k_0}{4\beta^3} + k_0 \right) dz . \qquad (10.8)$$

For simplifying the further calculations, we limit ourselves to the case of very high energy, supposing that $\beta = 1$, wherever possible, and using the above–mentioned undulatory coefficient. Then (10.8) yields the system of equations, already derived in Part II. This system describes induced phasing and radiation losses:

$$\frac{d\gamma}{dz} = g \cos \varphi ; \quad \frac{d\varphi}{dz} = k_u - \frac{k_0}{2\gamma^2} (1 + \mathcal{K}^2) , \qquad (10.9)$$

Here

$$g = \frac{qE_0\mathcal{K}}{2^{3/2}\gamma mc^2} \left[J_0 \left(\frac{\mathcal{K}^2 k_0}{4k_u\gamma^2} \right) - J_1 \left(\frac{\mathcal{K}^2 k_0}{4k_u\gamma^2} \right) \right] ; \quad \mathcal{K}^2 = \frac{\beta_\perp^2 \gamma^2}{2} . \qquad (10.10)$$

In the case of a helical undulator and a circularly polarized wave, the calculations are somewhat simpler because the oscillator longitudinal velocity is not modulated. As before, the equations of phasing preserve the form of (10.9). However, now

$$g = \frac{qE_0\mathcal{K}}{2\gamma mc^2} ; \quad \mathcal{K}^2 = \beta_\perp^2 \gamma^2 . \qquad (10.11)$$

The physical reason of this difference is clear. Really, the undulatory coefficient is prescribed by the particle mean–square transverse velocity. In the case of a helical motion, this parameter is determined by the particle highest transverse velocity. If the particle motion is plane, the particle mean–square transverse velocity is $\sqrt{2}$ times lower.

As compared with the estimations of radiation emitted by a moving harmonic oscillator, systematic and oscillating variations in the particle longitudinal velocity are taken into account in (10.10) and (10.11). These changes

10.1 FEL–oscillators. Low-Gain Regime

are conditioned by the finiteness of the mean–square transverse velocity $\beta_\perp^2$. In particular, one can see that striving for heightening the radiation power by heightening the amplitude of the transverse oscillations is unexpectedly hampered by the limitation:

$$\mathcal{K} \ll \gamma . \tag{10.12}$$

Starting from $\mathcal{K} \approx 1$, the violation of this condition causes a strong decrease in the resonant frequency:

$$k_0 \approx \frac{2\gamma^2 k_\mathrm{u}}{1 + \mathcal{K}^2 + \gamma^2 \vartheta^2} . \tag{10.13}$$

Thus, the angle made by the particle trajectory with the direction of propagation of emitted radiation should not exceed γ^{-1} (it is the natural angle of the spontaneous radiation emission). From this viewpoint, this limitation can be hardly regarded as unexpected one.

One can suppose that the presence of the small factor $\mathcal{K}/\gamma$ in the expressions for g essentially diminishes the efficiency of the wave–particle interaction in comparison with the case of the Cherenkov radiation emission. However, the reader should keep in mind that the free wave transverse electric field can substantially exceed the longitudinal field in a slow wave structure—to say nothing that in the case of ultrashort waves the latter can hardly be realized.

The amplification factor is small even notwithstanding the fact that the interaction length can be relatively large. This is conditioned by the existence of distance along the undulator where a longitudinal modulation of the beam particle density develops. As regards the mathematical model, this fact is related to the smallness of the coefficient $\alpha \approx \gamma^{-3}$. From the viewpoint of physics, it is conditioned by very slow bunching of the relativistic particles. In practice, their longitudinal velocity is equal to the velocity of light c and weakly depends on the electron energy. An original device that permits to overcome this drawback is called an optical klystron [69].

The undulator of the optical klystron is divided into two parts, optically connected. They are the modulating and emitting ones. A strongly dispersive magnetic insert is placed between them. The time of flight of an electron through this insert is prescribed by the particle energy rather by its velocity because the particles of different energies fly through this insert along essentially different trajectories. As a result, the bunching area length becomes substantially shorter. If the undulator length remains the same, the coefficient of amplification increases essentially. As usual, a drawback follows from an advantage. This scheme is very sensitive to the beam energy deviations. However, under the specific conditions of a storage electron ring, the optical klystron permits exciting the wavelengths of the ultraviolet range.

10.1.3 The Self–Excitation Threshold

To describe qualitatively the process of the self–excitation of oscillations, let us suppose that the light beam modes characterized by different k_0 are excited independently from one another. Generally speaking, this statement is justified only in the case when the initial signal represents a rather broadband noise, characterized by the spectrum width, which substantially exceeds the difference in frequencies of the neighboring modes (c/L_c). Under these conditions, phases of the excited modes are statistically independent. The total energy stored in the cavity is equal to the sum of energies of individual modes.

Here we will not reiterate the calculations performed in Chap. 6. We just submit the result in the corresponding interpretation. The small–signal theory does not take into account the interaction of different modes. In a unit of time, each mode receives from the beam the energy

$$-\frac{Imc^2}{q} \langle \gamma(L) - \gamma_i \rangle = g^2 I \frac{\alpha k L^3 mc^2}{4q} \frac{\mathrm{d}}{\mathrm{d}\mu} \left(\frac{\sin \mu/2}{\mu/2} \right)^2 . \quad (10.14)$$

Only the modes of frequencies lower than the resonance frequency (10.13) are amplified. The mode with $\mu \approx 2.6$ is subjected to the maximal amplification.

Neglecting some transverse and longitudinal inhomogeneities of the field, let us suppose that the stored–in–cavity energy of the mode, characterized by the electric field amplitude E_0, is equal to the cavity volume, multiplied by $|E_0|^2/8\pi$. Making use of the relation of E_0 to g, the energy balance for each mode may be written as

$$\frac{\partial W}{\partial t} = AF(\mu)IW - \frac{Wc}{QL_c} . \quad (10.15)$$

Here

$$A = \frac{q\alpha k L^3 \mathcal{K}^2}{2a^2 L_c \gamma^2 mc^2} .$$

In the neighborhood of its maximum, the function

$$F(\mu) = -\frac{\mathrm{d}}{\mathrm{d}\mu} \left(\frac{\sin \mu/2}{\mu/2} \right)^2$$

can be approximated by a parabola

$$F(\mu) \approx 0.27 - 0.04 \left(\mu - 2.6\right)^2 .$$

The first term on the right-hand side of (10.15) describes the beam energy losses for the stimulated radiation emission. The second term characterizes diffraction and dissipation losses during the wave double passage along the cavity. The parameter Q is the mode quality in units of L_c/λ. These units are convenient for comparing this magnitude with the wave energy relative absorption ν when the wave is reflected from the mirror. Really, if one neglects the diffraction losses,

$$Q^{-1} = 2\nu + \nu' . \quad (10.16)$$

Here ν' is the part of the wave energy extracted from the cavity (one talks about the wave falling on the output semitransparent mirror). According to its physical sense, Q characterizes the number of the light–pulse passages between the mirrors prescribed by the losses.

As it is easy to see, the system gets into the self–excitation regime if the current exceeds a threshold (or starting) value:

$$I_{\text{th}} \approx \frac{4c}{AQL_c} = \frac{4a^2\left(1+\mathcal{K}^2\right)}{Q\alpha k_u L^3 \mathcal{K}^2} I_0 \; ; \quad I_0 = \frac{mc^3}{q} \approx 17\,\text{kA}\,. \tag{10.17}$$

In principle, if the current in FEL were precisely equal to the threshold value, the device would operate in the single–mode regime. However, the time of stabilization of the oscillations would become infinitely long. Naturally, realization of this regime is senseless. If the current exceeds a threshold value, there arises a frequency band at the linear stage of the oscillation excitation. If this excess is small, the modes characterized by

$$\mu = 2.6 \pm 2.5\sqrt{(I - I_{\text{th}})/I_{\text{th}}}\,, \tag{10.18}$$

are exponentially amplified in time. If the excess is great, there takes place amplification of almost all modes within the frequency band $0 < \mu < \pi$ of stimulated radiation emitted.

The band of the relative width

$$\frac{\Delta k_0}{k_0} \approx \frac{\Delta \mu}{2\pi N} \ll 1 \tag{10.19}$$

corresponds to the band $\Delta \mu$. The band width conditionally determines the spectral composition of the radiation emitted by a FEL–oscillator. In this context, the term "conditionally" implies that within the framework of the small–signal theory the modes characterized by different μ are exponentially increasing with different time constants. The width of the real frequency band (e.g., at half of the maximum intensity) is somewhat varying in the process of the oscillation excitation. Besides, the frequency band width also depends on the mirror selective reflectivity (the so–called mode selection). Anyway, if all other conditions are identical, the spectrum is broadening as the current is increasing.

10.1.4 Steady–State Oscillations and Output Power

As regards the radiation instability, the small–signal theory predicts the threshold and increment of the wave exponential increase. However, the output power in the steady–state regime of the excited oscillations cannot be estimated within the framework of the theory. In practice, limitations on the output power are conditioned by many factors – to start from the finite duration of the current pulse which limits establishment of oscillations in a high–quality cavity and up to purely technological reasons (e.g., resistive losses in the mirrors). Here we will discuss only one of these power–limiting factors. Being important from the viewpoint of physics, it is inherent in all mechanisms of interaction between a beam and a traveling wave. That is, one talks about nonlinear saturation of the interaction caused by capture of particles by the wave.

Within the small–signal theory, the power of stimulated radiation is regarded as proportional to the field energy stored in the cavity. In essence, within this approximation the radiation reaction is taken into account just with respect to the beam phasing but not with respect to changes in the beam energy. In other words, the small–signal theory is true only when lengths of the beam–particle interaction are smaller than a certain saturation length, inversely proportional to the field amplitude, because the maximal change in the electron energy is crucial in this case.

To find the steady-state regime and the saturation length, one must take into account the interaction of modes. The methods of numerical simulations must be used for this purpose. One can obtain the rough estimates under the supposition that the saturation is determined by the decay of the mode when in the phase plane of Fig. 6.2 the first "fold" arises, that is, when the particle energy becomes ambiguously related to the phase. If the undulator length exceeds the saturation length in the single – mode regime, there arises a deep modulation of the oscillations, the modulation frequency being equal by the order of magnitude to the frequency of synchrotron oscillations. If there exist many modes – and also as a result of a nonlinear decay of a single mode – these oscillations are smoothed away. The beam energy transferred to the wave reaches then a constant level (see Fig. 10.3).

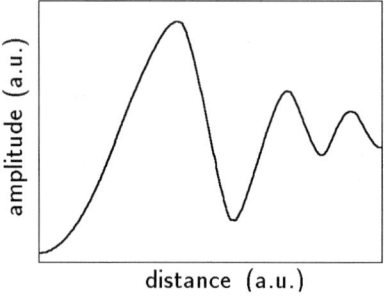

Fig. 10.3. Saturation of the amplification and nonlinear beats of the amplitude

Proceeding from these considerations, let us suppose that the amplitude exponentially increases in time until the saturation length becomes equal to the interaction length, which prescribes the output level of power. We now make use of (6.27), which describes the dependence of the particle phase on the traversed path and on the initial conditions:

$$\varphi = \varphi_i + \zeta + \frac{g}{\alpha k \delta_i^2} [\cos \varphi_i - \cos(\varphi_i + \zeta) - \zeta \sin \varphi_i] . \tag{10.20}$$

Supposing that $\partial \varphi / \partial \varphi_i = 0$, one gets the equation which determines the distance $z_0(\varphi_i)$, where there arises the fold under a fixed value of φ_i:

$$\alpha k \delta_i^2 + g [\sin(\varphi_i + \alpha k \delta_i z_0) - \alpha k \delta_i z_0 \cos \varphi_i - \sin \varphi_i] = 0 . \tag{10.21}$$

Minimizing this distance with respect to φ_i, one gets the saturation length $L_s = z_{0\,\mathrm{min}}$. The minimizing value of φ_i is determined by the relations:

$$\sin \varphi_i = \frac{1 - \cos \zeta}{\sqrt{(1 - \cos \zeta)^2 + (\zeta - \sin \zeta)^2}} ;$$

$$\cos \varphi_i = \frac{\zeta - \sin \zeta}{\sqrt{(1 - \cos \zeta)^2 + (\zeta - \sin \zeta)^2}} , \tag{10.22}$$

where $\zeta = \alpha k \delta_i L_s$.

Together with (10.21), this system permits finding L_s under a prescribed value of the field amplitude. However, now we are interested in determining the field amplitude value under which the saturation length is equal to the interaction length, that is, $\zeta = \mu$:

$$g_s = \frac{\alpha k \delta_i^2}{\sqrt{(1 - \cos \mu)^2 + (\mu - \sin \mu)^2}} . \tag{10.23}$$

The average energy loss at this distance is equal to

$$\langle \delta - \delta_i \rangle_{\mathrm{max}} = -\delta_i \frac{1 - \cos \mu - (\mu/2) \sin \mu}{(1 - \cos \mu)^2 + (\mu - \sin \mu)^2}$$

$$= -\frac{\mu}{\alpha k L} \frac{1 - \cos \mu - (\mu/2) \sin \mu}{(1 - \cos \mu)^2 + (\mu - \sin \mu)^2} . \tag{10.24}$$

For the optimal mode ($\mu = 2.6$) and $\alpha \approx \gamma^{-3}$, this yields the output power and efficiency:

$$P_{\mathrm{out}}\,[\mathrm{MW}] \approx 0.1 \left(1 + K^2\right) \frac{\gamma}{N} I\,[\mathrm{A}] ; \quad \eta \approx 0.2 \frac{1 + K^2}{N} . \tag{10.25}$$

There is another approach to the qualitative estimation of the saturation length. This notion can imply the path where occurs one forth of a linear synchrotron oscillation for a particle inside the separatrix. This length is equal to $\pi/2\alpha k$ so that the corresponding estimations coincide with the ones submitted above up to a numerical coefficients. However, both the estimations are rather ambiguous. In the first case, the estimation is made at the limit of applicability of the small–signal theory. In the second case, the particles are supposed to be completely trapped, which hardly corresponds to the regime with the optimal $\mu \approx 2.6$. One can just state that the efficiency is equal to $(2N)^{-1}$ by the order of magnitude (N is the number of the undulator periods). The physical meaning of this estimation is quite clear: while emitting the radiation at the given frequency, the electron changes its energy values within the range where it still can stay in resonance with the wave field, that is,

$$\frac{\Delta \gamma}{\gamma} = \frac{\Delta k_0}{k_0} \approx \frac{1}{2N}.$$

10.1.5 Beam Quality

The model presented above is based on several idealized suppositions. In particular, the beam is regarded as initially monoenergetic. Besides, all its particles are considered to be moving along a prescribed undulating trajectory. Surely, both the suppositions are of the idealistic nature. The degree of the approximation should be specified separately.

First of all, the mechanism of the stimulated radiation emission in a beam system means that the energy spread of the beam particles must be at least smaller than the optimal detuning δ_i. Otherwise, almost half of the particles get into the absorption band according to the condition for their synchronism with the wave of a given frequency. This circumstance surely diminishes substantially the stimulated radiation intensity. Taking into account that $\delta_i/\gamma \approx (2N)^{-1}$, one must state that the demand for the beam to be monoenergetic is especially important when one deals with generating systems with long undulators.

Besides, as it follows from the reasoning presented above, the angular spread of beam particles influences substantially the effectiveness of the wave–particle interaction as well. Really, an angular spread of the particles that enter the undulator immediately results in the spread of trajectories characterized by different inclinations with respect to the direction of the wave propagation. Roughly speaking, the particles, for which this angle exceeds γ^{-1}, do not participate in the process of interaction because they are actually non-synchronous with the wave at the given frequency. Seemingly, the angular spread $\Delta \varphi$ could be diminished by the methods of electron optics. However, as well as in classic optics, this decrease is achievable only by enlarging the transverse positional spread Δr (the beam transverse size). Thus, the beam phase volume $\epsilon = \Delta \varphi \Delta r$ called "emittance" just does not increase in the best case. According to geometrical reasoning, the positional scattering of the particle trajectories must not exceed $2\pi/k_u \gamma$. This condition provides the overlap of the electron beam by the natural cone of the radiation emission – at least within one undulator period – for preserving the "transverse" coherence. Consequently, there arises a simple but rather unpleasant condition:

$$\epsilon < l/\gamma^2 \approx \lambda.$$

For instance, the beam emittance in the light range has to be smaller than one mm×mrad. This magnitude is by a couple of orders smaller than the value typical for linear accelerators operating in the range 10–100 MeV. One should keep in mind that smallness of the emittance must be provided under the condition of large current, which brings additional complications. More complete information on the subject of the influence of various imperfections of the beam and the FEL electron–optical system is available in the monograph [61].

In this connection, we are just going to mention the special role played by proper matching of the phase volume occupied by the beam and the undulator acceptance. According to Liouville theorem, the emittance remains invariable under any of the beam transformations. Strictly speaking, this is correct for the six–dimensional phase space including the longitudinal extension of the beam and its energy spread. In many cases, the most strict conditions are imposed either on the transverse or on the longitudinal degrees of freedom. The phase volume then can be optimized by introducing the corresponding correlation between these degrees of freedom. Sometimes this results in obtaining a rather impressive increase in the gain (the so–called "beam conditioning").

Surely, in addition to the demands enumerated above, one should regard the problems of putting the beam precisely into the required trajectory. Besides, the influence of the space charge can be great and cannot be neglected. These questions are not to be discussed here. However, one can state that from the viewpoint of the technical problems, additional equipment and, finally, expenses, the accelerator–injector and the systems controlling the beam make the most important problem in constructing a short–wave FEL.

We are also not going to discuss the problems of providing accuracy and stability of the optical cavity, the length of which is very large according to the laboratory norms (about scores of meters). It just should be emphasized that the transverse size of the beam of light must be as small as possible. This factor is important not only from the viewpoint of heightening the energy density of the radiation field, on which the coefficient of amplification depends. In addition, this is the way of diminishing the transverse size of the undulator working area very essential for practical realization.

10.2 FEL–Amplifier: High-Gain Regime

If the beam current is so large and/or the undulator is so long that the gain substantially exceeds unit, the energy density of the stimulating field already cannot be regarded as constant all along the interaction region. The group velocity of the wave emitted is very close to the velocity of particles. Therefore, in the absence of the cavity, the radiation power has to increase exponentially in space remaining constant in time in the steady-state regime. This mode of operation in the absence of a feedback can be called the regime of the direct spatial amplification. For its realization, surely, an initial input signal is necessary.

First of all, the regime of direct amplification is attractive from the viewpoint of obtaining very high powers when the mirror stability becomes an essentially limiting factor. And what is more, up to now this regime is the only one which can be realized in the ranges of vacuum ultraviolet and soft x–rays, where no mirrors reflecting effectively have yet been elaborated.

10.2.1 Dispersion Relation

To calculate the gain, that is, to derive and to solve a dispersion equation, certain features of eigenwaves in a uniform longitudinal magnetic field of Sect. 7.1.3 are to be revised. A beam in an undulator moves across the magnetic field in a system with periodic parameters. Hence, the eigenwaves, strictly speaking, cannot be harmonic functions of z. Moreover, they cannot be classified as TE or TM modes so that electromagnetic waves are accompanied by space-charge waves even in an uniform one-dimensional flow. Note that the one-dimensional model is valid if the beam transverse size essentially exceeds the wavelength but remains smaller than the undulator period. The last condition is necessary to treat the undulator magnetic field as transversely homogenous one.

In the one-dimensional approximation, the nonlinear fluid equation of motion has a simple solution. The transverse momentum (in units of mc) equals to
$$\mathbf{p}_\perp = -\frac{q}{mc^2}\mathbf{A}_\perp, \tag{10.26}$$
where $\mathbf{A}_\perp = \mathbf{A}_u + \mathbf{A}_w$ is a vector potential including vector potentials of the undulator field $\mathbf{A}_u$ and of the wave $\mathbf{A}_w$. The solution can be readily checked by straightforward substitution the following relations being taken into account:
$$\mathbf{B}_\perp = \mathrm{rot}_\perp \mathbf{A} = \frac{\partial}{\partial z}[\mathbf{e}\times\mathbf{A}_\perp]\,; \quad \mathbf{E}_\perp = -\frac{1}{c}\frac{\partial}{\partial t}\mathbf{A}_\perp.$$
Obviously, it follows directly from the generalized momentum conservation law.

If the transverse velocities driven by the wave and undulator fields are small as compared with longitudinal one, it follows from (10.26) that
$$\mathbf{v}_{\perp w} = -\frac{q}{mc\gamma}\mathbf{A}_w\,; \quad \mathbf{v}_{\perp u} = -\frac{q}{mc\gamma}\mathbf{A}_u\,. \tag{10.27}$$
In the same approximation, the longitudinal component of the equation of motion reads:
$$\left(\frac{\partial}{\partial t} + \bar{\beta}c\frac{\partial}{\partial z}\right)v_\| = \frac{q}{m\gamma^3}\left(E_\| + \frac{1}{c}\mathbf{e}\left[(\mathbf{v}_w+\mathbf{v}_u)\times\nabla(\mathbf{A}_w+\mathbf{A}_u)\right]\right). \tag{10.28}$$

Taking into account the transverse integral of motion, one gets

$$\left(\frac{\partial}{\partial t} + \overline{\beta}c\frac{\partial}{\partial z}\right) v_{\parallel} = \tag{10.29}$$

$$\frac{q}{m\gamma^3}\left(E_{\parallel} - \frac{q}{mc^2\gamma}\frac{\partial}{\partial z}\mathbf{A}_{\mathrm{u}}\mathbf{A}_{\mathrm{w}}\right).$$

We neglected in (10.28) and (10.29) longitudinal velocity variations proportional to A_{u}^2, using an averaged value $\overline{\beta}c$. In a relativistic case, the deviations of β from $\overline{\beta}$ are negligible except of the expression $1 - \overline{\beta}$, which differs from $1 - \beta$ by the factor $1 + \mathcal{K}^2$. Here $\mathcal{K}^2 = \beta_{\perp\mathrm{u}}^2\gamma^2$ is the undulator coefficient squared.

From the linearized continuity equation

$$\frac{\partial}{\partial t}\rho_{\mathrm{w}} + \frac{\partial}{\partial z}\left(\rho_{\mathrm{w}}\overline{\beta}c + \rho_0 v_{\parallel}\right) = 0$$

we have for small deviations of density ρ_{w} from the equilibrium value: ρ_0:

$$4\pi\frac{\partial}{\partial z}\rho_0 v_{\parallel} = -4\pi\left(\frac{\partial}{\partial t} + \overline{\beta}c\frac{\partial}{\partial z}\right)\rho_{\mathrm{w}} = -\frac{\partial}{\partial z}\left(\frac{\partial}{\partial t} + \overline{\beta}c\frac{\partial}{\partial z}\right) E_{\parallel} \tag{10.30}$$

or

$$4\pi\rho_0 v_{\parallel} = -\left(\frac{\partial}{\partial t} + \overline{\beta}c\frac{\partial}{\partial z}\right) E_{\parallel}. \tag{10.31}$$

Now, let us take into account that the equilibrium density ρ_0 in a uniform beam is constant. In a modulated beam, it depends on the difference $z - \overline{\beta}ct$. Hence, in both cases the operator $\partial/\partial t + \overline{\beta}c\,\partial/\partial z$ does not act on ρ_0. Applying it to (10.31) and using (10.29), we obtain

$$\left(\frac{\partial}{\partial t} + \overline{\beta}c\frac{\partial}{\partial z}\right)^2 E_{\parallel} = -\omega_{\mathrm{p}}^{*2} E_{\parallel} + \frac{q}{mc^2\gamma}\omega_{\mathrm{p}}^{*2}\frac{\partial}{\partial z}\mathbf{A}_{\mathrm{u}}\mathbf{A}_{\mathrm{w}}. \tag{10.32}$$

One can see, in particular, that longitudinal bunching is driven by two physically different factors [62]. First, this is a longitudinal restoring force associated with space charge waves. Second, a pondermotive force appears (the second term in right-hand side of (10.32)), which is proportional to the longitudinal velocity and the transverse magnetic field product,[6] exactly as it happens in the case of the single particle motion.

One more link between longitudinal and transverse electric field components follows from Maxwell equations. The transverse vector potential obeys the wave equation:

$$\left[\frac{\partial^2}{\partial z^2} - \frac{1}{c^2}\frac{\partial^2}{\partial t^2}\right]\mathbf{A}_{\mathrm{w}} = -\frac{4\pi}{c}\left(\rho_{\mathrm{w}}\mathbf{v}_{\mathrm{u}} + \rho_0\mathbf{v}_{\mathrm{w}}\right). \tag{10.33}$$

---

[6] This is a reason to treat the combination $q\mathbf{A}_{\mathrm{u}}\mathbf{A}_{\mathrm{w}}/\gamma mc^2$ as a ponderomotive potential.

Bearing in mind the transverse integral of motion (10.27) and the relation $4\pi\rho_{\rm w} = \partial E_{\|}/\partial z$, it can be written as

$$\left[\frac{\partial^2}{\partial z^2} - \frac{1}{c^2}\frac{\partial^2}{\partial t^2} - \frac{\omega_{\rm p}^2}{c^2}\right]{\bf A}_{\rm w} = \frac{q}{mc^2\gamma}{\bf A}_{\rm u}\frac{\partial E_{\|}}{\partial z}. \qquad (10.34)$$

In absence of the undulator field, Eqs. (10.32) and (10.34) describe four partial waves. Two of them are transverse electromagnetic waves propagating in opposite directions. Another two are longitudinal space-charge waves with phase velocities close to the average particle velocity $\overline{\beta}c$. The undulator field provides their coupling, which takes place in spite of differences in phase velocities. The transverse electric field excites not only transverse current but the longitudinal one as well because of the ponderomotive force. For the same reason, a perturbation of the longitudinal velocity produces a transverse current. As far as the periodic undulator field includes harmonics with wavenumbers multiple of $k_{\rm u}$, the wavenumber of the ponderomotive force is shifted with respect to electromagnetic one by $k_{\rm u}$. Hence, the simplest condition of synchronism between the direct electromagnetic wave of frequency ω and space-charge waves looks like

$$k_{\rm em} \equiv \frac{1}{c}\sqrt{\omega^2 + \omega_{\rm p}^2} = k_{\rm u} + (\omega \pm \omega_{\rm p}^*)/c\overline{\beta}; \quad \omega_{\rm p}^{*2} = \frac{\omega_{\rm p}^2}{\gamma^2} = \frac{4\pi\varrho_0 q}{m\gamma^3}. \qquad (10.35)$$

Strictly speaking, these arguments are valid only for a longitudinally uniform beam. Nevertheless, they are qualitatively applicable to amplification in a long (in a wavelength scale) electron bunch. Certain features of this regime will be considered in Sect. 10.3. We omit also exotic resonances with higher harmonics of the undulator field and influence of the beam dielectric permeability.

Normal eigenwaves of the system generally depend on z in a nonharmonic way (this is why we did not substitute up to now ik and $-i\omega$ for the operators $\partial/\partial z$ and $\partial/\partial t$ correspondingly). However, if coupling near the resonance (10.35) is weak, one can neglect the counterpropagating electromagnetic wave and nonresonant harmonics of the undulator field. Then

$$\begin{aligned}{\bf A}_{\rm w} &= {\bf A}^+ \exp\left[i\omega\left(z/c - t\right) + i\delta z\right]; \\ {\bf A}_{\rm u} &= {\bf A}_{\rm u}^+ \exp\left(ik_{\rm u}z\right) + {\bf A}_{\rm u}^- \exp\left(-ik_{\rm u}z\right),\end{aligned} \qquad (10.36)$$

where

$$\delta = \frac{\omega}{c}\frac{1-\overline{\beta}}{\overline{\beta}} - k_{\rm u} \approx k_0\frac{1+\mathcal{K}^2}{2\gamma^2} - k_{\rm u} \qquad (10.37)$$

defines detuning, that is, a small deviation of the single particle motion from synchronism. Amplitudes of the direct electromagnetic wave $A^+(z)$ and of the waves of the longitudinal electric field

$$E_{\|} = E(z)\exp\left[i\omega\left(z/\overline{\beta}c - t\right)\right] \qquad (10.38)$$

can be treated as slow functions of z.

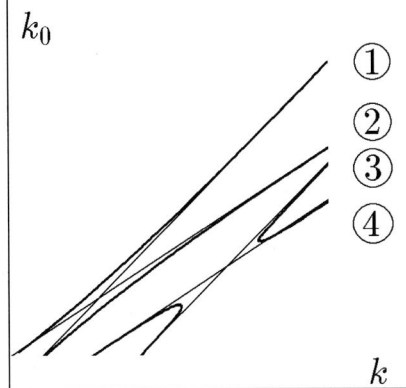

Fig. 10.4. The dispersion diagram depicts coupling between the electromagnetic wave and space charge waves: 1 is the direct electromagnetic wave $k_0 = k_{\rm em}(k)$; 2 is the space charge fast wave $k_0 = k\overline{\beta} + k_{\rm p}^*$; 3 is the wave of the ponderomotive force; and 4 is the space charge slow wave of negative energy $k_0 = k\overline{\beta} - k_{\rm p}^*$

In (10.34) and (10.32), we now neglect the derivatives of the second order and/or the terms which for sure are nonresonant. That yields the following system:

$$\frac{\partial \mathbf{A}_\perp^+}{\partial z} = -\mathrm{i}\frac{2\pi q \mathbf{A}_{\rm u}^-}{mc^3 \gamma k_0 k_{\rm em}\overline{\beta}}$$
$$\times \left[j^+ \left(k_0 + k_{\rm p}^* \right) \mathrm{e}^{\mathrm{i}z\left(\delta + k_{\rm p}^*/\overline{\beta}\right)} + j^- \left(k_0 - k_{\rm p}^* \right) \mathrm{e}^{\mathrm{i}z\left(\delta - k_{\rm p}^*/\overline{\beta}\right)} \right] ;$$
$$\frac{\partial j^\pm}{\partial z} = \pm \mathrm{i}\frac{k_0^2 \rho_0 c}{2\overline{\beta}^2 k_{\rm p}^*} \left(\frac{q}{mc^2\gamma^2}\right)^2 \mathbf{A}_{\rm u}^+ \mathbf{A}_\perp^+ \mathrm{e}^{-\mathrm{i}z\left(\delta \pm k_{\rm p}^*/\overline{\beta}\right)} . \qquad (10.39)$$

Here the parameter $k_{\rm p}^* = \omega_{\rm p}^*/c$.

As it is easy to demonstrate, the system (10.39) admits harmonic solutions

$$\mathbf{A}_\perp^+ \propto \exp\left[\mathrm{i}\left(\mu + \delta\right) z\right] ; \quad j^\pm \propto \exp\left[\mathrm{i}z\left(\mu \mp k_{\rm p}^*/\overline{\beta}\right)\right] ,$$

if μ satisfies the dispersion relation

$$(\mu + \delta)\left(\mu + k_{\rm p}^*/\overline{\beta}\right)\left(\mu - k_{\rm p}^*/\overline{\beta}\right) = -\frac{K_{\rm u} k_{\rm p}^{*2}}{\overline{\beta}^2} . \qquad (10.40)$$

The first factor on the right-hand side of (10.40) corresponds to the electromagnetic wave, the second and the third ones describe the fast and slow space charge waves, respectively. The negative coupling constant is proportional to the beam density and to the coefficient

$$K_{\rm u} \approx \frac{k_0^2}{2k_{\rm em}\overline{\beta}^2}\left(\frac{q}{mc^2\gamma}\right)^2 |A_{\rm u}^\pm|^2 = \frac{k_0^2 \overline{\beta_{{\rm u}\perp}^2}}{4k_{\rm em}\overline{\beta}^2} \approx \frac{k_{\rm u}}{2}\frac{\mathcal{K}^2}{1+\mathcal{K}^2} , \qquad (10.41)$$

On the right–hand sides of these expressions, the coefficient K_u is expressed via the particle transverse velocity induced by the undulator field and via the undulator coefficient.

Surely, one is mostly interested in the range of parameters of the cubic dispersion relation where it has two complex–conjugated roots. One of them, characterized by a negative imaginary component, describes an increment of spatial amplification of the normal proper wave. This increment depends on the deviation from the condition of the wave–particle synchronism and on the beam density. The standard procedure of determination of roots of the cubic equation (10.40) yields the increment as a function of detuning with the beam density as a parameter. The corresponding graphs are presented in Fig. 10.5.

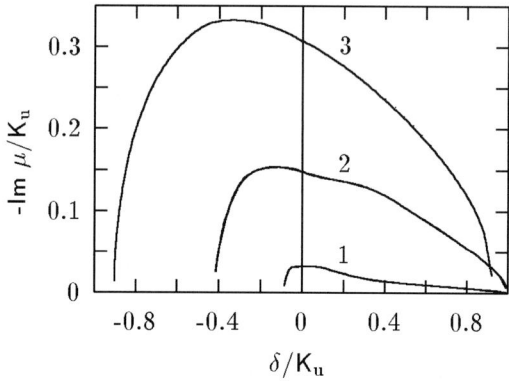

Fig. 10.5. The increment as a function of detuning for various beam densities. $1 : k_p^{*2}/\bar{\beta}^2 K_u^2 = 10^{-4}$; $2 : 10^{-2}$; $3 : 10^{-1}$

Above we have derived the expression for the increment of the wave spatial increase $\Gamma = |\text{Im}\,\mu|$, or the radiation length $L_r = \Gamma^{-1}$. They are the principal characteristics of the high-gain regime – surely, under the condition that the total interaction length substantially exceeds L_r. At the same time, L_r has to remain smaller than the saturation length. In particular, the tolerances estimated above and the expressions for the efficiency and the amplification band must contain not N but N_r (it is the number of the undulator periods within the radiation length). However, these considerations concern the asymptotic exponential regime only. In practice, before establishment of this regime, there exists a rather long section of the undulator, where the beam becomes bunched and modulated. At the same time, there the radiation field is still weak and its increase is much slower than exponential. So, a high–power FEL should be constructed on the basis of a two–cascade scheme with two undulators. The first one is optimized for obtaining a high gain. The output power and the efficiency could be low. The second undulator produces high power whereas the gain is relatively low.

10.2 FEL–Amplifier: High-Gain Regime

For small and large beam densities, the gain depends on the beam current in qualitatively different ways corresponding to two different modes of operation.

Compton Regime

It takes place for small density beams when the spatial increment is smaller than the space-charge parameter $k_\mathrm{p}^*/\bar{\beta}$. Under this condition, the space charge waves (fast and slow) have approximately equal phase velocities and the roots of the dispersion equation (10.40) can be written as

$$\mu_1 = -\delta - \frac{G^3}{\delta^2}; \quad \mu_{2,3} = \pm\sqrt{-\frac{G^3}{\delta}} \quad \text{for} \quad |\delta| \gg G$$
$$\mu_{1,2,3}^3 = -G^3 \quad \text{for} \quad |\delta| \ll G, \quad (10.42)$$

where

$$G^3 = K_\mathrm{u} k_\mathrm{p}^{*2}/\bar{\beta}^2 \,.$$

One can easily see that the amplification band is located mainly at positive values of δ but the maximal increment is reached at $\delta = 0$. For zero detuning, it is equal to

$$\Gamma \equiv |\mathrm{Im}\,\mu| = \frac{\sqrt{3}}{2} G, \quad (10.43)$$

and is proportional to the cubic root of the beam current. Note that this result has the same explanation as the Cherenkov radiation instability considered in Part II in the single particle approximation. In both the cases, individual particles interact only via radiation field because the plasma frequency is too small to play an essential role. This mode of operation is called traditionally as a Compton regime. The name came from the interpretation of the undulator field as an incoming wave reflected by a fast electron with large transformation of its frequency. Frankly speaking, this notation can hardly be considered as a good one because the involved physics has nothing common with the well-known quantum Compton effect.

Raman Regime

Calling the opposite case a Raman regime is more justified. The radiation emission is accompanied then by plasma oscillations or, to be more precise, by excitation of a negative energy slow space charge wave. The fast waves are not exited because their frequency detuning exceeds essentially the increment. For this reason, the second multiplier in the left-hand side of (10.40) can be replaced by $2k_\mathrm{p}^*/\bar{\beta}$, reducing the dispersion equation to a quadratic (two-wave) one:

$$(\mu+\delta)\left(\mu-k_{\rm p}^*/\overline{\beta}\right)=-\frac{K_{\rm u}k_{\rm p}^*}{2\overline{\beta}}.$$

Its solution

$$\mu=\frac{1}{2}\left(\delta-\frac{k_{\rm p}^*}{\overline{\beta}}\right)\pm\frac{1}{2}\sqrt{\left(\delta-\frac{k_{\rm p}^*}{\overline{\beta}}\right)^2-\frac{2K_{\rm u}k_{\rm p}^*}{\overline{\beta}}}. \qquad (10.44)$$

shows an amplification band

$$-\sqrt{2K_{\rm u}k_{\rm p}^*/\overline{\beta}}+k_{\rm p}^*/\overline{\beta}<\delta<\sqrt{2K_{\rm u}k_{\rm p}^*/\overline{\beta}}+k_{\rm p}^*/\overline{\beta} \qquad (10.45)$$

shifted to negative values of δ. The maximal increment

$$\Gamma=\sqrt{K_{\rm u}k_{\rm p}^*/2\overline{\beta}},$$

is reached in middle of the band. So, in the Raman regime the increment is proportional to the square root of the beam current as it is shown in Fig. 10.6.

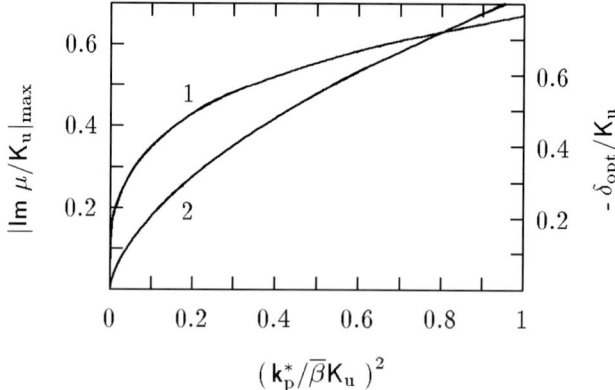

Fig. 10.6. Maximal increment (line 1) and maximizing detuning (line 2) versus beam density

Coming back to the unequality separating Compton and Raman regimes, one should keep in mind that this is the first one that is typical for short-wave FELs. Really, the value of $k_{\rm p}^{*2}$ drops with an increase in γ – as γ^{-3} for a fixed current or as γ^{-4} for a fixed power. The Raman regime takes place actually at small energies typical for submillimeter free electron masers.

Power and Efficiency

The FEL's output power depends, of course, on a great number of factors including technical ones. We consider here only nonlinear saturation related to particles capturing by the wave.

10.2 FEL–Amplifier: High-Gain Regime

As it has been already noted, the length of a FEL-amplifier in the linear regime must be larger than the radiation length but smaller than the saturation distance where nonlinear effects appear. The main nonlinear effect is capturing of particles by the growing wave and the related slipping of the bunch toward accelerating phases. In other words, it means appearance of coherent synchrotron oscillations. First of all, they manifest themselves as satellite lines in the radiation spectrum shifted from the basic line by a multiple number of the synchrotron oscillations frequency. They call this manifestation as a side-band instability leading to broadening of the spectrum and, hence, to a decrease in gain. In spite of some possible methods of the instability suppression, the efficiency of a FEL in a linear regime turns out to be rather low.

An estimation of the saturation length is even less reliable than it was for generators because of the field variation with distance. One can assume that at saturation majority of particles are trapped in the stability region of the synchrotron oscilations. Nevertheless, the corresponding length has to be computed. As regards the saturation power, one can expect it to be defined by the minimal particles energy achieved in the process of the oscillations:

$$\delta_{\min} = \delta_i - \sqrt{2g_{\max}/\alpha k}, \quad \text{where} \quad g_{\max} = \frac{qE_{0_{\max}}\mathcal{K}}{mc^2\gamma},$$

(for a helical undulator). At this distance the maximal field amplitude $E_{0_{\max}}$ determines the average power flux density $P = cE_{0_{\max}}^2/8\pi$, while the power supplied by the beam is equal $mc^2\left(\delta_i - \delta_{\min}\right)/q$. Combining these relations and putting $\alpha = \gamma^{-3}$ yields the following estimation:

$$P = \left(\frac{2\mathcal{K}^2\left(1+\mathcal{K}^2\right)^2}{\pi}\right)^{1/3} \frac{P_0}{l_u^2} \left(\frac{jl_u^2}{I_A}\right)^{4/3}, \tag{10.46}$$

where

$$P_0 = \frac{m^2c^5}{q^2} \approx 8.7\,\text{GWt}\,; \qquad I_A = \frac{mc^3}{q} \approx 17\,\text{kA}.$$

Of course, the numerical coefficients in this expression cannot be trusted. Moreover, the one-dimensional model used above includes only the current and power densities (by the way, an effect of so-called beam guiding considered below makes the light beam cross section of the same order of magnitude as that of the electron beam). Nevertheless, one can guess that the power can be rather high, especially bearing in mind that linear induction accelerators are quite capable to produce beam currents exceeding 17 kA.

Starting from the considerations above, the efficiency of the beam-light energy transfer can be estimated as

$$\eta = \left(\frac{2\mathcal{K}^2\left(1+\mathcal{K}^2\right)^2}{\pi}\right)^{1/3} \frac{1}{\gamma} \left(\frac{jl_u^2}{I_A}\right)^{1/3}, \tag{10.47}$$

We will discuss this expression later on. As regards the saturation in a FEL – amplifier, it is more convenient to estimate an input signal limiting power density that provides the saturation at the fixed system length L:

$$P_\mathrm{i} = P_\mathrm{max} \exp\left(-\Gamma L\right).$$

However, one should bear in mind that all these arguments are concerned with the asymptotic exponential regime. In practice, a long part of an undulator precedes where the the beam is modulated and bunched while the radiation field increase is rather low.

In a FEL oscillator, the low efficiency of the energy transfer is an unpleasant but still tolerable shortcoming. For an amplifier with a high current it can be fatal to some extent. The total energy can be saved using different schemes of its recovery, but the output power limitations still take place. It is worth to add that an effective energy recovery can hardly be realized in the case of inductive accelerators which are suitable for the high-current injection.

The efficiency estimated above can be improved essentially by using so-called tampered undulators with parameters (mainly a period) varying along the system [63, 64]. This device permits to keep the wave–particle synchronism in spite of energy changing. If the synchronous energy variations $\gamma_\mathrm{s}(z) = \sqrt{l_\mathrm{u}\left(1+\mathcal{K}^2\right)/2\lambda}$ are within certain reasonable limits the synchronism can be kept at a fixed frequency. To get this, the parameter $l_\mathrm{u}\left(1+\mathcal{K}^2\right)$ should be a definite function of z to hold the synchronous particle at a definite phase. Nonsynchronous particles would then oscillate in energy and phase according to the phase stability mechanism being bunched near the resonant one. Hence, they lose their energy in the average following the prescribed low. In essence, this is just a reversed mechanism of phase stability in linear accelerators, which can be called an enforced resonant deceleration. In this case, the saturation limitations can be eliminated. Note that the particles are to be preliminary trapped as it is shown in Fig. 10.7.

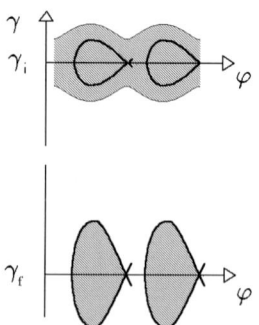

Fig. 10.7. Resonant particles deceleration in the phase plane

The scheme of the resonant deceleration has, of course, its own intrinsic disadvantages. First, a number of the trapped particles are limited by the phase volume of the stability region (acceptance) so that the field level is to be large enough. Second, the prescribed energy change rate, that is, the synchronous particle energy losses, should be supported for more or less definite number of trapped particles. Third, the undulator period must decrease with distance while its minimal value is a rather tough parameter. Nevertheless, the scheme is quite reliable and adequate to high-power amplifiers. Its certain potentialities are illustrated in Fig. 10.8 taken from [62]. It is worth to mention that the method gave the record result – efficiency about 30% for amplification of CO_2 laser radiation.

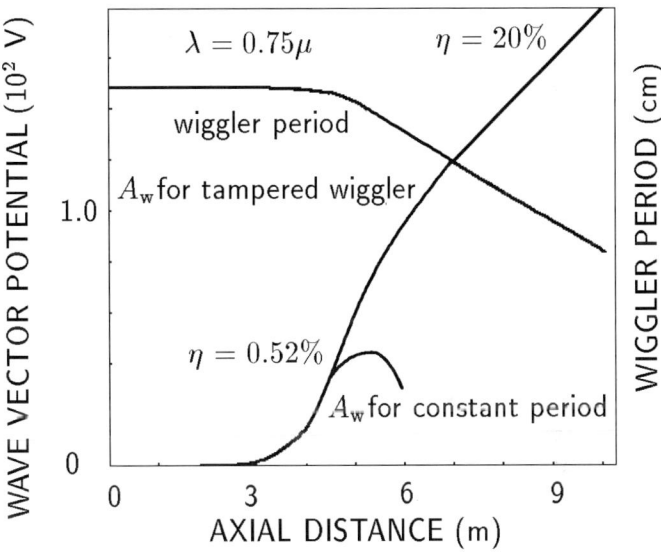

Fig. 10.8. Efficiency increase in a tampered undulator

10.2.2 Two-Dimensional Effects

The theory in the previous sections was based on a one-dimensional model when the wave was supposed plane and propagating strictly along the beam. As it has already been claimed, the first assumption is justified for wavelengths essentially smaller than the beam radius. As a rule, this is fulfilled in practice. Nevertheless, for a beam of a finite radius a boundary diffractive effects must exist and the wave direction is not properly defined. This angular uncertainty of order of λ/a can easily be larger than the beam angular aperture a/L, especially if the system is long enough and $L \gg a^2/\lambda$. Then the cross section of the light beam is larger than that of the electron beam and the gain has to decrease. The one-dimensional theory, of course, does not take this into account. However, the requirements to the beam quality discussed above say that the beam radius should not exceed $\gamma\lambda$ so that $a^2/\lambda \approx \gamma^2\lambda$. For this reason, the diffraction effects are worth of attention already for $L \gg l_u$, that is, practically always. Of course, in a high-gain system the radiation length should be substituted for L but that actually does not matter because N_r is always much larger than unity.

To avoid misunderstanding, note that the diffraction pattern for large angles is not of interest. A pencil beam emission at large angles is small and the first interference maximum being of importance overlaps the electron beam as a whole. For a FEL-oscillator, this overlapping is determined by the optical cavity caustics, but for an amplifier the diffraction effects are necessarily involved.

The two-dimensional theory, not mentioning the three-dimensional one, is rather bulky and requires investigations of high-order transverse modes. Up to our knowledge, these problems are more or less fully discussed in the monography [61], where the results below are taken from.

Beam Guiding

Here we limit ourselves by the so-called beam guiding effect [65], which is of a general physical interest and has no analogy in other devices. Basically, it says that the first interference maximum in a high-gain system is essentially enforced as compared with that in a system of fixed beam modulation. The radiation field turns out to be concentrated in the beam vicinity as if the beam were a waveguide. This is the analogy that gave the name to the effect.

First of all let us recall the statement that an electron beam creates not only active (positive or negative) load for the wave but a reactive one as well. That is, a change in the amplitude is always accompanied with changes in the wave phase velocity. It can be seen directly from the solution of the dispersion relation (10.42) for $\delta = 0$, where

$$\mu^3 = -G^3 = -\frac{K_u k_p^{*2}}{\bar{\beta}^2}$$

and, hence,

$$\mu_{1,2,3} = G \exp\left[i\frac{\pi}{3}(1+2s)\right]; \quad s = 0, \pm 1. \tag{10.48}$$

In particular for an exponentially growing wave ($s = -1$), which is of the main interest at the moment

$$\Gamma \equiv -\mathrm{Im}\,\mu = iG\sin\frac{\pi}{3}; \quad \mathrm{Re}\,\mu = G\cos\frac{\pi}{3} = \frac{\Gamma}{\sqrt{3}}. \tag{10.49}$$

As $\Gamma > 0$, $\mathrm{Re}\,\mu$ is positive as well. So, the plane wave interaction with a beam at a fixed frequency increases the longitudinal wavenumber, that is, decreases

10.2 FEL–Amplifier: High-Gain Regime

the phase velocity. Note that this effect takes place under amplification only and is not related to the nonresonant beam permeability, that is, to the deviation of $k_{\rm em}$ from k_0. (Moreover, these two factors counteract each other so that the phase velocity decreases if the gain is sufficiently high.)

In a plane uniform flow the decrease in the phase velocity would not be essential. However, the radiation field of a bounded beam increases with z not only inside it. Outside (in vacuum) the phase velocity of a free wave along z must be at least equal to c. This is not correct for a wave driven by the beam because at a fixed frequency the longitudinal wave number is prescribed by the condition of synchronism and by beam reactive loading:

$$k = k_{\rm em} + k_{\rm u} + {\rm Re}\,\mu = \sqrt{k_0^2 - k_{\rm p}^2} + k_{\rm u} + \Gamma/\sqrt{3}\,.$$

The dispersion relation in vacuum

$$k_0^2 = k^2 + \kappa^2$$

must be fulfilled as well, so one has to admit that the transverse wavenumber

$$\kappa = \sqrt{k_0^2 - k^2} \approx \sqrt{k_{\rm p}^2 - 2k_0\Gamma/\sqrt{3}}$$

is imaginary if the gain is high enough. This condition is readily achieved for Compton mode of operation when

$$\frac{\sqrt{3}k_{\rm p}^2}{2k_0\Gamma} \approx \frac{K^2}{2}\left(\frac{\bar{\beta}k_{\rm p}^{*2}}{K_{\rm u}^2}\right)^{1/3} \ll 1\,.$$

Outside the beam the field behaves as $H_n^{(1)}(\kappa r)$. As far as κ has a positive imaginary part the radiation field asymptotically drops exponentially with radius as

$$r^{-1/2}\exp\left(-r\sqrt{\frac{2}{3^{1/2}}k_0\Gamma}\right)\,.$$

Of course, the arguments above just indicate the possibility of beam guiding and say nothing on the field pattern in the beam vicinity where the asymptotic expressions fail. It should be added that even inside a transversely uniform beam of a finite radius the wave is not exactly plane, not speaking about real beams.

For a quantitative description we consider, as above, a plane undulator and the lowest modes. Note, first of all, that the transverse currents of slow and fast space charge waves are bound with the field by the second equation (10.37). The transverse coordinate r enters this equation only algebraically via dependence $k_{\rm p}^*(r)$. Hence, to modify the system (10.37), one should just add the transverse Laplace operator to the left-hand side of the first equation. However, this simple modification complicates essentially the problem reducing it to a partial derivative equation of a high order:

$$\left[\left(\frac{\partial^2}{\partial z^2}+\frac{k_\mathrm{p}^{*2}}{\bar{\beta}^2}\right)\left(2ik_0\frac{\partial}{\partial z}+\Delta_\perp-2k_0\delta\right)+2K_\mathrm{u}k_0\frac{k_\mathrm{p}^{*2}}{\bar{\beta}^2}\right]\mathbf{A}_\perp=\mathbf{0}. \quad (10.50)$$

Strictly speaking, this is based on the parabolic approximation of the wave equation. As was mentioned in sect. 10.1.1, this is correct only for sharply directed radiation. This is exactly what has to be expected in the high gain and optical guiding regime. In other words, the problem can be reduced to searching of a beam-like partial solution satisfying reasonable boundary conditions at $z=0$. It is worth to emphasize a certain difference of the problem comparing the low-gain regime. In the last case, a choice of a partial solution was governed by the cavity geometry providing a high-quality factor for the first Gauss–Laguerre beam (or several such beams). For this reason the pattern of the field was practically independent of the beam density transverse distribution. The electron beam just played the role of an energy source for certain modes of the field. Instead of such external mode selection in the high-gain regime, the light beam configuration is provided by the sharply selective amplification. In general, characteristics of such beams differ from those in vacuum.

Let us consider now the eigenfunctions of (10.50), using Fourier transformation with respect to z:

$$\mathbf{A}_\perp^\dagger=\int_{-\infty}^{+\infty}\mathbf{A}_\nu(\mathbf{r})\exp\left(i\nu z\right)\mathrm{d}z. \quad (10.51)$$

The Fourier transforms $\mathbf{A}_\nu(\mathbf{r})$ have to satisfy the ordinary derivative equation

$$\left[\frac{1}{r}\frac{\mathrm{d}}{\mathrm{d}r}r\frac{\mathrm{d}}{\mathrm{d}r}-2k_0\left(\nu+\delta\right)+\frac{2k_0k_\mathrm{p}^{*2}K_\mathrm{u}}{k_\mathrm{p}^{*2}-\nu^2\bar{\beta}^2}\right]\mathbf{A}_\nu=\mathbf{0} \quad (10.52)$$

being finite at $r=0$ and vanishing for $r\to\infty$. Nontrivial solutions of this equation are possible for a set of discrete values of the parameter ν corresponding to various light beams. The spectrum of these values and corresponding field distributions within a beam depend, of course, on the electron beam density profile, that is, on the function k_p^*. It is worth to remind that these functions do not form an orthogonal system. By the way, this is a general feature of the beam-like solutions.

As an example of such approach we consider an electron beam of density uniform inside a cylinder of radius a. The continuous solution for a corresponding vector potential component is

$$A_\nu=\begin{cases}J_0\left(\kappa r/a\right)/J_0\left(\kappa\right) & \text{for}\quad r<a;\\ K_0\left(\varphi r/a\right)/K_0\left(\varphi\right) & \text{for}\quad r>a,\end{cases} \quad (10.53)$$

where

$$\kappa^2(\nu) = -2k_0 a^2 (\nu + \delta) + \frac{2B^2 \overline{\beta}^2}{a^4 k_0^2 \left(k_p^{*2} - \nu^2 \overline{\beta}^2\right)};$$

$$\varphi^2(\nu) = 2k_0 a^2 (\nu + \delta).$$

The signs of κ and φ are chosen to provide $\operatorname{Re}\kappa > 0$ $\operatorname{Im}\varphi > 0$. The value

$$B = \left(k_0^3 a^6 G^3\right)^{1/2} \tag{10.54}$$

is a so-called diffraction parameter suggested in [61].

For operation in Compton regime, one can neglect $k_p^{*2}/\overline{\beta}^2$ in the denominator. Then

$$\kappa^2 + \varphi^2 = -8B^2/\varphi^4. \tag{10.55}$$

Being combined with the condition of the magnetic field continuity at the beam boundary:

$$\kappa J_1(\kappa) K_0(\varphi) = \varphi J_0(\kappa) K_1(\varphi) \tag{10.56}$$

the relation above determines the spectrum of φ and, hence, the gain in various light beams $\Gamma = -\operatorname{Im}\varphi^2/2k_0 a^2$. By the way, $-\operatorname{Im}\varphi^2$ represents the ratio of Rayleigh length to the radiation one.

Starting from this interpretation, one can expect large values of φ for $B \gg 1$ so that $\kappa J_1(\kappa) \approx \varphi J_0(\kappa)$. But for the main (low-order) beams where the field has no zeroes at $r < a$ the value of κ should be of order of unity. That can be provided with $\kappa \approx \kappa_i$ only when κ_i is one of the first roots of the Bessel function. In particular, for the first beam mode $\kappa \approx \kappa_1 \approx 2.405$ and

$$\varphi^2 \approx 2(-B)^{2/3} - \kappa_1^2 + \cdots. \tag{10.57}$$

Then the increment $\Gamma \approx \sqrt{3} B^{2/3}/4 k_0 a^2$ is close to the value of Γ_0 calculated for a plane flow of the same density. It is easy to see that the field distributed as the Bessel function inside the beam is comparatively small outside and provides a low-intensity halo at large distances.

In the opposite limiting case of small B, the value of κ differs essentially from κ_1. So, the light beam is much wider than the electron one and has an intensive halo (see Fig. 10.9). The corresponding asymptotics of the Macdonald function yields a transcendent equation, which cannot be solved in terms of elementary functions. Moreover, optimal detuning providing the maximal increment turns out to be dependent on B. For these reasons, we just cite the numerical solution [61] for $\Gamma_{max}(B)$, which overlaps both limiting cases of small and large B (see Fig. 10.9).

Figures 10.9 and 10.10 demonstrate a surprising insensitivity of the beam guiding effect to density distribution variations (of course, for a fixed total current). It should be underlined once more that this is the main mode (with the maximal increment), which is a subject of beam guiding. Hence, the angular distribution of radiation exceeds that of a "pencil" beam of a fixed density modulation. One more sequence of beam guiding is a decrease of diffraction

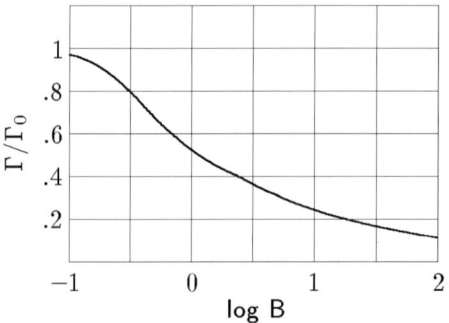

Fig. 10.9. Relative decrease in the increment for a fixed uniform density versus diffraction parameter B. Γ_0 is the increment from the one-dimensional theory

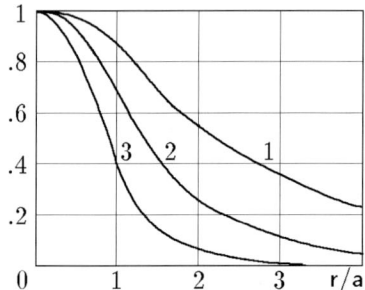

Fig. 10.10. Main mode field distribution in vicinity of a cylindrical beam. Gauss distribution with dispersion a. 1: $B = 0.1$; 2: $B = 1$; 3: $B = 10$

effects at the output of FELs where the light beam leaves the electron one. Even for a narrow electron beam, the light beam is insensitive to its liberation.

Initial Values Problem

The modes close to the optimal one have relatively large increments and play a dominant role in output radiation from a long undulator. However, this information is not sufficient for determination of the real gain, saturation length, efficiency, etc., when not only asymptotic behavior of various modes is of importance but their amplitudes and phases as well. The latter depend, of course, on input characteristics of the light and electron beams.

To find this dependence, one cannot be restricted to calculations of the dispersion equation roots but must solve the problem with initial conditions (meaning the point $z = 0$) taken into account. For this reason, Fourier transformation is to be replaced by Laplace transformation:

$$\mathcal{A} = p \int_0^\infty \mathbf{A}_\perp^+ \exp(-pz)\, \mathrm{d}z \ . \tag{10.58}$$

10.2 FEL–Amplifier: High-Gain Regime

Then we get the inhomogeneous equation for the Laplace transforms

$$\left[\frac{1}{r}\frac{d}{dr}r\frac{d}{dr} - 2k_0\left(ip + \delta\right)\frac{2k_0 k_p^{*2} K_u}{k_p^{*2} + p^2\overline{\beta}^2}\right]\mathcal{A} = 2ik_0 p \mathbf{A}_0, \tag{10.59}$$

where a distribution of the input signal $\mathbf{A}_0(r) = \mathbf{A}_\perp^+(r, z = 0)$ is presented. All other initial values, that is, amplitudes of direct and backward waves as well as those of space charge waves, we put equal to zero for the sake of simplicity.

For fixed functions $k_p^*(r)$ and $\mathbf{A}_0(r)$, a solution of the nonlinear inhomogeneous equation (10.59) should be found under conditions of finiteness at the axis and vanishing at infinity.

It is easy to see that the solution has simple poles in the complex plane of p at the points $p = -i\nu$, where ν are roots of the dispersion equation of the homogeneous problem. According to general rules, the contour of integration in the inverse Laplace transformation

$$\mathbf{A}_\perp^+ = \frac{1}{2\pi i}\int_{s-i\infty}^{s+i\infty} \mathcal{A}\exp(pz)\frac{dp}{p}$$

has to go round the poles from the right side. Shifting it to the left and closing counterclockwise yields the integral equal to the sum of residues at these points. The asymptotic behavior is determined, of course, by the pole ν_0 of the largest imaginary part as was predicted by Fourier transformation. However, now we get also the amplitude of the solution

$$\mathbf{A}_\perp^+(r, z) \asymp i\nu_0^{-1}\mathrm{Res}\,|_{p=-i\nu_0}\,\mathcal{A}\exp\left(-i\nu_0 z\right). \tag{10.60}$$

Of course, the calculation of the residue in (10.60) requires particular characteristics of the electron beam distribution.

This algorithm is illustrated below again with a cylindrical uniform beam of radius a. The solution of (10.59) satisfying the conditions above can be written as

$$\mathcal{A} = 2ik_0 p \times \tag{10.61}$$

$$\begin{cases} -\frac{\pi}{2}J_0(\kappa r/a)\int_0^r Y_0 r A_0 dr + \frac{\pi}{2}Y_0(\kappa r/a)\int_0^r J_0 r A_0 dr + C_1 J_1(\kappa r/a); & r < a \\[1ex] -I_0(\varphi r/a)\int_r^\infty K_0 r A_0 dr + K_0(\varphi r/a)\int_r^\infty I_0 r A_0 dr + C_2 K_0(\varphi r/a); & r > a. \end{cases}$$

The parameters φ and κ are defined in the same way as for Fourier transformation with substitution of ip instead of ν. The constants C_1 and C_2 are to be defined from sewing at the beam boundary

$$C_1 = -\frac{\pi}{2}\frac{\tilde{Y}_0 J_0 - \tilde{J}_0 Y_0}{J_0} - \frac{\tilde{J}_0 K_0 + \tilde{K}_0 J_0}{J_0 \Delta};$$

$$C_2 = \frac{\tilde{K}_0 I - \tilde{I}_0 K_0}{K_0} - \frac{\tilde{J}_0 K_0 + \tilde{K}_0 J_0}{K_0 \Delta}$$

$$\tag{10.62}$$

where
$$\Delta = \kappa K_0 J_0' - \varphi J_0 K_0' . \tag{10.63}$$

In what follows, the absence of a Bessel function argument means that $r = a$. The tilde sign means

$$\tilde{J}_0 = \int_0^a J_0(\kappa r'/a) r' A_0(r') dr' ; \quad \tilde{Y}_0 = \int_0^a Y_0(\kappa r'/a) r' A_0(r') dr' ;$$

$$\tilde{I}_0 = \int_a^\infty I_0(\kappa r'/a) r' A_0(r') dr' ; \quad \tilde{K}_0 = \int_a^\infty K_0(\kappa r'/a) r' A_0(r') dr' .$$

The inverse Laplace transforms will be defined by poles $\Delta = 0$, which correspond to the roots of the dispersion equation in the Fourier approach. For this reason the singular part only can be kept in the solution:

$$\frac{A}{2ik_0 p} \asymp -\frac{J_0 \tilde{K}_0 + K_0 \tilde{J}_0}{\Delta} \times \begin{cases} J_0(\kappa r/a)/J_0 & \text{for } r < a; \\ K_0(\varphi r/a)/K_0 & \text{for } r > a \end{cases} . \tag{10.64}$$

Now, the inverse Laplace transformation yields the asymptotics

$$A(r,z) \asymp -k_0 \exp\left[-i\nu_0 z\right] \frac{J_0^* \tilde{K}_0^* + K_0^* \tilde{J}_0^*}{(\partial \Delta/\partial p)_{p=-i\nu_0}} \tag{10.65}$$

$$\times \begin{cases} J_0^*(\kappa^* r/a)/J_0^* & \text{for } r < a \\ K_0^*(\varphi^* r/a)/K_0^* & \text{for } r > a \end{cases} ,$$

where the star sign means that the corresponding values are taken at ν_0. We shall not adduce here a rather bulky expression for $\partial \Delta/\partial p$ as far as it can be easily obtained from (10.60) and (10.63).

The expression obtained above describes the field amplitude and phase distribution across the beam as well as coupling of the input signal with the beam. It is worth to note that the field asymptotic distribution inside and outside the beam does not depend on the input signal distribution and is determined by cooperative action of spatial amplification and diffraction. However, the asymptotic amplitude does depend on the input power profile as well as on the configuration of the input phase front. Not going into details, note that coupling usually turns out optimal if the cross section of the input light beam is close to the asymptotic one.

10.3 SASE Mode of Operation

Separation of FEL-oscillators from FEL-amplifiers accepted above is, in a way, conditional and based on realization of a controlled radiative instability of either absolute or convective type. Actually, they are distinguished by the radiation length to undulator length ratio and by existence of a determined input signal. However, it is obvious that an oscillator starts from field fluctuations and a role of an input signal in an amplifier can be played by beam noncoherent spontaneous radiation.

The last possibility is of a special interest for very short-wave UV and x-ray devices. For these frequencies, a deep feedback seems impossible because of the lack of adequate optics. Coherent tunable master oscillators are rather problematic as well in this frequency region.[7] The first proposal of self-amplification of spontaneous radiation appeared in 1980 [66] and is known now as a SASE-mode of operation (for Self-Amplification of Spontaneous Emission). To some extent, this regime is analogous to so-called superradiance in optics when the spontaneous emission from excited atoms appears as short flashes of quasi-coherent radiation [35]. Of course, peculiarities of an electron beam as an active media influence essentially characteristics of the phenomenon.

It is clear that the undulator part of a FEL-amplifier plays a role of an active band pass filter with a sharply nonuniform amplification coefficient. Hence, a spectral density of a signal should vary along z its bandwidth being decreased. In other words, the radiation field becomes more and more coherent and its spatial–temporal characteristics tend to the single-mode ones. However, this development of coherency appears only at the length essentially larger than the radiation one but smaller than the saturation length. At least two important questions then arise:

– How coherency changes with distance?
– How the spatial–temporal structure of the output signal looks like and how large is the output power?

10.3.1 Amplification of Spontaneous Radiation in Uniform Flow

Let us consider, first of all, the input noise origination. Stochastic fields at the undulator entrance can be caused by various factors starting from thermal field fluctuations and finishing with quantum effects due to emission of separate photons. Of course, the last effect is negligible and is of a theoretical interest only. The quantum features of emission can be of importance only for a small number of photons per one elementary cell in the phase space or per one mode. For all realized or discussed FEL projects, this parameter exceeds unity by several orders of magnitude and the classic description is perfectly appropriate.

The main source of the stochastic signal in devices of the traveling wave tube type is a discrete distribution of the beam electrons. In the case of the SASE regime, it plays a positive role originating the output quasi-coherent

---

[7] Certain ideas on amplification of spectral lines of heavy elements hot plasmas are worth to mention in this respect.

radiation due to the very selective large amplification within a narrow passband. However, one essential feature of FELs should be emphasized which differs the phenomenon from the TWT proper noise or from the optical superradiance. The working medium density in FEls is very low. To be amplified up to the saturation level, an initial density fluctuation must be large enough; that is, a large number of electrons must be involved. This "cooperative number" n_c can be estimated as a number of radiators in the λ^3-volume (in the rest frame). In a TWT, it is large even for millimeter long waves but the SASE mode can hardly be expected there because of relatively small interaction length and gain. For optical superradiance in the visible light region, the wavelength smallness is compensated by large medium densities. As FELs are concerned, a simple estimation gives by order of magnitude in the rest frame (denoted by primes) $n_c \approx \rho_0' \lambda'^3/q$. In the laboratory frame

$$n_c \approx \rho_0 \lambda^2 l_u / q \approx j_0 \lambda^2 l_u / qc ,$$

where it is taken into account that $\rho_0' = \rho_0/\gamma$; $\lambda' = \lambda\gamma$; $l_u \approx \lambda\gamma^2$ and that n_c is a relativistic invariant. One can easily see that under the practical limitation $l_u > 1\,\text{cm}$ the condition $n_c \gg 1$ is rather ambiguous in the x-ray region.

In principle, the smallness of the cooperative number might be compensated by increasing the interaction length. However, beside of practical reliability this would make ambiguous the considerations above even for $n_c \approx 1$. Really, for particles obeying Poisson statistics relative fluctuations of the number of particles in the cooperative volume λ'^3 go as $n_c^{-1/2}$. Hence, the small signal amplification theory above is not valid for small n_c because of very large relative fluctuations of density. For this reason we restrict ourselves by the case of $n_c \gg 1$ when the fluctuations in the input current can be treated as perturbations.

Coupling between fluctuations of the input current and excited electromagnetic fields is an important parameter in this approach. Practically, one has to solve the main equations (10.39), where the current amplitudes $j^\pm(z=0)$ play the role of initial conditions instead of the input signal $\mathbf{A}_\perp^+(z=0)$. Note that these values are the amplitudes of real physical harmonic processes. In other words, the parameter k_0 remains free (although close to its resonant value). To simplify the mathematics, we restrict ourselves by the case of one-dimensional flow, a planar undulator, and a Compton mode of operation.

For linear amplification, the solution of (10.39) can be presented as a sum of three characteristic exponential functions:

$$A_\perp^+(z) = \sum_{j=1}^{3} A_j \exp\left[iz\left(\delta + \mu_j\right)\right] , \qquad (10.66)$$

where μ_j are complex roots of the dispersion equation for Compton regime:

$$(\mu + \delta)\mu^2 = -G^3 ; \quad G = \left(\frac{K_u k_p^{*2}}{\bar{\beta}^2}\right)^{1/3} . \qquad (10.67)$$

Then, according to the second of equations (10.39), the currents can be represented as

$$j^\pm = \pm \frac{k^2 \varrho_0 c A_u^+}{2\bar{\beta}^2 k_p^*} \left(\frac{q}{mc^2 \gamma}\right)^2 \tag{10.68}$$

$$\times \sum_{j=1}^{3} \frac{A_j}{\mu_j \mp k_p^*/\bar{\beta}} \exp\left[iz\left(k_0 \pm k_p^*/\bar{\beta} + \delta + \mu_j\right)\right].$$

Requiring the initial conditions, one gets

$$A_\perp^+(0) = 0; \quad j^\pm(0) = j_0/2. \tag{10.69}$$

Here $j_0(k_0)$ are Fourier harmonics of the input current. In a more general case, splitting of the space-charge waves taken into account, one should use independent initial conditions for $j^\pm$, that is, for possible plasma oscillations in the input beam. For Compton regime ($|\mu_j|^2 \gg k_p^{*2}$), they can be neglected. In the same approximation, we do not make a difference between k_{em} and k_0 putting

$$\delta = k_0 \frac{1-\bar{\beta}}{\bar{\beta}} - k_u. \tag{10.70}$$

The initial conditions above give an inhomogenious system of three linear equations for the amplitudes A_j:

$$\sum_{j=1}^{3} A_j = 0; \quad \sum_{j=1}^{3} A_j/\mu_j = 0; \quad \sum_{j=1}^{3} A_j/\mu_i^2 = \frac{2\bar{\beta}^6 A_u^-}{k_0 K_u} \frac{j_0(k_0)}{\bar{j}}, \tag{10.71}$$

where $\bar{j}$ is an average density of the input current. Solving it for A_j, we get

$$A_1 = \frac{2\bar{\beta}^6 A_u^-}{k_0 K_u} \frac{j_0(k_0)}{\bar{j}} \times \frac{\mu_i^2 \mu_2 \mu_3 (\mu_2 - \mu_3)}{\mu_1^2 (\mu_2 - \mu_3) + \mu_2^2 (\mu_3 - \mu_1) + \mu_3^2 (\mu_1 - \mu_2)}. \tag{10.72}$$

Analogous expressions for $A_{2,3}$ can be obtained by a circular variation of indices. In what follows the index 1 corresponds to the exponentially growing partial solution.

The gain at the middle of the passband is sufficient for our considerations. So, one has to put

$$\mu_j = G \exp\left[i\pi\left(1 - 2j\right)/3\right]; \quad G = \left(\frac{K_u k_p^{*2}}{\bar{\beta}^2}\right)^{1/3}. \tag{10.73}$$

Then

$$A_1 = M_1 \frac{j_0(k_0)}{\bar{j}} \tag{10.74}$$

where

$$M_1 = -\frac{\overline{\beta}^6 A_u^- (1+\sqrt{3})}{k_0 K_u} \frac{G^2}{3} = -\frac{4\overline{\beta}^6 A_u^- (1+\sqrt{3}) \Gamma^2}{9 k_0 K_0} . \tag{10.75}$$

The expression obtained above keeps the phase information as well.

So far as the input current is a random function of time, the current harmonics $j_0(k_0)$ as well as the amplitude of the vector potential A_1 and that of the electric field $E(k_0) = ik_0 A_1(k_0)$ are random functions of frequency. They can be characterized by corresponding spectral densities, that is, by the processes power per unit frequency interval. As it should be in a linear system the spectral density of output radiation $W_{\text{em}}(k_0)$ is to be proportional to the current spectral density $W_{\text{j}}(k_0)$:

$$W_{\text{em}}(k_0) = \left(\frac{k_0 c M_1}{4\pi}\right)^2 \frac{W_{\text{j}}(k_0)}{\overline{j}^2} \times \exp\left[-2\text{Im}\,\mu_1(k_0)z\right] . \tag{10.76}$$

Remember that the expression above ignores other branches of the dispersion relation within the passband and thus is valid only asymptotically.

We can discuss now the second of the questions above: how does the electromagnetic signal initiated by the input current noise change along the undulator. Qualitatively, one can represent the current noise as a random sequence of short spikes. So, we consider evolution of a signal generated at $t = 0$ by an infinitely short input current pulse

$$j_0(t) = \Delta j \delta(t); \qquad j_0(k_0) = \frac{\Delta j}{2\pi};$$

Using (10.70) and (10.74), we have

$$A_{\text{w}\perp}(z,t) \approx M_1 \frac{\Delta j}{2\pi \overline{j}} \int_{-\infty}^{+\infty} \exp\left[i(k_0 + \mu_1 + \delta)z - ik_0 ct\right] dk_0 \tag{10.77}$$

$$= \frac{M_1 \Delta j \overline{\beta}}{2\pi \overline{j} (1-\overline{\beta})} \exp\left[i\overline{k_0}(z-ct)\right] \times J_M ,$$

where

$$J_M(z,t) = \tag{10.78}$$
$$\int_{-\infty}^{+\infty} \exp\left[i\left(\mu_1(\delta) + \frac{\delta}{1-\overline{\beta}}\right)z - i\frac{\overline{\beta} ct \delta}{1-\overline{\beta}}\right] d\delta \qquad \overline{k_0} = \frac{\overline{\beta} k_u}{1-\overline{\beta}} .$$

The approximate equality sign stands here because only the "amplified" part of the signal with $\mu = \mu_1$ was taken into account. For this reason the expression (10.77) cannot describe fronts of the signal at low levels.[8] Bearing this in mind,

---

[8] For example, the total Fourier transformation shows the absence of the signal outside the light cone, that is, for $z > ct$.

the integral of (10.78) describes a modulated signal with the carrier frequency $c\overline{k}_0$.

To evaluate the integral note that for large z the main income comes from vicinity of the point $\delta = 0$, where $\operatorname{Im}\mu_1$ is close to its maximal value. Presenting $\mu_1(\delta)$ as an expansion over powers of δ/G

$$\mu_1 = G \exp\left(-i\frac{\pi}{3}\right) - \frac{\delta}{3} + \frac{\delta^2}{3^2 G}\exp\left(i\frac{\pi}{3}\right) + \cdots \tag{10.79}$$

up to quadratic terms, one gets the asymptotic behaviour for $Gz \gg 1$:

$$J_{\mathrm{M}} \asymp 3\sqrt{\frac{\pi G}{z}} \exp\left\{i\frac{\pi}{12} + 3G\zeta \exp(i\pi/6)\left(1 - \frac{3\zeta}{4z}\right)\right\}, \tag{10.80}$$

where

$$\zeta = \frac{\overline{\beta}}{1-\overline{\beta}}(ct - z)).$$

Information on phase of the carrier signal is irrelevant so we shall look directly for the power pulse profile (supposing $G^{-1} \ll z < ct!$):

$$|J_{\mathrm{M}}|^2 \asymp \frac{2\pi 3^{3/2}\Gamma}{z}\exp\left[6\Gamma\zeta\left(1 - \frac{3\zeta}{4z}\right)\right], \tag{10.81}$$

where $\Gamma = G\sqrt{3}/2$ is the amplitude increment for zero detuning.

The exponent in (10.81) describes the power pulse profile far away from the entrance. It is approximately gaussian. Its maximum arrives the point z at the moment $t = z(2 + \overline{\beta})/3\overline{\beta}c$ so it propagates with velocity

$$\beta_{\mathrm{e}} = \frac{3\overline{\beta}}{2 + \overline{\beta}}; \quad \overline{\beta} < \beta_{\mathrm{e}} < 1. \tag{10.82}$$

Note that this "energy propagation" velocity exceeds the beam speed by

$$\beta_{\mathrm{e}} - \overline{\beta} \approx \frac{1 + \mathcal{K}^2}{3\gamma^2} \tag{10.83}$$

but remains smaller than the group velocity. As a matter of fact in any medium with negative or positive absorption, a group velocity formally defined as $\partial k_0/\partial k$ is complex and means neither the energy flow velocity nor that of the signal. The latter is to be defined as a velocity of a fixed level of the signal, that is, from the relation $\zeta = \operatorname{const}$. If this level exceeds essentially the nonamplified input signal but is still smaller than the maximum, the velocities of the signal front and tail are

$$\beta_{\mathrm{f}} = 1; \quad \beta_{\mathrm{t}} = \overline{\beta} - \frac{1 + \mathcal{K}^2}{6\gamma^2}; \tag{10.84}$$

Hence, a pulse generated by a δ-fluctuation in the input current slowly goes forward with respect to beam particles and spreads.

If a sequence of the input current pulses is random and can be treated as a white noise within the amplification band, the electromagnetic field at a point $z \gg \Gamma^{-1}$ (for linear amplification) represents a stochastic signal with large correlation time $\tau \approx L_{\mathrm{r}}/\gamma^2$ and, correspondingly, with a narrow spectral bandwidth $\Delta\omega \approx \gamma^2 c/L_{\mathrm{r}} \approx k_0^* c/N_{\mathrm{r}}$, tending to that of quasi-coherent radiation.

From this point of view it is interesting to trace an evolution of an input chaotic sequence of random spikes and quasi-coherent radiation formation. Of course, we cannot ignore that phases are accidental because the s-th particle enters the undulator at the random moment of time t_s. Changing in (10.78) t for $t - t_s$ and summing over all particles which provide a signal at a fixed point z, one can get for $|J_{\mathrm{M}}|^2$:

$$|J_{\mathrm{M}}|^2 = \frac{9\pi G}{z} \left| \sum_s \exp\left(G\tau_s \exp\left(i\pi/6\right)\right) \right|^2, \qquad (10.85)$$

where

$$\tau_s = 3\zeta_s - \frac{9\zeta_s^2}{4z} = -\left(\frac{3\zeta_s}{2\sqrt{z}} - \sqrt{z}\right)^2 + z.$$

Let us suppose now that the moments t_s and, correspondingly, values ζ_s are independent, that is, that the particles obey Poisson distribution. Strictly speaking, this is justified for a steady state only. Moreover, proper collective degrees of freedom in the injected beams are neglected. For these reasons, the reliability of the model under consideration depends essentially on the type of the driving accelerator and on characteristics of beam transportation and focusing. Aside from that this "random noise" approximation seems adequate even to bunched beams if the bunch length exceeds the single particle signal length estimated above.

Under these conditions, the values ζ_s can be treated as statistically distributed uniformly over the interval $(0, 4z/3)$. Selecting terms with $s = p$ from the double sum (10.85) and bearing in mind that τ_s and τ_p are statistically independent for $s \neq p$, we get after averaging:

$$\left\langle |J_{\mathrm{M}}|^2 \right\rangle = \frac{9\pi}{z} \left[n \left\langle \exp G\tau_s \sqrt{3} \right\rangle + n(n-1) \left| \left\langle \exp G\tau_s e^{i\pi/6} \right\rangle \right|^2 \right], \qquad (10.86)$$

where n is a number of particles within the interval. For $Gz \gg 1$, which is necessary for applicability of the asymptotic formulae above, this averaging yields

$$\left\langle \exp G\tau_s e^{i\pi/6} \right\rangle = \sqrt{\frac{\pi}{4Gz}} \exp\left[-i\frac{\pi}{12} + Gze^{i\pi/6}\right] ;$$

(10.87)

$$\left\langle \exp\sqrt{3}G\tau_s \right\rangle = \sqrt{\frac{\pi}{4Gz\sqrt{3}}} \exp\sqrt{3}Gz \,.$$

Then

$$\left\langle |J_\mathrm{M}|^2 \right\rangle = \frac{3^{5/2}\pi^2}{8z^2\Gamma} \exp\left(2\Gamma z\right) \left[n\left(n-1\right) + n\sqrt{\frac{8\Gamma z}{3\pi}} \right]. \tag{10.88}$$

Concerning n, one can show that it is a number of particles passing the point z during the time interval $4z\left(1-\overline{\beta}\right)/3c\overline{\beta}$. Hence, it is equal to the number of particles per length $4z\left(1-\overline{\beta}\right)/3$. Expressing it via the average beam current I:

$$n = 4I\left(1-\overline{\beta}\right)z/3q\overline{\beta}c\,,$$

we can trace the radiation power evolution along the system. For $n \gg 1$

$$\left\langle |J_\mathrm{M}|^2 \right\rangle = \frac{2\pi^2 I^2 \left(1-\overline{\beta}\right)^2}{3^{1/2}\Gamma q^2 \overline{\beta}^2 c^2} \left[1 + \frac{q\overline{\beta}c}{I\left(1-\overline{\beta}\right)}\sqrt{\frac{3\Gamma}{2\pi z}} \right] \exp\left(2\Gamma z\right). \tag{10.89}$$

The absolute value of power can be evaluated as well but it is not of interest because our model of the input noise is somewhat artificial.

Beside of the natural exponential power increase with distance, the formula (10.89) reveals the development of coherency between separate spikes when the first term in the square brackets becomes larger than the second one. The distance where the input signal noise is smoothed can be estimated as

$$z_\mathrm{ot} \approx \frac{6\gamma^4 q^2 c^2}{\pi I^2 L_\mathrm{r}}\,. \tag{10.90}$$

10.3.2 Amplification and SASE Mode in Short Bunch

The SASE regime of operation considered above differs from its quantum analog because the initial energy distribution exhaustion occurs as a result of many quanta emission. That is why, it can be rather described as quasi-stationary amplification of the proper noise. Radiation spontaneously emitted at any region of the undulator propagates with the beam being amplified along the remaining distance and never reaching the head part of the electron bunch even if its length $l_\mathrm{b} \ll L_\mathrm{r}$ is relatively small. The bunch can be considered as a long one if its length exceeds the radiation length in the rest frame. In the laboratory frame the criteria look like

$$l_\mathrm{b} \gg L_\mathrm{r}/\gamma^2\,.$$

However, other beam structure are also intrinsic to some existing and foreseen FELs due to peculiarities of the driver accelerators. The general tendency of amplification decrease with the wavelength implies the necessity of large peak beam currents. With limited capabilities of beam sources, the latter can be achieved only with shortening of the bunch length. Moreover, the short bunches are required to keep the energy spread small enough while the energy itself must be large because of limitations of the undulator period. To make the long story short, an inverse condition of a "short" bunch

$$l_\text{b} \ll L_\text{r}/\gamma^2$$

is quite real for cases of practical interest.

Superradiance effect has been considered as a method of high-power short pulse generation in microwave devices [67]. The most interesting regime then occurs when the bunch length is comparable with the wavelength, that is, when the lowest modes are excited. However, even for FELs in extreme regimes picosecond bunches still are much longer than the wavelength. The radiation pulse looks like a wave packet with carrying frequency $k_\text{u}c/\left(1-\overline{\beta}\right)$ and small duration of order of $L_\text{r}/\gamma^2 c$. For this reason, the results obtained above for a narrow band signal are to be reconsidered.

Nevertheless, the main equations (10.32) and (10.34) are still reliable so far as the equilibrium density $\rho_0(\zeta)$ was supposed to be dependent on the combination $\zeta = (z - \overline{\beta}ct)/(1-\overline{\beta})$ and had not to be constant. However, the problem is not stationary anymore and must be solved under initial and boundary conditions. It is convenient now to use the variables (z, ζ) instead of (z,t) and to look for a solution in the form:

$$\begin{aligned}
\mathbf{A}_\text{w} &= \mathbf{A}^- \mathbf{A}(z, \zeta) \exp\left[\mathrm{i}\left(k_\text{u} + \delta/\overline{\beta}\right)(z-ct)\right] \\
&= \mathbf{A}^- \mathbf{A}(z, \zeta) \exp\left[\mathrm{i}\left(k_\text{u} + \delta/\overline{\beta}\right)(\zeta - z)\right] ; \quad (10.91) \\
E_\parallel &= E(z, \zeta) \exp\left[\mathrm{i}\left(k_\text{u} + \delta/\overline{\beta}\right)\zeta\right] ,
\end{aligned}$$

where $\mathbf{A}(z,\zeta)$ and $E(z,\zeta)$ are slow functions of their arguments. Then we have the system

$$\left[\frac{\partial}{\partial \zeta} + \frac{\partial}{\partial z} + \mathrm{i}\delta/\overline{\beta}\right] A(z,\zeta) = \frac{q}{2mc^2\gamma\overline{\beta}} E(z,\zeta) ;$$

$$\left[c^2\overline{\beta}^2 \frac{\partial^2}{\partial z^2} + \omega_\text{p}^{*2}\right] E(z,\zeta) = \mathrm{i}\frac{qk_\text{u}}{mc^2\gamma\left(1-\overline{\beta}\right)} \omega_\text{p}^{*2} A(z,\zeta) .$$

(10.92)

Equations (10.92) indicate existence of two partial wave subsystems coupled via the undulator field. The first one corresponds to plasma oscillations (space charge waves) carried with the average velocity $\overline{\beta}c$ and locked in the bunch as in a cavity. In absence of coupling, their amplitude can be an arbitrary function of the coordinate ζ. The second one represents electromagnetic waves traveling with the phase velocity c (the beam dielectric constant being neglected). One can easily see that in a homogeneous beam where the plasma frequency ω_p^* is independent of ζ the complex amplitudes $A(z, \zeta)$ and $E(z, \zeta)$ can be presented as $A \propto E \propto \exp[i\mu z]$, where μ is defined by the dispersion relation (10.40).

In a nonhomogeneous beam, the wave inside the bunch has no definite wavenumber. Moreover, a solution has to obey certain initial and boundary conditions so far as it cannot be steady in the laboratory frame. In particular, even linear amplification of an external monochromatic signal must be accompanied by generation of additional harmonics and by corresponding frequency band spreading.

Under initial conditions $A(0, \zeta)$; $E(0, \zeta)$; $\dot{E}(0, \zeta)$, where dots are for derivatives over z, equations (10.92) for fixed ζ shrink to the uniform linear first-order equation for the Laplace transform $\mathcal{A}(p, \zeta)$:

$$\left[\frac{d}{d\zeta} + D(p, \zeta)\right] \mathcal{A}(p, \zeta) = p\mathcal{F}(p, \zeta), \quad (10.93)$$

where

$$\mathcal{F}(p, \zeta) = A(0, \zeta) + \frac{q\overline{\beta}}{2m\gamma} \frac{\dot{E}(0, \zeta) + pE(0, \zeta)}{c^2 p^2 \overline{\beta}^2 + \omega_p^{*2}} \quad (10.94)$$

and

$$D(p, \zeta) = p + i\delta/\overline{\beta} - iK_u \frac{\omega_p^{*2}}{c^2 p^2 \overline{\beta}^2 + \omega_p^{*2}}. \quad (10.95)$$

Bearing in mind that the field vanishes at $\zeta \to -\infty$ the solution of (10.93) can be written as

$$\mathcal{A}(p, \zeta) = p \int_{-\infty}^{\zeta} \mathcal{F}(p, \zeta') \exp\left[\int_{\zeta}^{\zeta'} D(p, \zeta'') \, d\zeta''\right] d\zeta'. \quad (10.96)$$

To find the time and space dependencies of the field, one should substitute a definite density profile $\omega_p^{*2}(\zeta)$ into (10.94) and (10.96) and perform the inverse Laplace transformation.

To be more precise with initial conditions, more details should be taken into account including a stray magnetic field configuration. Such problems can be essential for a particular device design. In what follows the coupling coefficient K_u is "instantly switched on" at $z = 0$ ignoring details of the output radiation pulse shape over time intervals smaller than the transition time.

To simplify arithmetic, we restrict ourselves by Compton regime, which is typical for short-wave FELs and implies that no plasmon has time enough to be emitted during the flight through the undulator. Then the value of ω_p^{*2} can be neglected in the expression for D and

$$\int_{\zeta'}^{\zeta} D(p, \zeta'') \, d\zeta'' \approx \left[(p + i\delta/\overline{\beta})(\zeta' - \zeta) - i\frac{G^3}{p^2}\right], \quad (10.97)$$

where
$$G^3(\zeta, \zeta') = K_u \int_{\zeta'}^{\zeta} \frac{\omega_p^{*2}(z'')}{c^2 \bar{\beta}^2} d\zeta''$$

is proportional to the charge contained between ζ and ζ'.

The Laplace transform $\mathcal{A}(\zeta, p)$ as a function of the complex variable p can be estimated for large z by the saddle point method. Really, it contains the integrals in the complex plane of the type

$$I_n = \int_{-i\infty + \sigma}^{+i\infty + \sigma} \exp\left[p\left(z + i\frac{\delta}{\bar{\beta}}(\zeta' - \zeta)\right) - i\frac{Q^3}{p^2}\right] p^n \, dp \quad n = 0 \pm 1. \quad (10.98)$$

Changing the variable
$$p = 2QZ^{-1/3}x \qquad Z = z + \zeta' - \zeta$$

brings them to the form convenient for asymptotic estimation:
$$I_n = \left(2QZ^{-1/3}\right)^{n+1} \int \exp\left[2QZ^{2/3}\left(x - \frac{i}{2x^2}\right)\right] x^{n+1} dx.$$

There are three saddle points x_s which are the roots of the cubic equation $x^3 = -i$, that is,
$$x_s = \exp\left[i\pi(2s/3 + 1/6)\right]; \qquad s = 0, 1, 2.$$

In their vicinity the exponential function can be presented as the expansion
$$\exp\left[3QZ^{3/2}x_s\left(1 + \frac{(x - x_s)^2}{x_s^2} + \cdots\right)\right]$$

so that
$$I_n = \sum_{s=0}^{3} \frac{i\sqrt{2\pi}}{\sqrt{3}Z^{(n+2)/3}} (2Qx_s)^{n+1/2} \exp\left(3QZ^{2/3}x_s\right). \quad (10.99)$$

This is $x_0 = \left(\sqrt{3} + i\right)/2$, which has a positive real part and, hence, plays the dominant role in the asymptotic expression (10.99).

To calculate an external signal amplification, it is natural to accept zero initial conditions for the longitudinal field and its derivative supposing that the initial modulation of the bunch current and density is absent. At the point $z = 0$, a constant harmonic signal of unit amplitude and fixed frequency $\omega = \bar{\beta}c(k_u + \delta)/(1 - \bar{\beta})$ is supported. In other words,

$$A(0, \zeta) = 1; \qquad \dot{E}(0, \zeta) = E(0, \zeta) = 0,$$

where a phaser of no importance is omitted.

10.3 SASE Mode of Operation   249

Further details would require more determined initial conditions and density distribution within the bunch. However, the general tendency is clear. First of all, the amplitude of the radiation field grows slower than exponentially with the distance passed by the bunch. Second, the field is sharply nonuniform inside the bunch because the Q value and, hence, the gain are small at its tail. Both effects are related to the "lethargy" phenomenon due to a slippage of the wave packet with respect to the bunch. By the way, a small controlled slippage could be used for formation of ultra short pulses interesting for certain applications.

Amplification of the proper noise differs from the problem above in initial conditions and in a statistical character of interpretation. At $z = 0$ there are random fluctuations of density and velocity so that the longitudinal field $E(\zeta, 0)$ and its derivative $\dot{E}(\zeta, 0)$ are not equal to zero. Beside, an initial transverse field $A(\zeta, 0)$ exists because of spontaneous radiation at the undulator entrance.

To avoid the misunderstanding, note that the parameter δ in (10.98) is not fixed anymore and characterizes Fourier harmonics of the noise. So the following statistical averaging should relate the spectral density of the output signal to that of the noise (as for the uniform beam case above).

The proper noise can be treated as a perturbation with spectral density uniform within the amplification band if the cooperative number of electrons is large. More realistic statistics of proper noises should include correlations between $A(0, \zeta)$, $E(0, \zeta)$, and $\dot{E}(0, \zeta)$, and require special considerations being dependent on driver parameters as well as on conditions of transportation and undulator entrance.

Two features of amplification in a short bunch are worth to be mentioned. First, there are two additive terms in the expression for the output power. The first one originated by $A(0, \zeta)$ is due to the amplified spontaneous emission from particles which were not initially independent because they belonged to the same bounded bunch. This term can be identified as partly coherent spontaneous radiation [68]. The second term is directly related to the stimulated emission.

Second, for small slippage one can expect in this model a spiky structure of the output signal which could be of special interest for certain experiments with high temporal resolution. However, in our opinion, the model itself still needs additional justification.

In conclusion we would like to emphasize that the schemes above are not related to FELs only. They reveal also interesting physics not typical for traditional electronics. Perhaps, this could justify the volume of this chapter, which still cannot pretend to be a total description of FELs.

11

Blowup Effect in Linear Accelerators

As has been mentioned occasionally, the stimulated emission could be treated as a radiative beam instability. As a matter of fact, the problem of electromagnetic waves generation and amplification can be reduced to the provocation of the controlled instability in a desirable frequency region.

The previous sections were devoted to numerous difficulties of the task. However, as it usually happens, instabilities are easily self-excited when and where they are not desirable or even harmful. According to well-known Murphy low, damping of these parasitic instabilities requires sometimes even more efforts than exciting the desirable ones.

The theory of collective instabilities in such complicated systems as high current particle accelerators deserves a special book and, in any case, is outside of our scope. Nevertheless, one example is worth to be mentioned here briefly.

We mean a so-called blowup effect experimentally found out in large linear accelerators at currents exceeding a rather low threshold value of order of several tens of milliamperes. The accelerated current pulse with typical duration of 2–3 ms was found out to shorten sharply. Increase in the injection current shortened the pulse even more so that the total accelerated charge remained the same or decreased. At the same time, hard x-ray radiation appeared indicating high-energy electrons bombarding the chamber walls. These effects were accompanied by electromagnetic radiation with frequency exceeding 1.5–2 times the frequency of the main accelerating mode.

The last obviously indicated a parasitic mode self-excitation, that is, the coherent radiation emission in a higher propagation band.[1] The electron bombardment proved that the excited mode had transverse components at the axis and was axially nonsymmetric.

Transverse focusing taken into account, one can consider each electron as an oscillator moving with a relativistic velocity in a system permitting prop-

---

[1] Remind that the dispersion characteristic of a linear accelerator's waveguide consists of bands of transparency. Certain spatial Fourier harmonics of propagating modes have phase velocity lesser than that of light.

agation of slow waves. Hence, in our conception we can talk about radiation under conditions of anomalous Doppler effect when growing of oscillations can be expected. In this short chapter, we pay attention to this effect because the negative energy waves had been considered above only as longitudinal space charge ones. In the present case, self-excitation and phasing of transverse displacement waves are of interest.

In a linear accelerator, the beam looks like a train of short bunches separated by the accelerating wave length which is not an integer number of the excited wave one. For this reason the microwave equilibrium structure of the beam is not of importance for self-excitation but gives a possibility to consider each bunch as an individual point-like particle.

A structure of a nonsymmetric wave in a periodic waveguide is rather complicated even if the waveguide itself is symmetric. Opposite to uniform systems, only axially symmetric modes belong to definite E or M types. In general, the proper waves have all six components and for this reason are called HEM-waves (Hybrid ElectroMagnetic). However, only quasi-synchronous harmonics with wavenumbers $k \approx \omega/\overline{\beta}c$ and phase velocities $\beta_\mathrm{p} \approx \overline{\beta}$ are of importance for interaction with a particle moving along z with a practically constant velocity $\overline{\beta}$. To avoid misunderstanding, note that the phase velocity $\beta_\mathrm{p} = \omega/kc$ should be considered in our case as a fixed parameter. Boundary conditions in a waveguide of period l can be provided only by cooperation of harmonics shifted in wavenumbers by multiples of $2\pi/l$ and not taking part in the synchronous interaction.

The field of the lowest synchronous harmonic with one variation over azimuth can be expressed in cylindrical coordinates via three components of the vector-potential. For a wave linearly polarized in a $x = r \cos \theta$ plane

$$A_\mathrm{r} = I_2 \left(kr\sqrt{1-\beta_\mathrm{p}^2}\right) \cos \theta;$$
$$A_\theta = I_2 \left(kr\sqrt{1-\beta_\mathrm{p}^2}\right) \sin \theta; \qquad (11.1)$$
$$A_z = \mathrm{i}\sqrt{1-\beta_\mathrm{p}^2} I_1 \left(kr\sqrt{1-\beta_\mathrm{p}^2}\right) \cos \theta,$$

where I_n is a Bessel function of an imaginary argument. Standard calculations yield for the field components:

$$E_\mathrm{x} = -\mathrm{i}k\beta_\mathrm{p}\sqrt{1-\beta_\mathrm{p}^2} I_2 \cos 2\theta;$$
$$E_\mathrm{y} = -\mathrm{i}k\beta_\mathrm{p} I_2 \sin 2\theta;$$
$$E_z = k\beta_\mathrm{p}\sqrt{1-\beta_\mathrm{p}^2} I_1 \cos \theta;$$

$$\qquad (11.2)$$

$$B_\mathrm{x} = -\frac{\mathrm{i}}{2}k\sqrt{1-\beta_\mathrm{p}^2}\left(1+\beta_\mathrm{p}^2\right) I_2 \sin 2\theta;$$
$$B_\mathrm{y} = -\frac{\mathrm{i}}{2}k\sqrt{1-\beta_\mathrm{p}^2}\left[\left(1-\beta_\mathrm{p}^2\right) I_0 - \left(1+\beta_\mathrm{p}^2\right) I_2\right] \cos 2\theta;$$

$$B_z = k\sqrt{1-\beta_p^2}\,I_1 \sin\theta\,.$$

In particular, considering the polarization plane, the components

$$E_z = \frac{1}{2}k^2 x \beta_p \left(1-\beta_p^2\right)\,; \qquad B_y = -\frac{i}{2}k\left(1-\beta_p^2\right)^{3/2}\,. \tag{11.3}$$

do not vanish at the axis. A schematic structure of the force lines in the paraxial region is presented in Fig. 11.1

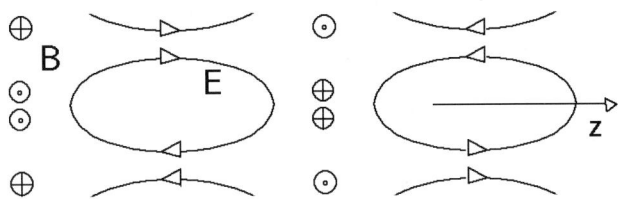

Fig. 11.1. Force lines of a HEM wave in the paraxial region

Two main deductions should be made. First of all, a particle travelling along the z-axis cannot radiate a HEM wave but experiences a deflecting Lorentz force. As a result, it is shifted in x-direction to the domain of possible emission and/or absorption. One can easily see that waves slightly faster than the particle accomplish negative work and, hence, are amplified. Correspondingly, slow waves are absorbed. This mechanism of the stimulated emission is in somewhat more complicated than the longitudinal phasing above. The latter effect also exists in our case but plays a secondary role.

In reality, the length of a single section is too small for developing of an absolute instability due to induced radiation.[2] But in a chain of many sections the instability occurs in spite of their electrodynamic independency. The necessary coupling takes place because information about a transverse displacement in a certain section is transported by the beam to all following ones.

To simplify the description, we neglect all transient effect at the ends of the sections constituting a waveguide of a large linear accelerator. Their independency can be imitated by putting zero the group velocity of the HEM wave. Radiative processes inside a single section cannot be considered in this model. We also suppose for simplicity the rigid structure of the beam bunched in the accelerating field. By the way this makes impossible the longitudinal bunching in the comparatively weak HEM wave.

Consider a sequence of particles (bunches) deflected by the synchronous HEM wave proportional to $\exp(ikz - i\omega t)$. When passing a point z the s-th particle is under action of the same force as the $(s-1)$-th one but shifted in phase by $-\omega T$ where T is a time interval between the particles. Besides,

---

[2] However, in industrial high-current accelerator, it can happen.

an additional force acts because of the $(s-1)$-th particle radiation. This is proportional to the particle deviation from the axis with a certain complex coefficient Z. For our purposes, an obvious fact is sufficient that Z value is proportional to the particle charge, that is, to the beam average current I. (Of course, calculations of the instability threshold would require the exact value of Z as well as the group velocity and damping constant of the HEM mode.) Now the equation for the transverse deviation of the s-th particle can be written in the form:

$$\left[\frac{d^2 x_s}{dz^2} + \nu^2 x_s\right] \exp(i\omega T) = \frac{d^2 x_{s-1}}{dz^2} + \nu^2 x_{s-1} + Z x_{s-1}. \tag{11.4}$$

Here ν^2 describes a possible external focusing and the phasor $\exp(i\omega T)$ reflects the phase shift of the radiation field during the interval between particles.

The transverse deflections of two successive particles also have a phase shift of ωT. In any case, a formal substitution

$$x_s = X_s \exp(-i\omega T s)$$

excludes the exponential factor from (11.4) and gives the following equation for slowly varying amplitudes:

$$\left[\frac{d^2}{dz^2} + \nu^2\right](X_s - X_{s-1}) = Z X_{s-1}. \tag{11.5}$$

Let us suppose now that particle-to-particle variations of the amplitude are small so that the index s can be considered as a continuous variable. As was mentioned above, it does not mean necessarily that the beam itself is continuous (the accepted model might fail only if the phase shift ωT is a multiple integer of 2π, i.e., if the HEM wave and the accelerating wave are coherent). In this approximation, Eq. (11.5) looks like

$$\frac{\partial}{\partial s}\left[\frac{\partial^2 X(z,s)}{\partial z^2} + \nu^2 X(z,s)\right] = Z X(z,s), \tag{11.6}$$

where the constant Z differs in somewhat from that of (11.4) and s can be treated now as time accounted from the moment when the head of the train passed the point z.

A solution of (11.6) depends on initial and boundary conditions, in particular on initial amplitude of the wave and on initial beam displacement. However, if time and distance are large enough, the amplitude asymptotic behavior is independent of initial conditions.

As far as the field vanishes ahead of the train Laplace transformation of (11.6) gives the second-order homogenous equation

$$\frac{d^2 X(z,p)}{dz^2} + \left(\nu^2 - \frac{Z}{p}\right) X(z,p) = 0 \tag{11.7}$$

with a general solution

$$X(z,p) = L^+(p)\exp(\Gamma z) + L^-(p)\exp(-\Gamma z) ,\qquad (11.8)$$

where

$$\Gamma(p) = (Z/p - \nu^2)^{1/2} ; \quad \mathrm{Re}\,\Gamma > 0;\qquad (11.9)$$

$$L^\pm = \left[X(z,p) \pm \Gamma^{-1}\frac{\partial X(z,p)}{\partial z}\right]_{z=0}.$$

The asymptotic behavior of (11.8) for $z \to \infty$ is obviously determined by the first term in the right-hand side. The inverse Laplace transformation then yields

$$X(z,s) \asymp \frac{1}{2\pi i}\int_{-\infty+i0}^{+\infty+i0} L^+(p)\exp\left[ps + \Gamma(p)z\right] dp.\qquad (11.10)$$

For large z and s, the integral value is determined by saddle points p_0 in the complex plane of p, which are the roots of the equation:

$$s + z\frac{d\Gamma}{dp} = 0 \quad \text{or} \quad p^2\sqrt{Z/p - \nu^2} = \frac{Zz}{2s}.\qquad (11.11)$$

The root of interest corresponds to the maximal real part[3] of the exponent argument in (11.10). Passing the integration contour through the point and expanding the argument over powers of $p - p_0$:

$$ps + \Gamma(p)z \approx p_0 s + \Gamma(p_0)z + \frac{s^3 p_0^3}{Z^2 z^2}\left(3\frac{Z}{p} - 4\nu^2\right)(p - p_0)^2\qquad (11.12)$$

we obtain

$$X(z,s) \asymp \qquad (11.13)$$

$$\frac{L^+(p_0)}{2\pi i}\exp\left[p_0 s + \frac{Zz^2}{2sp_0^2}\right]\int_{-\infty+i0}^{+\infty+i0}\exp\left[\frac{s^3 p_0^3}{Z^2 z^2}\left(3\frac{Z}{p_0} - 4\nu^2\right)(p - p_0)^2\right]dp.$$

The substitution

$$u = (p - p_0)\left(4\nu^2 - 3\frac{Z}{p_0}\right)^{1/2}\frac{(sp_0)^{3/2}}{Zz}$$

gives for the integral (11.13):

$$X(z,s) \asymp -\frac{L^+(p_0)\,Zz}{2\sqrt{\pi}\,(sp_0)^{3/2}(3Z/p_0 - 4\nu^2)^{1/2}}\exp\left[p_0 s + \frac{Zz^2}{2sp_0^2}\right].\qquad (11.14)$$

---

[3] There can be two such roots but it does not make an essential difference.

In the limiting case of zero focusing when $\nu \ll |2sZ/z|^{1/3}$

$$p_0 = Z^{1/3} \left(\frac{z}{2s}\right)^{2/3}.$$

The branch of the cubic root is chosen to maximize the real part. The asymptotic estimate then reads

$$X(z,s) \asymp -\frac{L^+(p_0)}{\sqrt{3\pi}} \left(\frac{2}{Z^{1/2} z s^{1/2}}\right)^{1/3} \exp 3\left(Zz^2 s/4\right)^{1/3}. \qquad (11.15)$$

In the opposite case of large $\nu \gg s/z$ valid for particles lagging the train head closer than the betatron oscillations wavelength $p_0 = \pm (iZz/2s\nu)^{1/2}$

$$X(z,s) \asymp L^+(p_0) \left(\frac{2\nu}{Zzs^2}\right)^{1/2} \exp\left[\pm i\frac{\pi}{4} \mp iz\nu + \left(\pm i\frac{Zzs}{2\nu}\right)^{1/2}\right] \qquad (11.16)$$

(the sign has to be chosen here to correspond to the amplitude growing with z and s). One can see that the external focusing retards the instability but cannot eliminate it. The latter should be expected treating the instability as a result of interaction of positive and negative energy waves. Actually, focusing means that two waves of transverse displacement with frequencies $(k\pm\nu)\bar{\beta}c$ may propagate along the beam. The slow one has a negative energy. The crossing of its dispersion curve with that of the HEM wave provides the simultaneous growth of amplitudes along the beam. In the limiting case of $\nu \to 0$ when the proper waves splitting becomes zero the two-wave interaction takes place. This reminds to certain extent the transition between Raman and Compton regimes in the case of longitudinal space charge waves.

So far as these analogies are the main purpose of this chapter, we will not calculate here the "impedance" Z. If necessary, this can be done in the same way as in Part I. We would like just to emphasize once more that the possibility of coupling between HEM modes and beam transverse displacements is originated from the synchronous interaction within a single section. The key role here is played by the harmonic with $\beta_p \approx \bar{\beta}$ and by linear dependence of the HEM wave longitudinal electric field on transverse coordinates.

References

1. L. Landau and E. Lifshitz: *The Classical Theory of Fields*. (Pergamon Press, Oxford, 1968)
2. A. Einstein: Verhandl Dtsch.Ph.Ges. B.**18**(1916) 318
3. W. Heitler: *The Quantum Theory of Radiation*. (Clarendon Press, Oxford, 1936)
4. W.M. Fine, Ya.I. Khanin: *Quantum Electronics*. (MIT Press, Cambridge, MA, 1968)
5. V.L. Ginzburg: *Theoretical Physics and Astrophysics*. (Pergamon Press, Oxford, 1979)
6. E. Fermi: Phys. Rev. **57**(1940)485
7. G.N. Watson: *Bessel Functions and Their Applications*. (Cambridge University Press, Cambridge, 1922)
8. H. Bateman: *Higher Transcedental Functions, 2nd edn*. (McGraw–Hill, New York, 1953)
9. M. Born, E. Wolf: *Principles of Optics*. (Pergamon Press, London, 1959)
10. L. Landau and E. Lifshitz: *Electrodynamics of Continuous Media*. (Pergamon Press, Oxford, 1982)
11. V.L. Ginzburg: Sov. Phys. Doklady. **56**(1947)145
12. Ya. Fainberg and A. Yegorov: Plasma Phys. Rep. **12**(1999)123
13. M.V. Kuzelev, A.A. Rukhadze, P.S. Strelkov: *Relyativistskaya plazmennaya elektronika (Plasma Relativistic Electronics)*. (Moscow, 2002)
14. W.P. Allis, S.J. Buchsbaum, and A. Bers: *Waves in Anisotropic Plasma*. (MIT Press, Camridge, 1963)
15. N.A. Kroll and A.W.Trivelpiece: *Principles of Plasma Physics*. (McGraw–Hill, New York, 1973)
16. J.V. Jelley: *Cherenkov Radiation and Its Applications*. (Pergamon Press, London, 1958)
17. B.M. Bolotovsky: *Oliver Heavyside* (Nauka, Moscow, 1985)
18. I. Frank and V. Ginzburg: Sov. Phys. JETP **16**(1946)15
19. V.L. Ginzburg and V.N. Tsytovich: *Transition Radiation and Transition Scattering*. (Gordon & Breach, London, 1990)
20. B.M. Bolotovsky and G.V. Voskresensky: Sov. Phys. Uspekhi **9**(1966) 73
21. L.A. Vainshtein: *Teoria difrakcii i metod faktorizacii (Diffraction Theory and Factorization Method)*. (Sov. Radio, Moscow, 1966)

22. B. Noble: *Methods Based on Wiener–Hopf Technique for the Solution of Partial Differential Equations.* (London, 1958)
23. S. Smith, E. Purcell: Phys.Rev. **92**(1953)1069
24. P. Morse and H. Feshbach: *Methods of Theoretical Physics.* (McGraw–Hill, New York, 1953)
25. Ya. Fainberg and N. Khizhnyak: Sov. Phys. JETP **32**(1957)32
26. M.V. Kuzelev, A.A. Rukhadze: *Electrodinamika plotnykh puchkov v plazme (Electrodynamics of Dense Beams in Plasma).* (Nauka, Moscow, 1990)
27. I.S. Gradstein, I.M. Ryzhik: *Tables of Integrals, Sums, and Products.* (Academic Press, New York, 1980)
28. A.A. Kolomensky, A.N. Lebedev: Sov. Phys. Doklady **7** (1962) 492
29. V.Ya. Davydovsky: Sov. Phys. JETP, **16** (1963) 629
30. H. Motz: JAP **22**(1951)527
31. G. Goldstein: *Classical Mechanics* (Addison–Wesley Press, Cambridge, 1974)
32. G.I. Budker in: *Proc. Int. Conf. on Storage.* (Orsay, 1966)
33. A.A. Kolomensky and A.N. Lebedev: *Theory of Cyclic Accelerators.* (N.-H., Amsterdam, 1966)
34. V.I. Kurilko and Yu.V. Tkach: Physics – Uspekhi. **38** (1995) 231
35. R.H. Dicke: Phys. Rev. **93** (1954) 99
36. A.N. Kolmogorov: Sov. Phys. Doklady **59** (1954) 527; V.I. Arnold: Russ. Math. Surv. Sov. Math. **18** (1963) 9,85; J. Moser: Machr. Acad. Wiss. Gottingen. Math. Phys. Kl. **IIa** (1962) 1
37. A.J. Lichtenberg, M.A. Lieberman: *Regular and Stochastic Motion.* (Springer–Verlag, New York, 1983)
38. B. Chirikov: Phys. Reports **52** (1979) 265
39. V. Arnold: *Matematicheskie metody v klassicheskoy mekhanike (Mathematical Methods of the Classical Mechanics).* (Nauka, Moscow 1977)
40. *Basic Plasma Physics.* Edit. A. Galeev, R. Sudan et al. (N.-H. Co, Amsterdam, 1984)
41. E. Lifshitz, L. Pitayevsky: *Physical Kinetics.* (Pergamon Press, Oxford, 1981)
42. L. Landau, E. Lifshitz: *Fluid Mechanics.* (Pergamon Press, Oxford, 1987)
43. A.I. Akhiezer, Ya.B. Fainberg: Sov. Phys. Doklady **69** (1949) 525
44. L.D. Landau: J. Phys. USSR **10** (1946) 26
45. D.D. Ryutov: Plasma Phys. Control. Fusion **41** (1999) A1–A12
46. M.A. Evgrafov. *Asimptoticheskie ocenki (Asymptotic estimations).* (GITTL, Moscow 1957); N.McLachlan: *Complex Variable and Operational Calculus with Technical Applications* (University Press, Cambridge, 1946)
47. A.V. Gaponov, A.L Goldenberg, D.P. Grigoryev, M.I. Petelin: JETP Lett. **2**(1965)430; V. A. Flyagin, A.V. Gaponov , M.I. Petelin, V.K. Yulpatov: IEEE Trans. **MTT-25** (1977)514
48. V.A. Balakirev et al: Sov.Phys.JETP **84** (1983) 507
49. M.J. Clauser, M.A. Sweeney: *Proc. 1st Inter. Topical Confer. Electron Beam Research and Technol.* (Albuquerque 1976) **1** 135
50. R.B. Miller: *Intense Charged Particle Beams.* (Plenum Press, New York, 1983)
51. C. Nielsen, A. Sessler: Rev. Scient. Instrum, **30** (1959) 80
52. A.A. Kolomensky, A.N. Lebedev: Atomnaya energiya (Atomic Energy) **7** (1959) 549
53. Ya.B. Zeldovich: Sov. Phys. Uspekhi **18** (1975) 79
54. V.L. Bratman, N.S. Ginzburg, G.S. Nusinovich: Journ Techn. Phys. Lett. **3** (1977) 395

55. G.R. Smith, A.N. Kaufman: Phys. Fluids. **21** (1978) 2230–2241
56. Y. Gell, R. Nansch: Phys. Fluids. **23** (1980) 1646–1655
57. D.R. Shklyar: Sov. Phys. JETP **53** (1981) 1187
58. A.N. Antonov, V.A. Buts, O.F. Kovpik, et al.: JETP Lett. **69** (1999) 851
59. V.A. Buts, O.V. Manuilenko, Yu.A. Turkin: Plasma Physics Reports **25** (1999) 737
60. A. Oraevsky: *Gaussian Beams and Optical Resonators* (Nova Science, New York, 1996)
61. E.L. Saldin, E.A. Schneidmiller and M.V. Yurkov: *The Physics of Free Electron Lasers.* (Springer-Verlag, 2000)
62. C.W. Roberson and P. Sprangle: *A Rewiew of Free Electron Lasers.* Physics of Fluids **B 1** (1989) 3–42
63. N.M. Krall, P.L. Morton, and M.N. Rosenbluth in: *Free Electron Generators of Coherent Radiation, Physics of Quantum Electronics.* (Addison-Wesley, Reading, MA, 1980)113; M.N. Rosenbluth, B.N. Moore, and H.W. Wong: IEEE J. Quantum Electron. **QE-21**, 1026 (1985).
64. P. Sprangle, C.M. Tang, and W.M. Manheimer: Phys. Rev. Lett. **43** (1979) 1932; Phys. Rev. **A 21** (1980) 302
65. G.T. Moore: Opt. Commun. **52** (1984) 46; E.T. Sharlemann, A.M. Sessler, and J.S. Wurtele: Phys. Rev. Lett. **54** (1985) 1925; NIM **A 239** (1985) 29
66. A.M. Kondratenko and E.L. Saldin: Part. Acc. **10** (1980) 207
67. N.S. Ginzburg: J. Techn. Phys. Lett. **14** (1988) 197; N. Ginzburg and Ju. Novozhilova: J. Techn. Phys. Lett. **15** (1989) 771
68. Zhirong Huang and Kwan-Je Kim: NIM **A 445** (2000) 105 (2000)
69. N. Vinokurov and A. Skrinsky in: *Proc. 10th Int. Conf. on High Energy Accelerators.* **2** (Serpukhov, 1977) 454

Index

Amplification
  Compton regime   227
  Raman regime   227
  spatial   112, 113, 209, 222
  temporal   217, 236
Autophasing   105
Autoresonance   67, 175, 189, 195, 200

Beam
  emittance   81, 220
  Gauss-Laguerre   212, 213, 234
  guiding   232
  instability in plasma   159
Bounce oscillations   107
Boundary Conditions   42, 53, 180
Bremsstrahlung   4

Causality Principle   45
Chaos   114, 118
Chaotic Dynamics   114, 119, 175, 201, 204
Chirikov Criterion   120
Coherence   3, 87, 89
  conditions   89
  factor   91
  in a train of particles   3, 6, 12, 14–16, 18, 36, 93
  lattices   92
  spatial   95
Cold System   5
Continuity Equation   131, 161, 223
Coulomb Fields   15, 17, 18, 20, 21, 24, 26, 33, 76, 98, 201, 204
Current   240, 242, 244

density   13, 22, 39, 76, 160, 177
Cyclotron
  autoresonance maser   67, 175, 205
  resonance   67, 174, 175
  waves   132, 134, 137–139, 143, 145, 149, 155, 174, 180

Deceleration by Radiation   4
Detuning   113, 163, 164, 166, 169, 224, 226, 227
  optimal   113, 164, 166, 219, 220
Diffraction   231
  parameter   235
Dispersion
  equation   48, 63, 64, 114, 173, 174, 188, 210
  periodic structures   28, 29
  plasma   163
  relations   47, 61, 125, 134, 136–139, 147, 151, 152, 154, 225, 240
  ribbon beam   178
  waveguide   46
Distribution Function   82, 169, 176, 201
Doppler Effect   7, 9

Efficiency   195, 215, 220, 229, 230
Eigenvalues   41
Elementary Emitter   3
Emission
  classical   102, 103
  quantum   98, 100
  spectral line   91, 99, 101
  spontaneous   4, 5

262 Index

stimulated 5, 98
zone 58, 59

Field Zone
  far 4
  near 38, 47, 50, 53
Filamentation 110
Free Electron Laser 87, 208

Gain 209, 241
Group Velocity 55, 110, 122, 163
Gyrotron 53, 175, 182, 195, 205

Harmonic Numbers 52, 56, 57
Helmholtz Equation 41, 42
Homocline Points 119

Impedance 35
  coupling 35
Increment 182, 183, 228, 235, 243
Instability
  absolute 162
  convective 167
  increment 113, 217, 226, 227
  local 9, 119, 120
  radiative 176

KAM Theorem 119

Landau Damping 171, 176
  negative 170
Langmuir Frequency 162
Lasing 102, 104
Lethargy 249
Lienard-Wichert Potentials 4, 38
Liouville Theorem 81, 82, 115, 221
Local Dispersion Relations 137
Lorentz Force 73, 76
Lyapunov Criterion 119

Multipoles 47

Negative Energy 140
Negative Mass Instability 183
Non-Linear
  parameter 227, 229
  resonance 115, 199
  saturation 218

Optical

  cavity 211
  klystron 215
Optimal Detuning 235

Parabolic Approximation 211, 234
Phase
  bunching 108, 109, 183
  dynamics 107, 111
  slippage 104, 106, 107, 121, 194, 196
  trajectory 80, 107
Polarization 52, 56, 214
  losses 16, 18
  oscillations 16
Potential
  ponderomotive 223, 224
  retarded 37, 38, 76, 79
Poynting Vector 44, 47, 54, 72, 79, 90

Radiation
  cyclotron 37, 53, 55, 68, 197
  friction 67
  instability 176
  length 112
  parametric transition 27
  reaction force 72, 76, 78, 79
  secondary 116
  spectral line 105
  synchrotron 37, 55, 65, 167
  transition 12, 22, 36, 48
  under uniform motion 8
Resonance 114
  cyclotron 83
  independent 189
  lines 67, 189, 192, 193
  overlapping 15
  secondary 116
  separation 114, 122
  width 114, 121, 122

SASE Regime 113
Saturation 114, 218, 219
Scattering
  by bound particles 65
  coherent 66, 80, 94
  Compton 60, 61
  cross section 80, 95
  Thomson 64
Spectral-Angular Distribution 57, 65, 113, 114, 218, 219

Stimulated Radiation 5
Superradiance 239, 240
Synchronism 10, 11, 27, 34, 54, 55, 93, 104, 159, 213
  for periodic motion 40
Synchrotron Oscillations 83, 107, 218, 229

Undulator 50, 69
  coefficient 51, 57, 214, 223, 226
  helical 55, 212, 214, 229

plane 212, 233, 240
radiation 37, 87, 209
tempered 230

Vlasov Equation 160, 176

Waves
  in free space 38, 48
  plasma 162, 163
  proper 38, 53
  space charge 132, 143

Printing: Krips bv, Meppel
Binding: Stürtz, Würzburg